Electronic Noses & Sensors for the Detection of Explosives

NATO Science Series

A Series presenting the results of scientific meetings supported under the NATO Science Programme.

The Series is published by IOS Press, Amsterdam, and Kluwer Academic Publishers in conjunction with the NATO Scientific Affairs Division

Sub-Series

I. Life and Behavioural Sciences	IOS Press
II. Mathematics, Physics and Chemistry	Kluwer Academic Publishers
III. Computer and Systems Science	IOS Press
IV. Earth and Environmental Sciences	Kluwer Academic Publishers
V. Science and Technology Policy	IOS Press

The NATO Science Series continues the series of books published formerly as the NATO ASI Series.

The NATO Science Programme offers support for collaboration in civil science between scientists of countries of the Euro-Atlantic Partnership Council. The types of scientific meeting generally supported are "Advanced Study Institutes" and "Advanced Research Workshops", although other types of meeting are supported from time to time. The NATO Science Series collects together the results of these meetings. The meetings are co-organized bij scientists from NATO countries and scientists from NATO's Partner countries – countries of the CIS and Central and Eastern Europe.

Advanced Study Institutes are high-level tutorial courses offering in-depth study of latest advances in a field.
Advanced Research Workshops are expert meetings aimed at critical assessment of a field, and identification of directions for future action.

As a consequence of the restructuring of the NATO Science Programme in 1999, the NATO Science Series has been re-organised and there are currently Five Sub-series as noted above. Please consult the following web sites for information on previous volumes published in the Series, as well as details of earlier Sub-series.

http://www.nato.int/science
http://www.wkap.nl
http://www.iospress.nl
http://www.wtv-books.de/nato-pco.htm

Series II: Mathematics, Physics and Chemistry – Vol. 159

Electronic Noses & Sensors for the Detection of Explosives

edited by

Julian W. Gardner

School of Engineering,
University of Warwick,
Coventry, United Kingdom

and

Jehuda Yinon

National Center for Forensic Science,
University of Central Florida,
Orlando, Florida, U.S.A.

Kluwer Academic Publishers

Dordrecht / Boston / London

Published in cooperation with NATO Scientific Affairs Division

Proceedings of the NATO Advanced Research Workshop on
Electronic Noses & Sensors for the Detection of Explosives
Warwick, Coventry, U.K.
30 September–3 October 2003

A C.I.P. Catalogue record for this book is available from the Library of Congress.

ISBN 1-4020-2317-0 (HB)
ISBN 1-4020-2319-7 (e-book)

Published by Kluwer Academic Publishers,
P.O. Box 17, 3300 AA Dordrecht, The Netherlands.

Sold and distributed in North, Central and South America
by Kluwer Academic Publishers,
101 Philip Drive, Norwell, MA 02061, U.S.A.

In all other countries, sold and distributed
by Kluwer Academic Publishers,
P.O. Box 322, 3300 AH Dordrecht, The Netherlands.

Printed on acid-free paper

Printed in the Netherlands.

Dedication

This book is dedicated to my father, Dr. William Edward Gardner, who has both helped me with this book and inspired me all of my life.

Contents

Contributing Authors

Sergi Bermudez I Badia, Institute of Neuroinformatics, ETH-Universitat Zürich, Winterthurerstrasse 190, Zürich CH 8057, Switzerland.

Stewart Berry, Scintrex Trace Corporation, 300 Parkdale Ave, Ottawa, Ontario, Canada.

Charnjit Singh Bilkhu, Scintrex Trace Corporation, 300 Parkdale Ave, Ottawa, Ontario, Canada.

Mikael A. Carlsson, Department of Crop Science, Swedish University of Agricultural Sciences, PO Box 44, Sundsvagen 14, SE-230 53 Alnarp, Sweden.

Eric Chanie, Alpha MOS, 10 Av Didier Daurat, Z1 Montaudran, 31400 Toulouse, France.

Kwok Y. Chong, NeuroLab, Centre for Bioengineering, Department of Engineering, University of Leicester, University Road, Leicester LE1 7RH, UK.

David Christensen, Scintrex Trace Corporation, 300 Parkdale Ave, Ottawa, Ontario, Canada.

Colin Cumming, Nomadics Inc., Stillwater, Oklahoma, USA.

Mark Fisher, Nomadics Inc., Stillwater, Oklahoma, USA.

Julian W. Gardner, Smart Sensors and Devices Group, School of Engineering, University of Warwick, Coventry CV4 7AL, UK.

Bill S. Hansson, Department of Crop Science, Swedish University of Agricultural Sciences, PO Box 44, Sundsvagen 14, SE-230 53 Alnarp, Sweden.

Wen He, Scintrex Trace Corporation, 300 Parkdale Ave, Ottawa, Ontario, Canada.

Etienne Hugues, LORIA, Campus Scientifique, BP 239, Universite de Nancy, Vandoevre-Les-Nancy, France.

Phuong Huynh, Scintrex Trace Corporation, 300 Parkdale Ave, Ottawa, Ontario, Canada.

Maria Ivanovskaya, Research Institute for Physical Chemical Problems, Belarus State University, Leningradskaya 14, Minsk, Belarus.

Marc Kehm, Fraunhofer-Institut fur Chemische Technologie (ICT), Pfinztal-Berghausen, Germany.

Dzimtry Kotsikau, Research Institute for Physical Chemical Problems, Belarus State University, Leningradskaya 14, Minsk, Belarus.

Michael Krausa, Fraunhofer-Institut fur Chemische Technologie (ICT), Pfinztal-Berghausen, Germany.

Richard T. Lareau, Transportation Security Administration, Atlantic City International Airport, New Jersey, USA.

Shirley Locquiao, Scintrex Trace Corporation, 300 Parkdale Ave, Ottawa, Ontario, Canada.

Dominique Martinez, LORIA, Campus Scientifique, BP 239, Universite de Nancy, Vandoevre-Les-Nancy, France.

Dao Hinh Nguyen, Scintrex Trace Corporation, 300 Parkdale Ave, Ottawa, Ontario, Canada.

Vamsee K. Pamula, Duke University, Durham, North Carolina, USA.

Timothy C. Pearce, NeuroLab, Centre for Bioengineering, Department of Engineering, University of Leicester, University Road, Leicester LE1 7RH, UK.

Krishna C. Persaud, Department of Instrumentation and Analytical Science, University of Manchester Institute of Science and Technology. Manchester, UK.

Karsten Pinkwart, Fraunhofer-Institut fur Chemische Technologie (ICT), Pfinztal-Berghausen, Germany.

Lal A. Pinnaduwage, Life Sciences Division, Oak Ridge National Laboratory, Oak Ridge, Tennessee, USA.

Anna Maria Pisanelli, Department of Instrumentation and Analytical Science, University of Manchester Institute of Science and Technology. Manchester, UK.

Peter Rabenecker, Fraunhofer-Institut fur Chemische Technologie (ICT), Pfinztal-Berghausen, Germany.

Arunas Setkus, Semiconductor Physics Institute, A. Gostauto 11, Vilnius, Lithuania.

John Sikes, Nomadics Inc., Stillwater, Oklahoma, USA.

Edward J. Staples, Electronic Sensor Technology, 1077 Business Center Circle, Newbury Park, California, USA.

Tamar Sternfeld, Department of Chemistry, Tufts University, Medford, Massachusetts, USA.

Timothy M. Swager, Department of Chemistry and Institute for Soldier Nanotechnology, Massachusetts Institute of Technology, Cambridge, Massachusetts, USA.

Thomas Thundat, Life Sciences Division, Oak Ridge National Laboratory, Oak Ridge, Tennessee, USA.

William C. Trogler, University of California, San Diego, California, USA.

Paul F. M. Verschure, Institute of Neuroinformatics, ETH-Universitat Zürich, Winterthurerstrasse 190, Zürich CH 8057, Switzerland.

David Walt, Department of Chemistry, Tufts University, Medford, Massachusetts, USA.

Joseph Wang, Department of Chemistry, New Mexico State University, Las Cruces, New Mexico, USA.

Peter Wareham, Department of Instrumentation and Analytical Science, University of Manchester Institute of Science and Technology. Manchester, UK.

Jehuda Yinon, National Center for Forensic Science, University of Central Florida, Orlando, Florida, USA.

Lin Zhang, Scintrex Trace Corporation, 300 Parkdale Ave, Ottawa, Ontario, Canada.

Qiaoling Zhong, Scintrex Trace Corporation, 300 Parkdale Ave, Ottawa, Ontario, Canada.

Preface

This book examines both the potential application of electronic nose technology, and the current state of development of chemical sensors for the detection of vapours from explosives, such as those used in landmines. The two fields have developed, somewhat in parallel, over the past decade and so one of the purposes of this workshop, on which the book is based, was to bring together scientists from the two fields in order to challenge the two communities and, mutually, stimulate both fields.

The first chapter reviews the basic principles of an electronic nose and explores possible ways in which the detection limit of conventional electronic nose technology can be reduced to the level required for the trace levels observed for many explosive materials.

Chapters 2-6 describe the use of several different types of polymer-based sensors, i.e. chemiluminescent, fluorescent and optical, to detect explosive materials. In the optical case a large array of coated beads is employed to reduce the signal-to-noise ratio, and provide greater amplification of weak signals from trace levels of vapours.

Chapters 7-11 continue the theme and explore different types of chemical sensors. Chapter 7 describes the application of metal oxide semiconducting resistive sensors, and then Chapters 8-11 cover mainly recent developments of electrochemical sensors.

Part of Chapter 11 and Chapter 12 explore the use of, firstly, pattern recognition techniques to enhance the performance of the chemical sensors and secondly the modulation of the smell intensity to improve the signal processing.

Chapters 13 and 14 look at biological systems as a blue-print for chemical sensing. The biology is employed either to understand the way insects locate odorant sources (e.g. the pheromone mediated chemotactic search of the moth), or to understand the signal processing via a neural mode of the antennal lobe of the locust.

Chapters 15 and 16 continue with a discussion of some of the new types of electronic noses; namely a fast GC column with a SAW detector and secondly the use of micromechanical sensors.

Chapters 17 and 18 look at the importance of sampling technologies and the design of the microfluidic systems. In particular, the use of pre-concentrators and solid phase micro extractors to boost the vapour concentration before it is introduced to the chemical sensor or electronic nose.

Finally, in Chapter 19 there is an exploration of the next generation of trace detection sampling systems and Chapter 22 attempts to consider the two fields and answer the question as to whether they are producing complementary or replacement technologies.

As remarked above this book is based on material presented at a NATO Advanced Research Workshop held in Coventry (UK) in October 2003. The authors wish to express their gratitude to the NATO Science Committee for their financial support of the workshop and the preparation of this book, the University of Warwick for providing administrative support and to all the contributors. We would also like to thank the staff at the Westwood Training & Conference Centre, and single out Marie Bradley who worked tirelessly to ensure that the meeting ran smoothly.

Warwick University
March 2004

Julian W. Gardner
Jehuda Yinon

Foreword

   The contents of this book represent the views of the contributing authors and are not necessarily those of the co-editors. However, the adoption of the UK rather than US spelling of certain words in the main text, such as "odour" and "fibre", is the editors' fault! Finally, we provide below a group photograph of the participants at the NATO Advanced Research Workshop held in Coventry, UK.

NATO Workshop. 1-2 October 2003. Warwick University, UK

Chapter 1

REVIEW OF CONVENTIONAL ELECTRONIC NOSES AND THEIR POSSIBLE APPLICATION TO THE DETECTION OF EXPLOSIVES

Julian William Gardner
School of Engineering, Warwick University, Coventry CV4 7AL, UK

Abstract: During the past twenty years there has been enormous interest in the detection of simple and complex odours by means of electronic instrumentation. This has led to the commercialisation of so-called 'electronic noses' that typically comprise an array of partially selective sensors with suitable pattern recognition software; they have been applied to many different olfactory problems in industry, e.g. from the quality assurance of foodstuffs through to medical diagnostics. Here, the nature of an electronic nose is reviewed and various strategies are suggested to enhance their performance in terms of detection limit and specificity. The general problem of applying a sensor-based electronic nose to the detection of explosive materials is discussed and, the special requirement to detect very low vapour pressures of compounds, i.e. a limit of detection of at least parts per billion (10^9) in air and, in some cases, down to parts per trillion (10^{12}) in air. Although there is considerable potential to enhance the detection limit of sensor-based electronic noses, the technology is unlikely to displace in the near future the current 'gold standard' of ion mobility spectroscopy for explosives detection. Instead electronic noses could be regarded as a complementary technology used to screen for unknown compounds.

Key words: Electronic noses, artificial olfaction, chemical sensing.

1

J.W. Gardner and J.Yinon (eds.),
Electronic Noses & Sensors for the Detection of Explosives, 01-28.
© 2004 *Kluwer Academic Publishers. Printed in the Netherlands.*

1. INTRODUCTION

1.1 What is a smell?

A *simple* odour is usually a small, polar molecule that can become airborne, enter the nasal cavity and then be sensed by the mammalian olfactory system. There are many different functional groups found in odorous compounds; some of which are listed below in Table 1.

Table 1. Structure of some simple functional groups found in odorous compounds. (Table taken from reference [1], p13).

Functional group	Structure	Functional group	Structure
Alcohol	$R-O-H$	Lactone	(cyclic structure with O, n, O)
Ether	R_1-O-R_2	Amine	$R-N(H)(H)$
Aldehyde	$R-C(=O)H$	Nitrile	$R-C\equiv N$
Ketone	$R_1-C(=O)R_2$	Isonitrile	$R-N\equiv C$
Acid	$R-C(=O)O-H$	Thiol	$R-S-H$
Ester	$R_1-C(=O)O-R_2$	Amide	$R_1-C(=O)N(H)R_2$

Although there has been much study of the relationship between the structure and the sensory activity of odorous molecules, it frequently appears impossible to predict the odour from its chemical structure. In other words, very similar chemical structures can have quite different smells (e.g. the *para* versus the *meta* form, optical isomers). Contrarily, different compounds in terms of functional group, shape, etc can have similar smells [2].

The detection threshold of different molecules by the human olfactory system varies enormously from compound to compound. The human response to an odour, i.e. the intensity, generates a logistic-like function with intensity just perceivable at what is called the olfactory threshold and then saturating out above at a higher concentration. The olfactory (detection) threshold of some different odorant in air is given in Table 2 and shows that the detection threshold in air varies enormously from the mg/ml level for ether down to the ng/ml level for citral [1]. There are some compounds with an even lower detection threshold, such as musk xylene, that is detected by the human nose down at the pg/ml level. So the range of detection thresholds for the human nose varies by at least 9 orders of magnitude and depends on the compound!

Table 2. Detection thresholds for some typical odorants in air (From [1]).

Compound	Odour type	Threshold
Ether	ether	5.8 mg/ml
Limonene	lemon	0.1 mg/ml
Benzaldehyde	bitter almond	3.0 µg/ml
Butyric acid	rancid butter	9.0 µg/ml
Citral	lemon	3.0 ng/ml
Musk xylene	musky	0.8 pg/ml

Interestingly, the detection threshold of different flavour compounds in water also varies considerably with a grapefruit compound being an astonishing low value of 2 parts in 10^{14}. So the human sensory system has evolved a highly specialised sensitivity to, amongst other things, citrus compounds.

A *complex* odour (as opposed to a simple odour) is a collection of two or more different volatile chemical compounds that produces a smell. The number of different compounds in the headspace can vary from just a few to several thousand – each with a different concentration and each with a different detection threshold. The resultant smell may have various olfactory notes within it, such as green, floral, citrus etc.

The problem is further complicated by the fact that the detection threshold of a chemical compound in a complex odour can differ from that when it is on its own, i.e. as a simple odour. For example, compound A and B may not smell as individuals but when mixed together they do elicit an olfactory response. Thus, the structure-activity relationship may be straightforward and exist for simple odours but can be highly non-linear in mixtures. This makes it very difficult to understand and model the mechanisms by which the human olfactory system works and thus to develop an electronic analogue.

4

1.2   Conventional analytical headspace analysis

The most commonly used method to measure smells is with an organoleptic panel, i.e. a group of, typically, 4 to 8 people who sniff professionally the headspace of an odorant compound. In some cases, the headspace is first passed down a gas chromatography column and then the panel smells the separate compounds as they are slowly eluted from the end of the column – this is known as an *olfactometer*. However, organoleptic panels are expensive to train, take a considerable amount of time and effort to detect the compounds, and are subject to considerable variability – perhaps a factor of 3 or more from panel to panel. Consequently, there has been considerable effort to employ other headspace techniques that are well known in the field of analytical chemistry, such as:

- Gas chromatography (GC)
- Mass spectrometry (MS)
- Optical spectroscopy
- Combined GC-MS

The associated analytical instruments separate out some of the chemical compounds in a complex odour according to a physical property, for example, their solubility, mass, mass to charge ratio etc. Reference [1] provides an overview of these different analytical instruments. The disadvantages with these types of instruments for detecting odours are that they are relatively slow (minutes per reading), large, expensive and, more importantly, simply separate out the different constituents of a compound rather than provide a measure of their relative olfactory intensity. Indeed, the smell of a complex odour may well be dominated by a few key flavour compounds which occur below the detection threshold of even these analytical instruments – typically ng/ml.

1.3   Definition of an electronic nose

The definition of the term 'electronic nose' is itself a topic for debate. The term first began to be used by scientists in the mid to late eighties but the following definition was not published until 1994 [3]:

'An electronic nose is an instrument which comprises an array of electronic chemical sensors with partial sensitivity and an appropriate pattern recognition system capable of recognising simple or complex odours'

The basic architecture of an electronic nose is shown in Figure 1 with the signals from an array of chemical sensors being processed and the 'smell fingerprint' being identified against those fingerprints already held in a knowledge base (i.e. a database for odours).

The concept of an electronic nose has been around for many years but interest in the topic significantly increased after a Nature paper was published in 1982 on work carried out at Warwick University (UK) by chemists Persaud and Dodd. The first conference devoted to the topic of electronic noses was held in Iceland in 1991 and sponsored by NATO (see [4] for proceedings). In the 1990s hundreds of research papers were published on the topic of electronic noses, many review articles and two or three more books written.

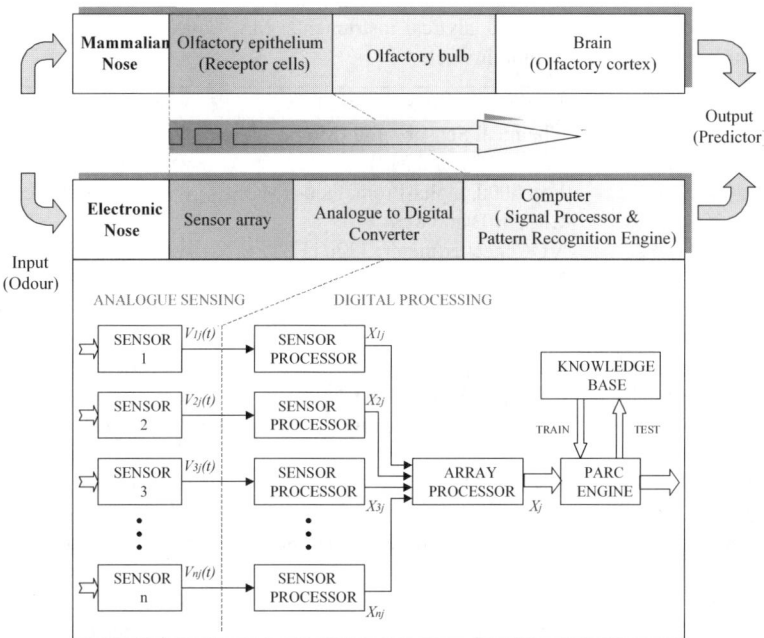

Figure 1. Schematic architecture of an electronic nose showing an array of chemical sensors, pre-processing, array processing and finally a supervised pattern recognition system. A crude analogue to the biological system is shown as well. See chapter 2 of reference [2] for details of the biology of olfaction.

A brief summary of electronic nose instruments currently available is now presented, followed by a discussion of strategies that may be or are being used to reduce the detection threshold of sensor-based electronic noses and hence increase their suitability to the ultimate detection of explosive materials. Readers interested in further details on electronic noses are directed towards a comprehensive book published in 2003 [5] that summarises all recent work on all types of electronic noses as well as 'machine olfaction'.

2.   ELECTRONIC NOSE TODAY

In 2003 there were at least 17 companies manufacturing and selling electronic nose instruments of various types and these companies are summarized in Table 3 together with the sensor technology employed and their application(s).

Table 3. Some commercial companies that manufactured sensor-based electronic noses and related instruments in 2003. Key to sensor technologies: MOS – metal oxide sensor, CP – conducting polymer, QMB – quartz crystal microbalance, FET – field effect transistor, SAW – surface acoustic wave; key to analytical instruments: MS – mass spectrometry, GC – gas chromatography, IMC – ion mobility cell

Manufacturer	Product (Technology used)	Website
Agilent	Chemical Sensor 4440 (MS)	www.agilent.com
Airsense Analytics	PEN (MOS and MS)	www.airsense.com
Alpha MOS	Fox 5000, alpha-Prometheus (MOS, CP, QMB and MS)	www.alpha-mos.com
Applied Sensor	VOCchecker/meter, 3320 (FET, MOS, QMB)	www.appliedsensor.com
Bloodhound Sensors Ltd	(CP)	www.bloodhound.co.uk/ bloodhound
Cyrano Science Inc.	C320 (CP)	www.cyranosciences.com
Environics Oy	ChemPro, M90, MGD-1 (IMS)	www.environics.fi
Electronic Sensor Technology	zNose model 4100/7100 (GC-SAW)	www.estcal.com
HKR Sensorsysteme	QMB 6, MS-Sensor	www.hkr-sensor.de
Illumina Inc.	SNP genotyping services (Optical bead array)	www.illumina.com
Lennartz Electronic	MOSES II (QMB, MOS, calorimetric)	www.lennartz-electronic.de
Marconi Applied Technologies	eNose (CP, MOS, QMB, SAW)	www.marconitech.com/che msens
Microsensor Systems Inc	Hazmatcad, VapourLab, Eagle (SAW or GC)	www.microsensorsystems. com
OligoSense	Organic semiconductor sensor modules	sch-www.uia.ac.be/struct /oligosense
Osmetech plc	OMA, Point of Care (CP)	www.osmetech.plc.uk
RST Rostock	SamSystem (QMB, SAW, MOS)	www.rst-rostock.de
SMart Nose	Smart Nose 300 (MS)	www.smartnose.com

These electronic noses employ a range of different classes of solid-state sensing materials (e.g. metal oxide, catalytic metal, polymer – both conducting and non-conducting), in a variety of different genii of transducers (e.g. resistive, potentiometric, amperometric). A detailed description of the different types of chemical sensors is provided in chapter 9 of [6]. Perhaps the most common type of sensor in commercial e-noses is the resistive sensor with the leading companies of Alpha MOS (France) principally employing tin oxide, Osmetech plc (UK) employing conducting polymer, and Cyrano Sciences (USA) employing carbon black/polymer composite. Figure 2 shows a discrete tin oxide resistive gas sensor and 32-element arrays of polymer resistive sensors.

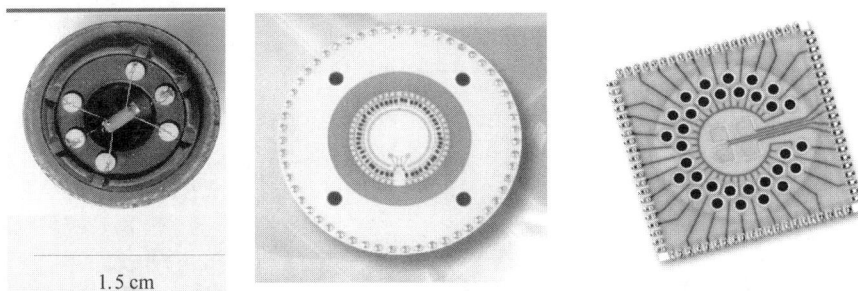

Figure 2. Examples of resistive gas sensors employed in commercial e-noses: Left - single MOS from Alpha MOS; Middle - 32 CP element array from Osmetech; Right 32 CBP element array from Cyrano Science (USA).

Large analytical e-nose instruments are commercially available and range in price from about €40k to €120k. Figure 3 shows the main instrument sold by the market leaders - the Fox - made by Alpha MOS in France and sold with different sensor technologies.

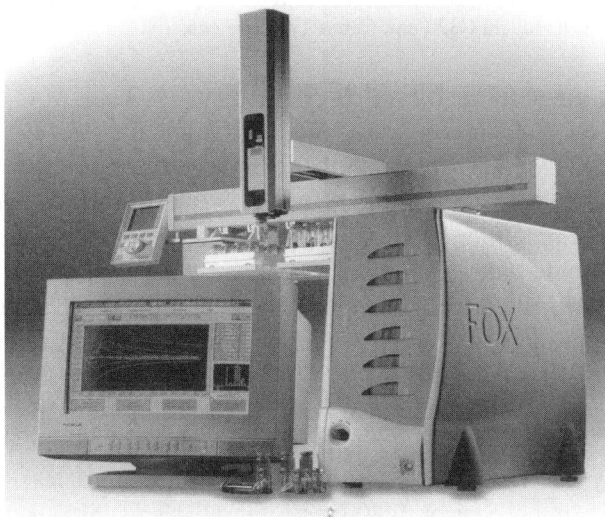

Figure 3. Fox electronic nose manufactured by Alpha MOS, France. This unit requires a PC (only VDU shown) to run the system.

The unit shown has an automated robotic headspace autosampler for the analysis of the headspace of 50 samples in small glass vials.

Smaller electronic noses are also available for process monitoring and for greater portability. Two examples are shown in Figure 4, the Alpha Centauri from Alpha MOS and the handheld C320 from Cyrano Sciences.

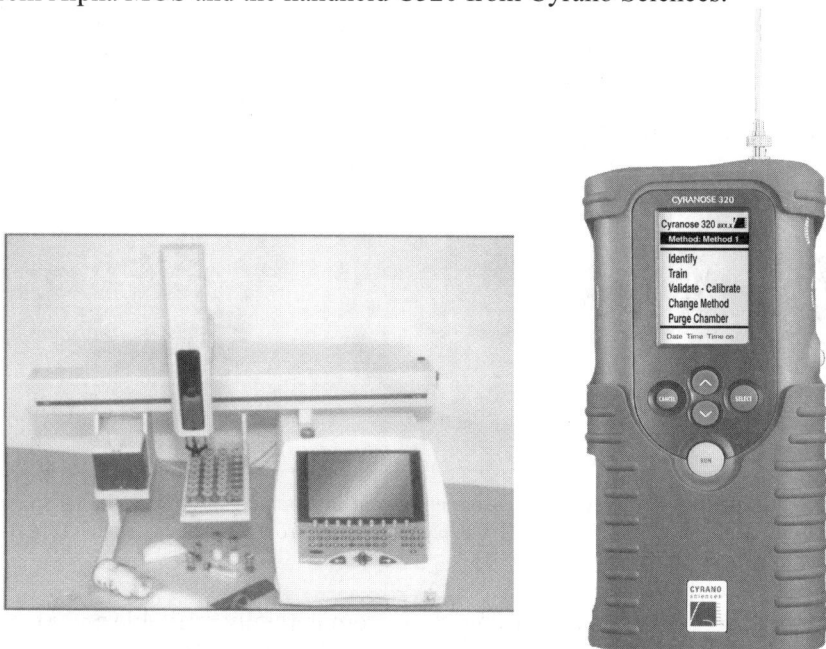

Figure 4. Examples of two portable electronic noses: Alpha Centauri from Alpha MOS and the handheld C320 unit from Cyrano Science Inc.

2.1 Quasi-sensor based technology

The initial electronic noses employed an array of non-specific solid-state sensors coupled with a pattern recognition system [1]. However, it has been recognized by practitioners that this type of instrument can have important limitations. Some units could only detect static headspace concentrations of certain odorants at, or around, the ppm level and would struggle to hold calibration and identify odours over a long period of time because of undesired drift in the sensor output. This led to the emergence of what will be referred to as 'quasi-electronic noses' bearing in mind the original 1994 definition [3] given above. Perhaps the first instrument of this type was launched by Hewlett-Packard (now Agilent Technologies) in 1998 and called the Chemical Sensor 4440 (see Figure 5). This instrument is simply a direct-injection quadrapole mass spectrometer; it is a conventional analytical instrument, but with reconfigured software to make the output look like an electronic nose system. The performance of this type of nose was attractive because the technology is mature, the limit of detection is down to the ppb level and there is practically no interference from ambient conditions, e.g. humidity, and airborne pollutants such as CO. Since then other companies have also employed mass spectrometers as commercial instruments for the recognition of smells, e.g. Alpha MOS, SMartNose.

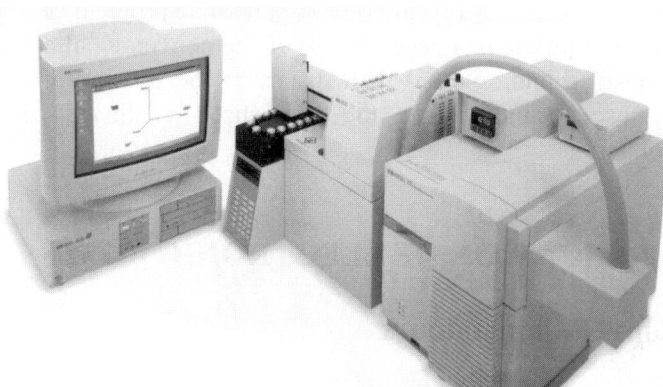

Figure 5. The Chemical Sensor 4440 manufactured by Agilent Technologies as a type of quasi-electronic nose based upon an automated headspace sampler and an MS.

Another example of a quasi-electronic nose is the use of a reconfigured GC column. One commercial example is the z-Nose which is a portable instrument based upon a short 1m GC column with an uncoated SAW detector. The instrument is calibrated using compounds similar to the target analyte and shows some promise in detecting explosives [7].

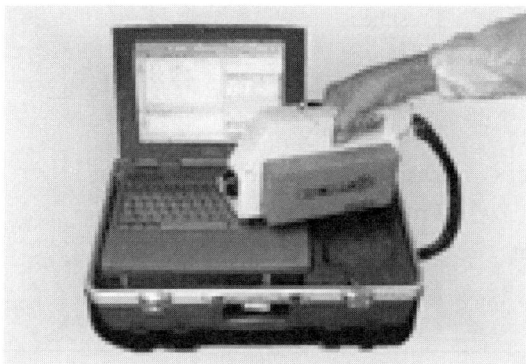

Figure 6. Quasi-electronic nose based upon a GC column: the z-Nose.

The discussion so far has reviewed the historical emergence of a range of different sensor-based electronic nose instruments employing a wide variety of sensing materials, transduction principles, and even different pattern recognition systems (see relevant chapters in references [1] and [5] for an overview). The situation has been further enriched by the more recent emergence of different conventional analytical instruments redesigned for the purpose of odour detection, such as mass spectrometers, gas chromatographs etc. However, the ultimate success of an electronic nose will depend solely upon its ability to solve a particular problem or application. The next section provides a brief review of the conventional application of electronic nose technology and then discusses the possibility of applying electronic nose technology to the specialised challenge of detecting explosive materials.

3. APPLICATION OF ELECTRONIC NOSES

3.1 Conventional odours

Electronic noses have been used to detect the odours of a wide variety of different products [1, 3-5]. For example, there has been considerable effort in the analysis of foodstuffs. Table 4 shows some of the wide variety of different foods that have been analysed by electronic noses and the purpose is often to determine the quality of food in terms of freshness, taints, malodours etc.

Table 4. Selected reports of e-noses applied in the food industries. Full references may be found in reference [1].

Food	Test	Sensor no/type
Seafood	Freshness	1/MOS
Fish (cod, haddock)	Freshness	4/MOS
Food Flavours	Flavour Identification	8/BAW
Wheats	Grade quality	4×4/EC
Ground Pork/Beef	Discriminate and effect of ageing	15/mixed
Fish	Freshness	1/ MOS
Cheese and wheat	Discriminate and ageing	20/CP
Fish (trout)	Freshness	8/EC
Grains	Classification	15/mixed

There have also been many reports of electronic noses used to detect the headspace of beverages and some of these are listed in Table 5.

Table 5. Selected reports of e-noses applied in the drinks industries. Full references are to be found in reference [1].

Drink	Test	No. of sensors /type
Coffees	Discriminate *C. arabica* and *C. robusta*	6/MOS
Various beverages	Discriminate different types of alcoholic drinks	6/MOS
Whiskies	Discriminate Japanese whiskies	8/BAW
Coffees	Discriminate different blends and roasts	12/MOS
Beers	Discriminate between lager and ales	12/CP
Liquors	Discriminate between brandy, gin and whisky	5/CP
Water	Taints in drinking water	4/MOS
Beers	Diactyl taint in synthetic beer	12/CP
Coffees	Discriminate between varieties	12/CP
Wines	Varieties and vintages of same wine	4/MOS
Colas	Discriminate between diet and normal colas	6/MOS

Perhaps a better way to discuss the application of electronic noses is in terms of the difficulty of the problem being addressed. This usually relates to the nature of the analyte and has been summarized below [1]:

Table 6. Different levels of difficulty in detecting odours in different applications.

Measurand	Reference odour?	Odour stability?	Vapour pressure	Number of components	Level of difficulty
Simple odours, e.g. ethanol	Yes	Good	High	One	Low
Solvents in polymers, paints, plastics etc.	Yes	Good	High	Several	Low
Perfumes/essential oils	Some	OK	High	Several	Medium
Food: Coffee quality	No	Poor	Medium	100's	Medium
Explosive materials (plastic)	Yes	Good	Very Low	Several	High
Human odours	No	Poor	Low	100's	High

It is immediately apparent that the detection of a large number of different materials all with a low vapour pressure is very difficult – just the case for detecting explosives. This may be regarded, at first sight, as an issue of having a suitably low detection limit – note that this term is often referred to as *sensitivity*. Strictly speaking, the term sensitivity is a measure of the change in output for a given change in input and so is the slope of the instrument's output.

The problem is further compounded when the analyte must be detected in conditions where the background odours vary considerably and are perhaps 9 orders of magnitude higher in concentration! The issue is now one of *selectivity* or specificity in which it is possible to recognize one type of odour (e.g. fruity) in a myriad of other odours (e.g. headspace of a red wine). This problem is referred to as one of segmentation in the field of cognitive science and is solved remarkably well by the mammalian olfactory system.

3.2   Detection of explosives

The detection of explosive materials is a very challenging one and chapter 23 of reference [5] is devoted to this issue. Figure 7 shows the concentrations of some different explosives and their different molecular weights. Some explosives have vapour pressures in the ppm range, such as dinitrotoluene (DNT), but others have vapour pressures in the ppt level, such as plastic explosives like RDX. The problem is that the molecular weight of many explosives is high being over 150 amu or Daltons and these materials can barely be described as volatile!

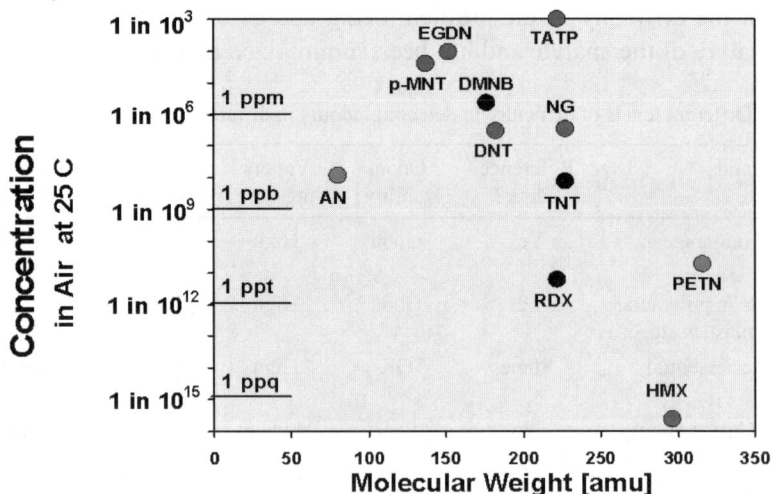

Figure 7. Typical concentrations in air of explosive compounds including di- and tri-nitrotoluene (DNT, TNT), nitroglycerine (NG) and plastic explosives like RDX with different atomic masses (data taken from US DOE study).

An important part of this application is to determine the precise location of explosive material. For example, millions of landmines have been laid in the past 50 years during many wars and conflicts. There is now the problem of detecting such mines after these wars in order to prevent further human suffering. These landmines have been laid in very different conditions and so the degree of selectivity required will vary from climate to climate. One can imagine that detecting landmines in a barren desert landscape (e.g. Gulf wars) is different from a rocky climate (e.g. Afghanistan war), and easier than jungle or swamp conditions (e.g. Korean and Vietnam).

4. STRATEGIES FOR ENHANCING THE PERFORMANCE OF ELECTRONIC NOSES

It was stated above that the detection threshold of odours for a typical sensor based electronic nose was around the ppm level and so the detection of very low concentrations (ppb or below) of an explosive is a considerable challenge. Nevertheless, this is not an unknown problem in the field of olfaction. Many of the key olfactory compounds of a product occur in its headspace at the ppb level or lower and so a number of strategies are emerging to enhance the performance of sensor-based electronic noses.

A discussion now follows that attempts to summarise the different ways in which electronic nose technology is moving towards this goal and some of the limitations of the methods chosen.

4.1 Selection of sensing material

The choice of sensing material in a sensor-based electronic nose is clearly of fundamental importance. The detection limit and specificity/selectivity of the sensing material S to the analyte A will depend upon the basic chemical reaction that takes place. In its simplest form the reaction can be defined as:

$$A + S \leftrightarrow AS \tag{1}$$

The equation here shows a basic reversible reaction because a sensor signal should return to its initial state after the analyte A is removed [6]. Irreversible reactions (such as catalysis) are also possible in which the analyte is consumed but the sensor signal should still return to its initial value.

The limit of detection of a chemical reaction is usually related to the strength of the binding constant. Weak types of interaction, such as hydrogen bonding, would be expected to lead to poorer detection limits. For example, polar odorant molecules, which are detected by conducting polymers, are generally attributed to hydrogen-bond type interactions and have sensitivities in the ppm level. In contrast a Lewis acid interaction, such as in a metallo-porphyrin, has a binding constant many orders of magnitude stronger and so can detect ppb or below of Lewis acid or base compounds.

Of course there are two downsides to this strategy. The first one is that the subsequent release of the analyte may be problematic. In other words the analyte may bind so strongly that the substrate is reluctant to release the analyte – thus there is a problem with reversibility of the sensors or, at the very least, a slow off-transient time. The other problem can be one of specificity. The larger the binding constant, then the greater the specificity to that specific analyte tends to be. So not only is a high detection limit desirable but also strong specificity. However, a high specificity can be an issue for electronic nose designers because the sensors will detect well one specific compound but will be blind to most of the other ones in a complex headspace. This may be regarded as sacrificing the bandwidth of an electronic nose for the sake of a high detection limit. Perhaps this can be expressed as a generic rule:

Fundamental rule: Bandwidth-selectivity product is a constant?

This is not a major issue when wanting to detect a known explosive but is an issue when wanting to detect a family of related explosive materials. So as a general strategy it has some value in this particular application:

Stratagem 1: Select a sensing material with the greatest binding constant.

The success of this strategy partly depends upon the nature of the target compound. If the explosive contains functional groups that are reactive (e.g. N-H), then there is considerable scope for employing existing, known substrates. However, if the explosive is a very inert substance, then it may require a synthetic chemist to design new sensing materials.

4.2   Principle of transducer

Equation (1) shows a basic reaction in which the target analyte A binds to the substrate material S. The progress of this reaction can be monitored in a wide variety of manners. The most common type of odour sensor is a resistive one, i.e. the bulk resistance (or conductance) of the substrate changes when the analyte is introduced. Examples of this are the tin oxide resistive sensor (Alpha MOS, France), the conducting polymer resistive sensor (Osmetech, UK), and the carbon black/polymer composite resistive sensor (Cyrano Sciences, USA). Yet there are many other types of sensing elements besides conductimetric, here are some of them [6]:

- Potentiometric - using catalytic gates on FETs (AppliedSensor)
- Amperometric - using selective coatings on electrochemical cells
- Piezoelectric - using stationary phase compounds on quartz crystal microbalances (AppliedSensor)
- Piezoelectric - using stationary phase compounds on SAW sensors (HKR)
- Resonant -   using coated microcantilever beams (IBM)

The sensitivity of the transducing element depends upon the effect of the analyte on the different properties of the substrate. For example, the partition coefficient of an analyte in to a polymer coating may be high, but if the molecular weight of the molecule is low then a gravimetric transducer (e.g. quartz crystal microbalance) is not ideal. Having said this, the potential to amplify at this stage (at reasonable cost) does vary from transducer to transducer. For example, resistance and frequency can be measured to high levels of precision using high-Q resonance circuits. So the strategy here is:

Stratagem 2: Select a transduction principle with the greatest amplification factor, i.e. gain.

The miniaturization of sensors is permitting the creation of very small nanomechanical sensors at IBM that could create very high sensitivities, e.g. the detection of a few molecules on the surface. Also, at the moment the use of an optical transducer looks promising. Optical sensors based on the physical vibration bands of molecules (e.g. infra-red spectrometers) are not particularly sensitive but combined with new materials the sensitivity can be impressive. For example, a recent report on colour-sensitive dyes based on Lewis acid/base reactions offers considerable promise [8]. Figure 8 shows an electronic nose based on a set of dyes that changes colour when the target analyte is introduced and this is detected using a CCD array chip. The unit has been shown to have ppb sensitivities to certain compounds.

16

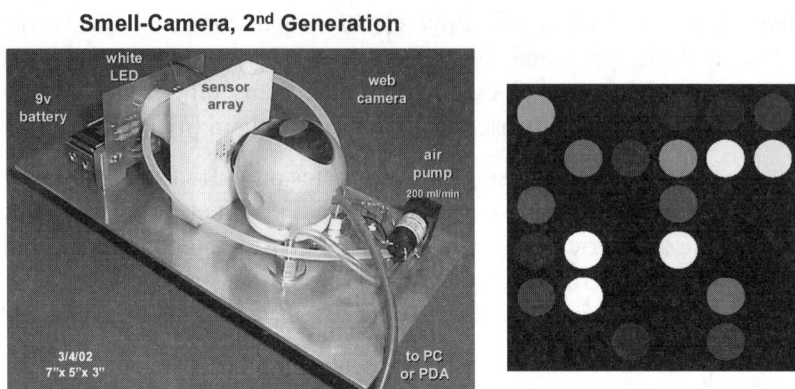

Smell-Camera, 2nd Generation

Figure 8. Smell camera from ChemSensing (USA) and a CCD colour image of 22 ppb of CH_3SH in air based on different substituted metalloporphyrins [8].

Another optical approach has been reported by Dickenson *et al.* [9] based on UV, rather than optical, absorption within the substrate material using optical fibres. Further details on this type of instrument to detect explosives can be found in reference [10].

An optical transducer is an attractive option in that selectivity can be enhanced through the use of optical filters to select specific wavelengths. When this is combined with a very specific optical sensing material, e.g. photo-luminescent quenching polymers, the possible limit of detection is extremely impressive (see chapter 23 of reference [5]) and [11].

4.3 Sensor arrays

In order to improve the detection limit of a sensor it is not only necessary to have a high gain but also to reduce the noise level and so provide a high signal to noise. Stratagem 2 is based on producing the highest gain but this is difficult when the sensor suffers from interfering signals, such as fluctuations in ambient temperature, humidity, pollution etc. Different hardware and software solutions have been proposed by e-nose designers to ameliorate these problems; see chapter 13 of reference [5].

One simple approach has been that of Cole *et al.* who propose the use of two sensors run in a *ratiometric* mode [12]. Figure 9 shows a CMOS chip that drives a pair of resistive odour sensors; one polymer sensor is active while the other is passivated. The output of the ratio of the resistances is amplified rather than the absolute resistance of just one sensor. This removes at the input stage of the amplifier any common mode signals, such as temperature drift.

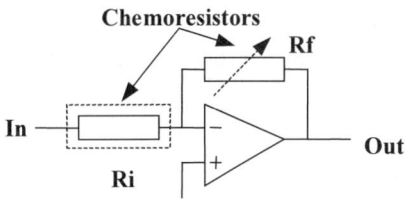

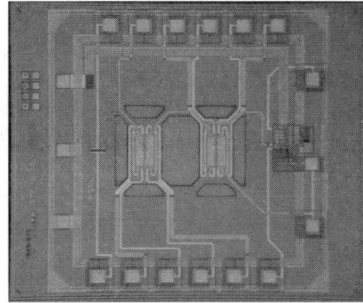

Figure 9. Principle of ratiometric CMOS odour sensor to enhance signal-to-noise ratio and photograph of realized CMOS chip [12].

Another approach is to construct a sensor chip with a large number of sensing elements. Figure 10 shows two examples of CMOS array chips for odour detection. The Warwick-Leicester-Edinburgh CMOS chip comprises an array of 5 different carbon black nanocomposite materials set in columns of 15 identical sensors making a total of 70 resistive sensing elements [13]. The Caltech CMOS chip consists of an array of 12 by 40 but in this case they are all of one type of polymer [14]. The large number of sensors can provide some level of redundancy in the system and so permit signal averaging to take place and thus enhance the signal to noise ratio. It can also be used for spatial as well as temporal signal processing to increase the discriminating power of the micro-nose.

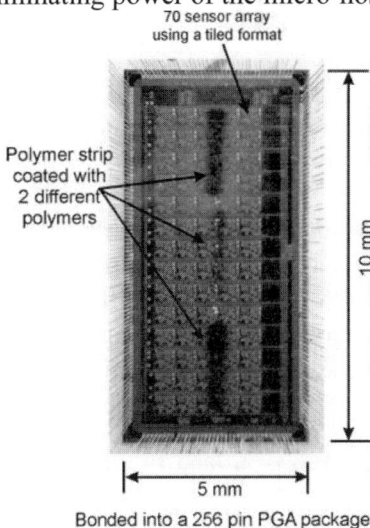

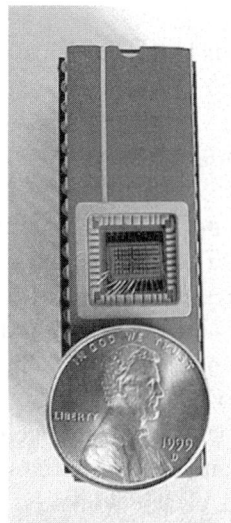

Figure 10. Array based CMOS odour chips: left - Warwick-Leicester-Edinburgh design [13] and right – smaller Caltech design [14].

The concept of large arrays is taken even further by Walt with the use of thousands of coated glass beads and then averaging the optical signals retrieved from them (see refs [9, 10] and chapter 6).

The root mean square amplitude of the signal noise should decrease according to the following equation:

$$Noise\ level \propto 1/\sqrt{n} \tag{2}$$

where n is the number of identical sensors in the array. Figure 11 illustrates an optical bead array for detecting odours and other chemicals [9].

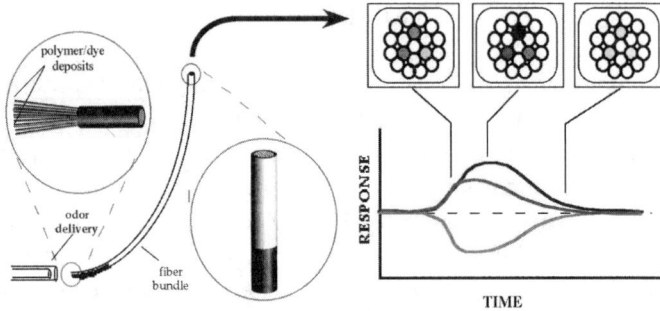

Figure 11. Large array of optical beads coating the end of an optical fibre with three primary tuning signals averaged to enhance detection limit.

This leads to the third possible stratagem:

Stratagem 3: Employ a large number of identical sensors to enhance the signal-to-noise level.

Employing a large number of identical sensors clearly has some drawbacks. The obvious ones are the increased cost of manufacturing large sensor arrays and then coating them with a number of different sensing materials. There is also the added problem of analysing the patterns in which the dimensionality is now high (known as the curse of dimensionality in pattern recognition). Yet the human biological system possesses some 50 million olfactory receptor cells and a large degree of convergence to the glomeruli and so this strategy, when combined with a suitable neural architecture, clearly works for nature.

4.4   Pattern recognition algorithms

The final stage in an electronic nose is the signal or pattern analyser. Readers are directed to reference [15] for an introduction to pattern recognition.

The various stages in the signal processing of an electronic nose system are illustrated in Figure 11 in which the signals generated from an array of sensors (or, for example, pseudo-sensors in a mass spectrometer) are first pre-processed and then fed in to a pattern analyser. In other words the n dimension vector in sensor space is transformed into feature space (a process called *feature extraction*) and then identified using some form of pattern classifier.

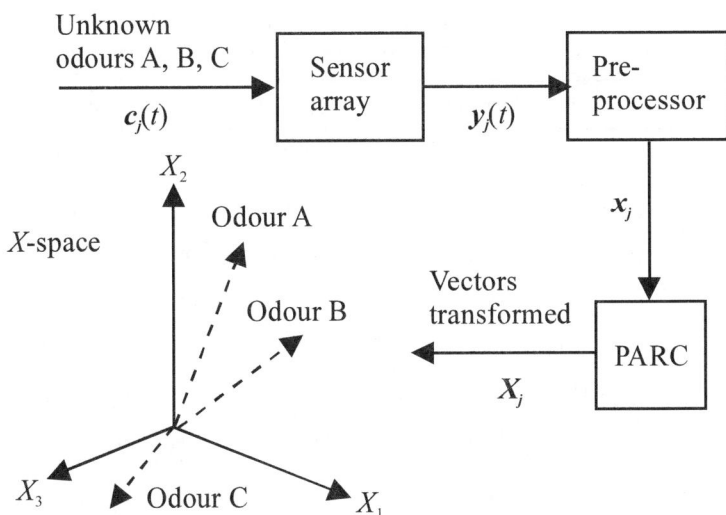

Figure 12. Overview of data processing within a sensor-based electronic nose.

Electronic nose designers have employed many different sensor pre-processing algorithms. The choice of pre-processing algorithms is large and could be:

- Absolute signal
- Differential signal
- Relative signal
- Log signal
- Normalised signal
- ....

Table 7 gives some of the different pre-processing algorithms recommended in various publications.

Pre-processing of sensor signals is an important procedure because it can help reduce the dimensionality of the problem to solve, reduce drift/noise and hence improve the pattern recognition success rate. However, the best pre-processing algorithm depends on the choice of sensor and the nature of the problem. So strategy 4 relates to sensor conditioning or pre-processing and expressed simply is:

Stratagem 4: Determine the optimal pre-processing algorithm for the sensor array and problem domain.

Table 7. Some different pre-processing algorithms used to generate static parameters x_{ij}; frequency (f), resistance (R), conductance (G), signal (s), baseline (0), sensor (i), odour (j).

Method	Generalised formulae	Sensor type	Specific formulae
Difference	$x_{ij} = (y_s - y_0)$	SAW	$x_{ij} = (f_s - f_0)$
Difference	$x_{ij} = (y_s - y_0)$	BAW	$x_{ij} = (f_s - f_0)$
Difference	$x_{ij} = (y_s - y_0)$	Metal oxide resistor	$x_{ij} = (R_s - R_0)$
Difference	$x_{ij} = (y_s - y_0)$	Metal oxide resistor	$x_{ij} = (G_s - G_0)$
Relative	$x_{ij} = y_s/y_0$	Metal oxide resistor	$x_{ij} = R_s/R_0$
Relative	$x_{ij} = y_s/y_0$	Polymer resistor	$x_{ij} = R_s/R_0$
Fractional change	$x_{ij} = (y_s - y_0)/y_0$	Polymer resistor	$x_{ij} = (R_s - R_0)/R_0$
Fractional change	$x_{ij} = (y_s - y_0)/y_0$	Metal oxide resistor	$x_{ij} = (G_s - G_0)/G_0$
Log parameter	$x_{ij} = \ln(y_s/y_0)$	Metal oxide resistor	$x_{ij} = \ln(R_s/R_0)$

There is a large number of pattern recognition techniques that have been applied to e-nose technology and reported. Some of these techniques are quantitative and others qualitative as illustrated in Figure 13.

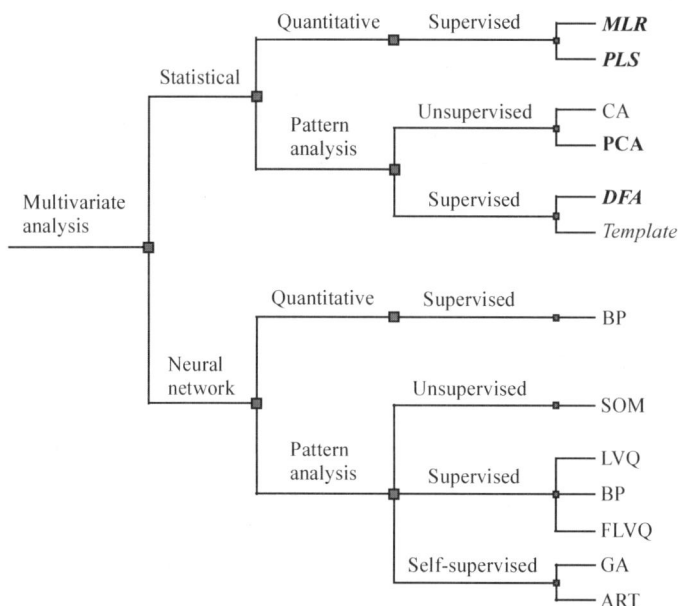

Figure 13. Multivariate data processing techniques employed by electronic noses. See reference [2] for definition of acronyms.

Many of these pattern analysis techniques have been applied in the field of electronic noses and Table 8, below, summarises some of them. Further reviews of the different techniques can be found in references [1], [4] and [5]. The performance of these static pattern analyzers again depends upon the nature of the sensors employed and the problem to be solved. The olfactory problem may be a linear one which means that the principle of linear addition applies and so simple linear pattern analyzers can be used as illustrated below:

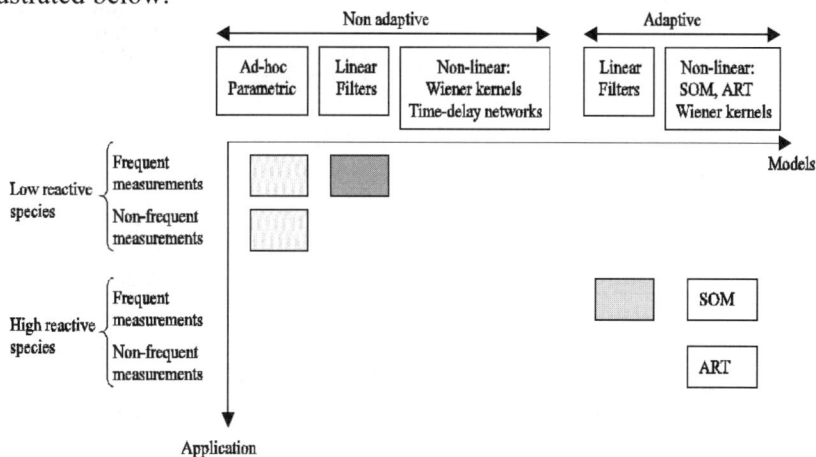

Figure 14. Selection chart for linear PARC methods [16]. Many explosive materials relate to the quadrant for low reactive species.

Table 8. Review of different pattern recognition techniques that have been applied to electronic nose data.

PARC method	Linear	Supervised learning	Number and type of sensors	Target odours
Statistical:				
PCA	Yes	No	8 MOS	VOCs
			12 MOS	Alcohols
			4 CP resistors	Spirits
			8 EC	Fish
			18 mixed	Paper type
			1 BAW (transient)	Wines
CA (Euclidean)	Yes	No	8 MOS	Notes
			12 MOS	Alcohols
			8 MOS	Coffees
			8 MOS	Whiskies
			2-18 EC	Grain quality
			CP resistors	Pig slurry
DFA (Linear)	Yes	Yes	8 MOS	Coffee
			12 MOS	Coffees blends
DFA (Quadratic)	No	Yes	12 MOS	Alcohols/coffees
Template matching	Yes	Yes	18 CP resistors	Lager beer taints
Neural networks:				
Hamming net	No	No	8 MOS	Alcohols/alkanes
SOM	No	No	12 MOS	Coffees
			2 MOS	Simulation
			6 MOS	Bacteria age
BP	No	Yes	6 QCM	Beverages
			6 QCM	Whiskies
			12 MOS	Alcohols
			6 QCM	Various notes Perfumes and flavours
			Mixed	Pork/beef meats
			18 CP resistors	Beer taints
			4 MOS	Bacteria
			1 MOS (transient)	Wines
FLVQ/Fuzzy	No	Yes	8 MOS	Whiskies
			3 MOS	Notes
GA	No	Yes	12 MOS	Alcohols/coffees
ART	No	Self	12 MOS	Alcohols/coffees

More often the problem is non-linear and there is a choice of non-linear methods as illustrated in Figure 15. The most commonly used approach to solve a non-linear problem, such as the ExOR one, is an artificial neural network.

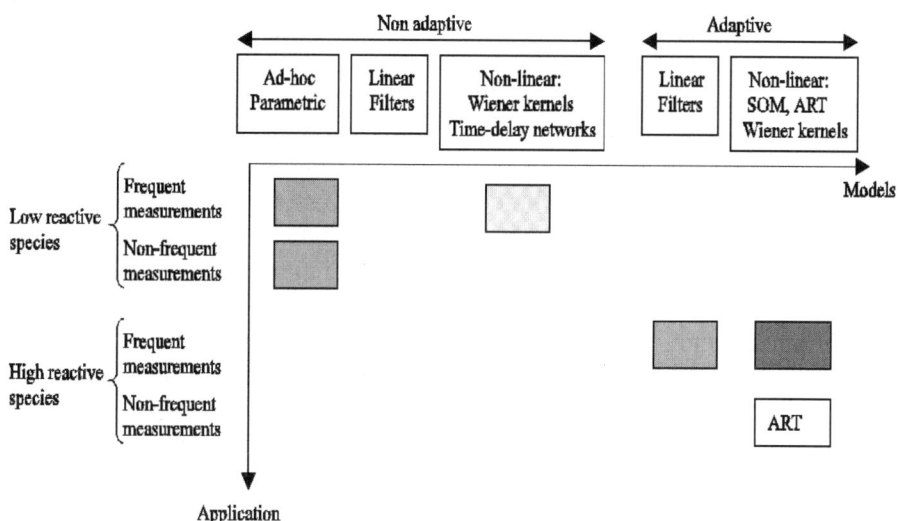

Figure 15. Selection chart for non-linear problems [16]. Many explosive materials relate to the quadrant for low reactive species

So the basic strategy proposed here is:

Stratagem 5: Determine whether the problem is linearly separable and employ the algorithm most appropriate to the nature of the problem.

A good electronic nose designer will be able to investigate a particular problem, and thus determine the optimal combination of pre-processing algorithm (e.g. relative) and pattern recognition (e.g. multilayer perceptron neural network). Nevertheless, a basic rule may be applied here and that is:

Fundamental rule: The choice of pattern classifier will at best increase the detection limit by a factor of 10?

4.5 Headspace pre-concentrators

Perhaps one of the most promising approaches to enhancing the performance of an electronic nose is based in the choice of sampling system. The concentration of an odour in a dynamic headspace will always be much lower than in the static headspace. Consequently, the use of a headspace pre-concentrator when trying to detect very low concentrations of a particular odour is often a logical decision. There are a variety of methods that can be used to enrich the headspace and some of these are now commonly used. The four main techniques are:

- Purge and trap (needs a vacuum system)
- Carbon absorber (e.g. Tenex)
- Solid phase micro extraction (SPME)
- Selective membrane

The different sampling techniques are summarized in Table 9 together with an estimate of their detection limit, precision, cost, time to use and simplicity.

Table 9. Some common sampling techniques and their characteristics. Adapted from [17].

Method	Detection limit (MS)	Precision (% RSD)	Expense	Time (min)	Simple
Saturated headspace	ppm	-	Low	30	Yes
Purge & Trap	ppb	1-30	High	30	No
SPME	ppb-ppt	<1-12	Low	2-30	Yes
Stripping	ppt	3-20	High	120	No
Liquid-liquid extraction	ppt	5-50	High	60	Yes
Solid phase extraction	ppt	7-15	Medium	30	Yes

Interested readers are referred to a recent conference on the topic of headspace sampling sponsored by an EU network on electronic noses (NOSE) and held in France [17], and a chapter here on micromachined pre-concentrators [18].

Headspace enrichment is a sensible strategy to adopt when designing a high sensitivity electronic nose and so the sixth stratagem is:

Stratagem 6: Employ a pre-concentrator as this can increase the detection limit by a factor of 10-100.

There is, as always, a down-side to most techniques and that is also true for pre-concentrators. Most methods involve the flowing of air containing the odour through the pre-concentrator for a fixed period of time after which the pre-concentrator is discharged (usually by heating) and the odour detected. This means that the sampling rate is much slower for an electronic nose that contains a pre-concentrator than one without. SPME is one of the fastest methods and sampling still needs to be for at least tens of seconds.

There is considerable work being carried out to make micro pre-concentrators and micro noses (e.g. see recent review from Sandia National Laboratories [19]), but this is still fundamentally a research program and micro nose systems are expensive to fabricate.

5.   CONCLUSIONS

Electronic nose technology is generally based upon an array of chemical sensors with partial specificity to both simple and complex odours. Its potential application to detecting low concentrations of explosive materials has been explored here. The inherent detection limit of sensor-based electronic noses and other instruments is depicted in Figure 16. The use of a pre-concentrator enhances the selectivity of a static headspace sensor array down from the ppm level to around the ppb level and thus approaching the limit of mass spectrometers. Various strategies have been discussed here that can be adopted to maximize the detection limit of an electronic nose to a specific molecule, such as an explosive material. However, even when all of the individual stages of an electronic nose have been optimized, it is difficult to see how the detection level can be reduced by much more than about 3 orders of magnitude, i.e. down to the parts per trillion (10^{12}) level. Consequently, the ion mobility spectrometer currently used is likely to remain the 'gold standard' for explosives detection for some years to come – not only does it have a ppt sensitivity but it responds in a fraction of a second. These have been manufactured by several companies, including Smiths Defence, UK.

The other technique that could challenge this method is that based on fluorescent polymers which are extremely specific to the target analyte and so blind to other compounds [20]. This type of instrument is therefore attractive when wanting to detect a known explosive but will be blind to any variations.

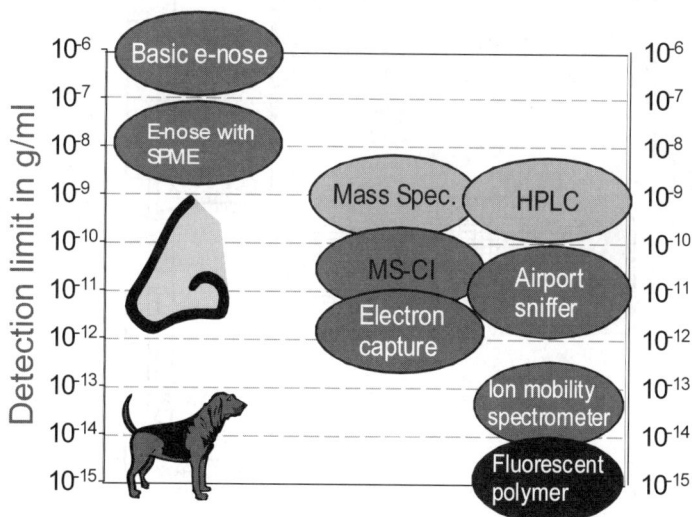

Figure 16. Review of detection limits of different instruments. The human nose typically detects odorant molecules in the ppm to ppt level, whereas the canine nose can work down to the PPQ level in which the snuffling of dust particles may aid this process.

26

One possible role of existing e-nose technology could be to scan a material to see if it contains a range of different classes of explosive materials rather than a specific compound. The increased breadth of the chemical scan will result in a poorer detection limit. Nevertheless it could still be regarded as a complementary tool to ion mobility spectroscopy when dealing with new and unknown types of explosives in the battle against terrorism.

Finally, electronic nose designers are currently developing a range of e-nose systems based upon spatio-temporal signal processing similar to that employed within the human olfactory system. A possible architecture for a neuromorphic spatio-temporal chip is illustrated below in Figure 17, which mimics the signal processing within the human olfactory system.

This type of neuromorphic chip, combined with sensors containing olfactory binding proteins, may be some way off but does hold out some promise for realizing electronic noses that start to approach the capability of the human and canine olfactory systems. The practical issues would then be ones of low cost manufacture, and long-term reliability. Olfactory receptor proteins provide high binding constants and so enhanced detection limits, however they only survive for about 15 days with nutrients [5] and so the continual replacement of biological sensors would be an essential feature of a biological micro-nose.

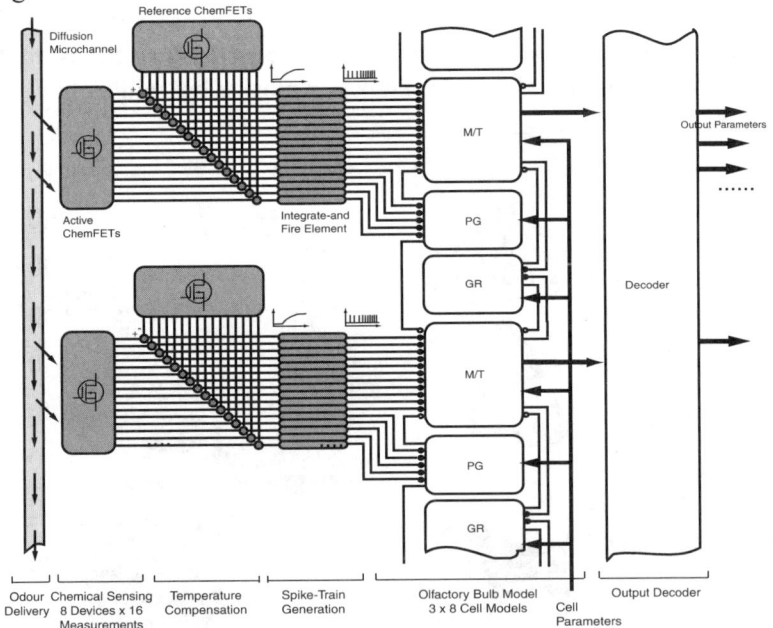

Figure 17. Concept of a spatio-temporal neuromorphic electronic nose chip [21]. The sensor stage has been fabricated [13] and the subsequent analogue VLSI implementation of the integrate-and-fire elements and neurons reported [21].

ACKNOWLEDGEMENTS

The author acknowledges here the contributions of many of his colleagues, and research students. In particular, he acknowledges Professor Philip Bartlett (Southampton University, UK) for the source of some of the material taken from reference [1]. The work has also been financially supported by the Engineering and Physical Sciences Research Council (UK), The Royal Academy of Engineering, and industry. The author also thanks the NATO Science Program for the financial support to host a NATO ARW on the topic of Electronic Noses/Sensors for Detection of Explosives near Warwick University, Coventry.

REFERENCES

1. J.W. Gardner and P.N. Bartlett, Electronic Noses: Principles and Applications, Oxford University Press, Oxford (1999) 245pp.
2. G. Ohloff, Scents and Fragrances, Springer-Verlag, Berlin (1990).
3. J.W. Gardner and P.N. Bartlett, A brief history of electronic noses, Sens. Actuators B 18 (1994) 211-220.
4. J.W. Gardner and P.N. Bartlett (eds), Sensors and Sensory Systems for an Electronic Nose, Kluwer Academic Publishers, NATO ASI series, Dordrecht (1991) 327pp.
5. T. Pearce, S. Schiffman, T. Nagle, J.W. Gardner (eds), Handbook of Machine Olfaction, Wiley-VCH: Weinheim, (2003), 592pp.
6. J.W. Gardner, Microsensors, Wiley, Chicester (1994) 331pp.
7. E.J. Staples, Detecting chemical vapours from explosives using the zNose, in Electronic Noses and Sensors for the Detection of Explosives, (eds JW Gardner and J Yinon), NATO ASI Series, Kluwer Academic Publishers, Dordrecht, 2004.
8. N.A. Rakow and K.S. Suslick, Colourimetric sensor array for odor visualisation, Nature 406 (2000) 710-714.
9. T.A. Dickenson, K, Michael, J.S. Kauer and D.R. Walt, Convergent self-encoded bead sensor arrays in the design of an artificial nose, Anal. Chem. 71 (1999) 2192-2198.
10. D. Walt, Optical sensor microarrays for explosives detection, in Electronic Noses and Sensors for the Detection of Explosives, (eds JW Gardner and J Yinon), NATO ASI Series, Kluwer Academic Publishers, Dordrecht, 2004.
11. C. Cumming, Amplifying fluorescent polymer arrays for chemical explosives detection, in Electronic Noses and Sensors for the Detection of Explosives, (eds JW Gardner and J Yinon), NATO ASI Series, Kluwer Academic Publishers, Dordrecht, 2004.
12. J. Garcia-Guzman, N. Ulivieri, M. Cole, and J.W. Gardner, Design and simulation of a smart ratiometric ASIC chip for VOC monitoring, Sensors and Actuators B, 95 (2003) 232-243.
13. J.A. Covington, S.L. Tan, A. Hamilton, T. Koickal, T.C. Pearce and J.W. Gardner, Combined smart chemoresistive/FET sensor array, Proc. IEEE Sensors 2003 Conference, Canada, 22-24 October 2003.
14. N. Lewis, Proc. IEEE Sensors 2002 Conference, Orlando, 12-14 June 2002.
15. A.R. Webb, Statistical Pattern Recognition, John Wiley & Sons Ltd, Chicester, (2002) 514pp.
16. E.L. Hines, E. Llobet and J.W. Gardner, Electronic noses: a review of signal processing techniques, Proc. IEE: Circuits, Systems and Devices 146 (1999) 297-310.

28

17. NOSE II Short Course on Sampling Techniques, France, 2003; K. Persaud, Solid phase microextraction methods for rapid pre-concentration and measurement of trace volatiles, in Electronic Noses and Sensors for the Detection of Explosives, (eds JW Gardner and J Yinon), NATO ASI Series, Kluwer Academic Publishers, Dordrecht, 2004.

18. A. McGill, A micromachined pre-concentrator for enhanced detection of explosives, in Electronic Noses and Sensors for the Detection of Explosives, (eds JW Gardner and J Yinon), NATO ASI Series, Kluwer Academic Publishers, Dordrecht, 2004.

19. S. Casalnuovo, Sensors at the interface: chemical and biological microsensors, Proc. IEEE Sensors Conference, Toronto, 21-24 October 2003.

20. W. Trogler, Luminescent inorganic polymer sensors for vapour phase and aqueous detection of TNT and other nitroaromatics, in Electronic Noses and Sensors for the Detection of Explosives, (eds JW Gardner and J Yinon), NATO ASI Series, Kluwer Academic Publishers, Dordrecht, 2004.

21. T.C. Pearce, T.J. Koickal, C. Fulvi-Mari, J.A. Covington, F.S. Tan, J.W. Gardner and A. Hamilton, Silicon-based neuromorphic olfactory pathway implementation, Proc. Brain Inspired Cognitive Systems, Stirling, UK, 29 Aug - 1 Sept, 2004.

Chapter 2

POLYMER ELECTRONICS FOR EXPLOSIVES DETECTION

Timothy M. Swager

Department of Chemistry and Institute for Soldier Nanotechnology, Massachusetts Institute of Technology, Cambridge, Massachusetts, USA

Abstract: The amplifying ability of semiconductive organic polymers in sensory schemes is described and its use for the detection of nitroaromatic explosives is reviewed. Semiconductive organic polymers serve as extremely efficient conduits for the transport of optically induced excitations and it is this transport property that allows for the high sensitivity of these materials to trinitrotoluene (TNT) and dinitrotoluene (DNT), the primary explosives used in landmines. Systematic molecular designs for the formation of improved sensitivity sensory materials are described.

Key words: Semiconductor, organic, fluorescence, excitons, sensors, explosives, TNT.

1.   INTRODUCTION

Research efforts to date for the detection of explosives and ordnances thereof have been substantial. Nevertheless a comprehensive solution has not been developed and current technologies are generally specific to a field of operation. Landmine detection ranks as perhaps the greatest detection challenge as a result of a highly variable environment presenting interfering signals and fluctuations in temperature heat, light, moisture, soil, and widely varying leak rates for various types of landmines. These factors have thwarted landmine detection technologies which rely on indirect physical signatures that lead to spectroscopic anomalies that are imaged (detected) by

J.W. Gardner and J.Yinon (eds.),
Electronic Noses & Sensors for the Detection of Explosives, 29-37.
© 2004 *Kluwer Academic Publishers. Printed in the Netherlands.*

electromagnetic radiation. Detection of explosive ordnances has been most often accomplished indirectly with metal detectors; however metal detectors suffer from lack of specificity and produce many false alarms even in uninhabited areas. Nuclear Quadrupole Resonance (NQR) spectroscopy provides for an explosive-specific signature due to the unique magnetic resonance of nitrogen nuclei in explosives. This method has recently emerged as a powerful method to detect bulk explosives, such as those found in landmines containing RDX and PETN [1]. Unfortunately, the nitrogen nuclei in TNT, the principal explosive in most landmines, have different spectroscopic characteristics that limit NQR detection. The NQR method, while having virtues for detection of bulk explosives, also has large power and electronics requirements that make it expensive, and thereby complicate the fielding of a portable (non-tethered) hand-held instrument.

The detection of trace explosives by vapour sampling represents the most versatile and comprehensive approach for pinpointing explosive sources and ordnances containing them. Indeed trained canines represent the best overall method for explosive detection and serve as the proof that indeed there is a reliable and specific vapour signature. This broad utilization of canines establishes that a direct vapour sampling method (sniffer) for the detection of explosives has the potential to be a comprehensive solution to both military and civilian needs. Indeed, security at airports and other installations would benefit greatly from a vapour sensor with performance similar or superior to that of canines. Commercial airport explosive detection systems typically require the collection of particles weighing multiple picograms to give a positive identification. Particle collection is not optimal as they can be removed by careful cleaning and do not necessarily map directly to a source of explosive. Put simply, the presence of explosive particles does not necessarily indicate a bulk source but rather indicates that an object has come in contact with explosives. For landmines that have been in the ground for an extended time, the soil around the landmine acts as a temporary repository and a natural pre-concentrator for explosive molecules. Collecting the concentrated explosives presents challenges. Liquid phase extraction of explosive molecules is time consuming and involves significant dilution, eliminating the benefits. These considerations unquestionably support my assertion that an explosives detection method based on vapour sampling with sufficient selectivity and sensitivity will be superior to conventional technologies in both military and civilian arenas.

As a result of their low vapour pressures, explosives detection in real-time using sniffer technologies presents significant detection challenges. A breakthrough that has enabled a paradigm shift toward explosive vapour detection came in 1998 when our group developed a sensory material with extraordinary sensitivity to TNT and DNT (trinitrotoluene and dinitrotoluene, respectively) [2]. The enabling feature in this technology is the use of organic semiconductive polymers that have the ability to self-amplify their response to the binding of TNT and DNT. These 'molecular wire amplification' methods [3] make use of the efficient transport of electronic excited states that allows for one molecule of TNT or DNT to extinguish the fluorescence from more than 10^4 chromophore (monomer) units. The extraordinary sensitivity provided by this amplification has now been validated by more groups and can be applied to a number of chemical and biological detection problems of national security as well as those related to the environment and health care.

2.   AMPLIFYING SEMICONDUCTIVE POLYMERS

Organic semiconducting polymers with extended electronic states are excellent conduits for the transport of excitons. This ability to transport optical excitations over large distances is the origin of the gain (amplification) in semiconducting fluorescence-based sensors. These polymers can be generally described as wide-band-gap (large E_g) semiconductors, where the molecular orbitals of the repeating units combine to form a continuum of orbitals (energy band). A schematic representation of this idea, and how it can be used to create amplification, is shown in Figure 1. A photon of energy $h\nu$ (must be larger than E_g) is absorbed by the polymer. This promotes an electron from occupied orbitals in the valence band to the unoccupied orbitals in the conduction band. The now high-energy electron is bound to the empty state (hole) in the valence band (shown as a '+') to produce what is referred to as an exciton. The molecular framework relaxes around the exciton and hence it is often convenient to consider it as an object (quasi-particle) that can move throughout the polymer. The sensory response occurs when the exciton interacts with an occupied receptor, in a manner like that observed for single-chromophore sensor schemes. SA results from the rapid migration of excitons through the material and the effective 'sampling' of many different receptor sites. This process increases the probability of finding an occupied binding site, and the sensor can be thought of as having an 'effective binding constant' that is equal to the product of individual binding constant for individual receptors and the number of sites the exciton visits [4].

The SA sensory polymer concept was first demonstrated in an analyte-induced quenching assay (Figure 1a) where the ligand to the receptor site was a quencher [4]. A key feature of the polymeric system was that the exciton migration caused greatly enhanced quenching relative to a chromophore bearing only one of the receptors. This is established by measuring what is referred to as a static Stern-Volmer constant for both the polyreceptor and the single-receptor systems. It was found that the exciton could sample about 130 of the individual phenylene ethyne units during its *one-dimensional random walk* along the polymer backbone. A related scheme involves the creation of a low-energy trap on ligand binding, arising either from a perturbation of the polymer's band structure as shown in Figure 1b [5] or from the positioning of another chromophore close to the polymer backbone [6]. These two mechanisms will both produce a shift in the wavelength of the polymer fluorescence to produce a new emission band.

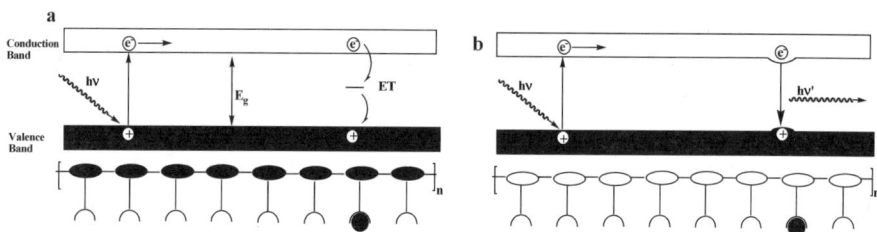

Figure 1. Schematic representation of energy migration in a conjugated polymer and how it produces amplification. The horizontal axis represents the distance along the polymer backbone and the semicircles and red balls represent receptors and ligands, respectively. (a) Excitons migrate along the polymer backbone and are quenched when they encounter a receptor that has complexed an acceptor ligand. (b) Excitons selectively recombine at trapping sites created by the occupation of a receptor with a non-quenching ligand, which produces a local minimum in the band gap.

3.   POLYMER STRUCTURES

To produce thin films of organic semiconductors with high sensitivity, stability, and reversibility we developed polymers shown in Figure 2 having rigid three-dimensional structures [7]. This rigid three dimensional structure prevents strong interchain electronic interactions that can increase the reactivity with electrophiles (neighbouring group participation) and also prevent reductions in the fluorescence intensity that often accompanies these processes.

Other polymers have been designed to exhibit improved recognition of TNT and DNT over potential interferents. A very successful approach is to work with polycyclic aromatics that have triphenylene residues. These structures present three fold symmetry sites that are complementary to the positive charge on DNT and TNT. This effect can be seen in Figure 3 where electrostatic potentials have been calculated for the fragments of two different polymers. Consistent with the expected improved binding of TNT to the polymer, we find that triphenylene polymer shown left displays greater sensitivity to TNT.

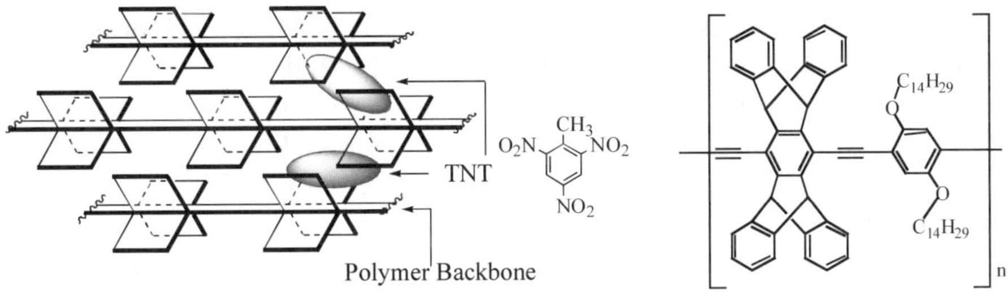

Figure 2. Highly stable sensor polymer designed to bind and detect TNT and DNT. The rigid three dimensional structure provides for interchain spacing that prevents fluorescence quenching and allows for the formation of galleries capable of binding nitroaromatics.

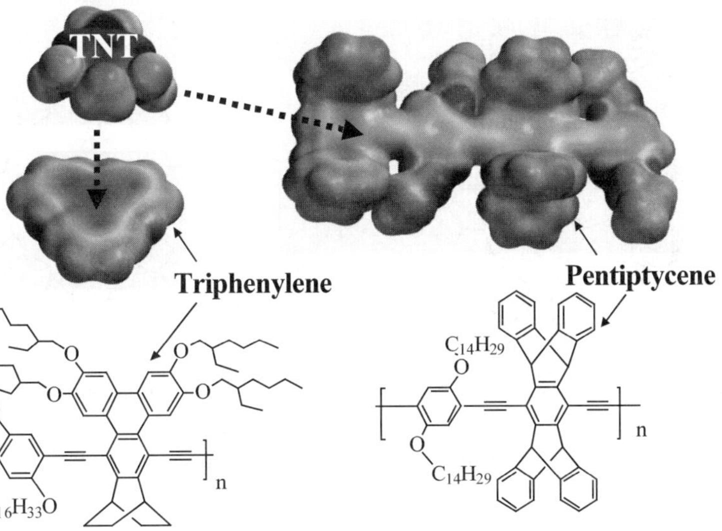

Figure 3. Surfaces of TNT and polymer structures, illustrating the shape and electrostatic complementarity between triphenylene and TNT. The lighter regions represent negative charge and the darker regions represent positive charge.

34

4.   ENERGY MIGRATION: MECHANISMS, DIMENSIONALITY, AND SUPERSTRUCTURE

There are two broad classes of mechanisms for the transfer of energy in these materials. The first is the Förster mechanism as wherein the polymer is thought of as a collection of chromophores. The high electronic coupling within semiconducting polymers also affords a second mechanism known as Dexter energy transfer. This mechanism requires direct wave function overlap. Detailed studies from our group demonstrate that the Dexter mechanism is dominant for energy transfer within our phenylene ethynylene SA polymers [8]. As a result, the semiconductive nature of the polymer is absolutely critical for achieving high degrees of amplification. A correlation between polymer lifetime and the diffusion length of excitons [8] has indicated new approaches to the enhancement of the amplification displayed by these polymers. This fact was revealed by the triphenylene polymers and creates a clear path toward increasing amplification by engineering the lifetimes of the polymers. These results led us to design polymers (Figure 4) that have long-lived excited states and enhanced sensitivity.

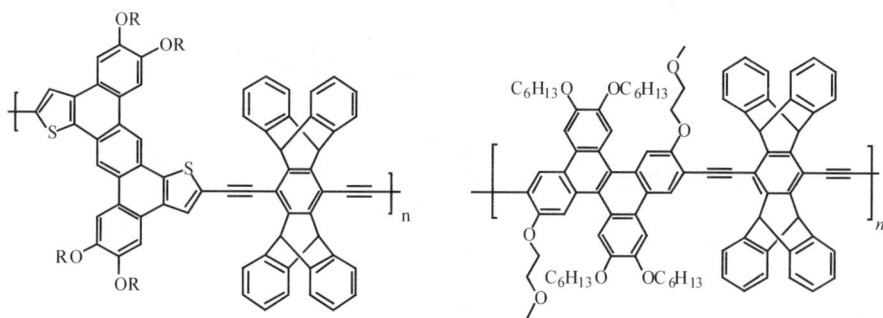

Figure 4. Polymers designed to have extended excited state lifetimes.

The first demonstrations of the utility of conjugated polymers in ultra-sensitive fluorescent chemosensors were performed in solution [4]. However, the amplification in solution studies represents the extreme lower limit for sensitivity enhancement that can be displayed by conjugated polymers because exciton migration through the dissolved polymer takes place in the form of a random walk (Figure 5). The isolated nature of the SA polymer chain in solution means that the random walk is restricted to one dimension and the excitons retrace the same segments of the polymer backbone multiple times. This inefficiency restricts the total number of receptors that an exciton can encounter during its lifetime. Sampling more of the polymer backbone increases the probability that an exciton will encounter an occupied receptor (or indicator element) and thereby provides greater amplification [9]. Excitons can sample 130 repeating units in one dimension and, assuming a random-walk mechanism (considering each hop to be one repeat unit), the exciton was required to make 130^2 hops. In solution, this is equivalent to an efficiency of less than 1%. As illustrated in Figure 5, most of the hops that the exciton makes in a random walk in thin films or particles of polymers are productive (the exciton samples new sites continually) and hence the amplification factor in a simple (un-optimized) one-component film is closer to 17,000 times that of a single, isolated chromophore-receptor combination. With improved electronic structures amplification factors exceeding 10^5 should be possible.

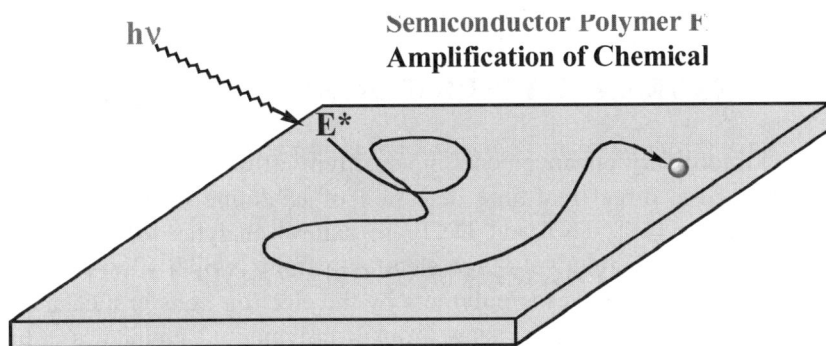

Figure 5. Schematic representation of a two-dimensional random walk of an exciton (E*). The amplification is directly proportional to the path length over which the exciton diffuses.

We continue to develop new polymers that have optimal transport properties. Our designs have focused on avoiding interchain interactions (3-D electronic interactions) with lead to self-quenching of fluorescence and inherent compromises in the mobility of charge and energy throughout the medium. This has restricted the dimensionality of the electronic delocalization and hence the performance of materials. To overcome this we have assembled these polymers into chiral 3D grids (Figure 6) [10]. We have further shown that these chiral grid polymer assemblies provide superior sensory properties for the detection of the explosive TNT [10].

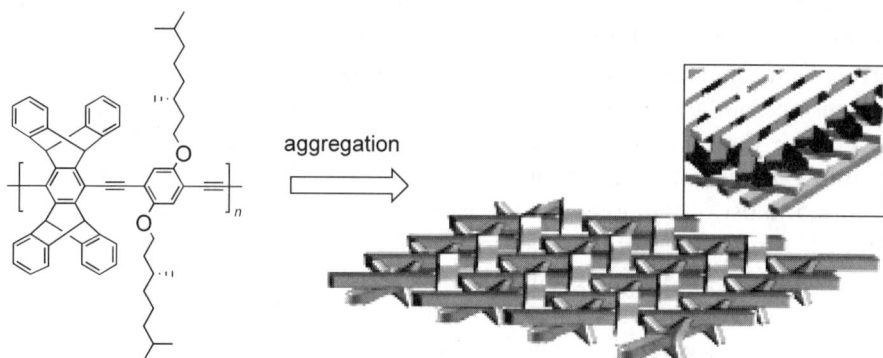

Figure 6. Aggregation of this chiral polymer creates grids that have 3-dimensional electronic structures and improved exciton and carrier mobility.

5. SENSORS AND FIELD TESTS

The extraordinary enhancements in sensitivity afforded by SA polymers have enabled the direct real-time detection of landmine using a chemical vapour sampling [11]. TNT and DNT are natural analytes for conjugated polymers due to their relatively high electron affinity, which allows them to quench the fluorescence of the polymers by the electron transfer mechanism shown in Figure 1a. Using our polymers, Nomadics Inc. (www.Nomadics.com), has built a series of completely portable instruments with femtogram detection limits of TNT, a detection limit that greatly exceeds other detection technologies. It is important to note that the TNT and DNT only weakly bind to these polymers and the response is rapid and reversible. Extension of these methods to the detection of other explosives including RDX and PETN is the subject of ongoing research.

ACKNOWLEDGEMENTS

TMS is grateful for the efforts of his many excellent coworkers. This research was sponsored by the Defense Advanced Project Agency, the Office of Naval Research, and the Army Research Office through the Tunable Optical Polymers MURI.

REFERENCES

1 A.N. Garroway, M.L. Buess, J.B. Miller, B.H. Suits, A.D. Hibbs, G.A. Barrall, R. Matthews and L. Burnett, J. IEEE Trans. on Geoscience and Remote Sensing 39 (2001) 1108-1118.
2 J.-S. Yang and T.M. Swager, Porous shape persistent fluorescent polymer films: an approach to TNT sensory materials, J. Am. Chem. Soc. 120 (1998) 5321-5322.
3 For a reviews see: (a) T.M. Swager, The molecular wire approach to sensory signal amplification, Accts. Chem. Res. 31 (1998) 201-7. (b) D.T. McQuade, A.E. Pullen and T.M. Swager, Conjugated polymer sensory materials, Chem. Rev. 100 (2000) 2537-2574.
4 (a) Q. Zhou and T.M. Swager, J. Am. Chem. Soc. 117 (1995) 7017-8. (b) Q. Zhou and T.M. Swager, J. Am. Chem. Soc. 117 (1995) 12593-12602.
5 T.-H. Kim, T.M. Swager, Angew. Chem. Int. Ed. 42 (2003) 4803-4806.
6 D.T. McQuade, A.H. Hegedus and T.M. Swager, J. Am. Chem Soc. 122 (2000) 12389-90.
7 J.-S. Yang and T.M. Swager, J. Am. Chem. Soc. 120 (1998) 11864.
8 A. Rose, C.G. Lugmair and T.M. Swager, J. Am. Chem. Soc. 123 (2001) 11298-11299.
9 I.A. Levitsky, J. Kim and T.M. Swager, J. Am. Chem. Soc. 121 (1999) 1466-1472.
10 S. Zahn and T.M. Swager, Three dimensional electronic delocalization in chiral conjugated polymers, Angew. Chem. Int. Ed. Engl. 41 (2002) 4225-423.
11 J.C. Cumming, C. Aker, M. Fisher, M. Fox, M.J. la Grone, D. Reust, M.G. Rockley, T.M. Swager, E. Towers and V. Williams, IEEE Trans. on Geoscience and Remote Sensing 39 (2001) 1119-1128.

Chapter 3

LUMINESCENT INORGANIC POLYMER SENSORS FOR VAPOUR PHASE AND AQUEOUS DETECTION OF TNT

William C. Trogler

Department of Chemistry, University of California at San Diego, La Jolla, California, USA

Abstract: Photoluminescent conjugated organic polymers have been applied to the detection of nitroaromatic explosives, such as TNT. The low energy unoccupied π^* orbitals in nitroaromatics can accept an electron from the excited state of luminescent polymers. This electron transfer quenching of luminescence provides detection limits as low as the parts per trillion (ppt) range. Photoluminescent organometallic polymetalloles and metallole copolymers have been synthesized and may also be used for the detection of nitroaromatic explosives such as picric acid, trinitrotoluene (TNT), 2,4-dinitrotoluene (DNT), and nitrobenzene. These soluble polymers are extended oligomers with a degree of polymerization of about 10 to 20 metallole units, and show high sensitivity for detecting nitroaromatics in solution or the vapour phase. The efficiency of photoluminescence quenching follows the order TNT > DNT > nitrobenzene, which is correlated with their reduction potentials. Quenching of photoluminescence is primarily attributable to electron transfer from the lowest excited state of metallole polymers to the π^* LUMO of the nitroaromatic analyte. Quenching of photoluminescence is a static process, since the excited state lifetime is invariant with varying quencher concentration. Each metallole polymer has a unique ratio of quenching efficiency to the corresponding analyte and each analyte has a variety of different responses to different metallole polymers, which could be used to specify the analyte by pattern recognition methods. These inorganic polymers are robust and insensitive to common interferents, such as organic solvents and inorganic acids.

Key words: TNT, explosives, silole, germole, polysilane, polygermole, polysilole, sensors, luminescence, fluorescence, nitroaromatics, quenching, inorganic polymers

J.W. Gardner and J.Yinon (eds.),
Electronic Noses & Sensors for the Detection of Explosives, 39-52.
© 2004 *Kluwer Academic Publishers. Printed in the Netherlands.*

1.   INTRODUCTION

Chemical sensors for the rapid detection of ultra-trace analytes from explosives are important because explosives detection is important to military applications, such as demining, as well as to environmental applications, such as munitions site remediation [1]. Both military and homeland security applications are attracting increased research because of aviation security concerns, suicide bombings, and other acts of terrorism. Even nuclear and radiological weapons require a conventional primary explosive as the initiator. In the Unites States, the volume of checked aviation baggage (~8 million), imported shipping containers by sea (~6 million), and individuals clearing customs (~500 million at 300 locations) pose tremendous challenges for explosives detection [2]. Control of access to facilities, such as nuclear power plants, chemical companies, oil refineries, gas storage areas, prisons, embassies, federal buildings, courts, corporate headquarters, banks, tunnels, Olympic venues, railway, bus, and subway terminals, underground parking areas, police stations, post offices, mailboxes, schools, and lockers are examples where explosive detection equipment could prove valuable. There are more than 500,000 federal facilities in the US and overseas, which are potential terrorist targets. In addition to these applications, explosive detection is crucial to forensic investigations, such as post-blast residue determinations [3,4]. Metal detectors, which are widely used as portable instruments for explosives detection, cannot locate the plastic casing of modern land mines. Trained dogs, often regarded as the gold standard for a mobile explosives detector, are expensive, are difficult to maintain, and are easily fatigued [5].

Physical detection methods for explosives include gas chromatography coupled with a mass spectrometer [6], surface-enhanced Raman spectroscopy [7], nuclear quadrupole resonance [8], energy-dispersive X-ray diffraction [9], neutron activation analysis, electron capture detection [1], and cyclic voltammetry [10]. These techniques exhibit varying degrees of selectivity. Some are expensive and others are not easily fielded in a small, low-power package. Most trace detection methods for explosives are only applicable to vapour samples because of interference problems encountered in aquatic media. Sensing TNT and picric acid in groundwater or seawater is important for the detection of buried unexploded ordnance and for mine sweeping [11-13]. There are important environmental applications for characterizing soil and groundwater contaminated with toxic nitro-based explosives at military bases and munitions production and distribution facilities [14].

Structures of Nitroaromatics Used in High Explosives

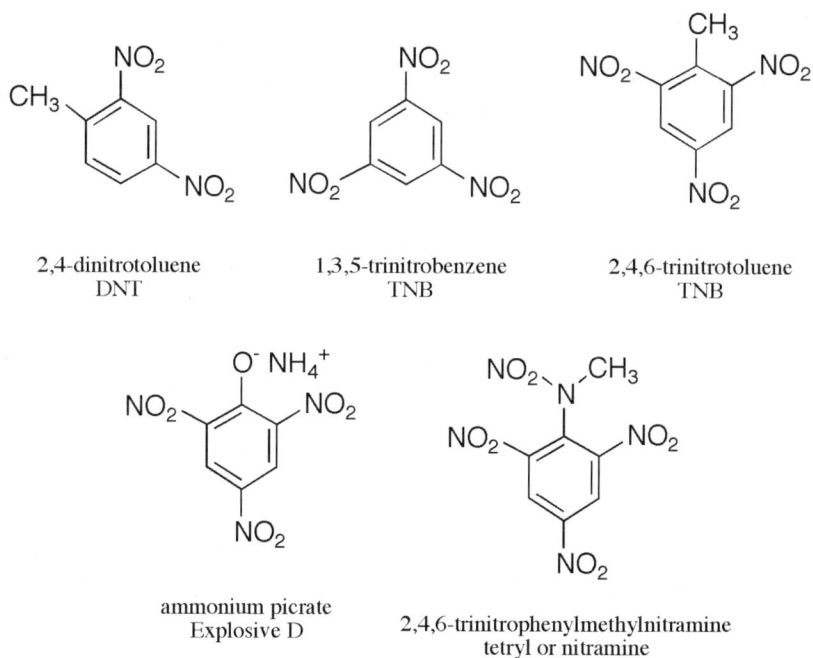

2,4-dinitrotoluene DNT	1,3,5-trinitrobenzene TNB	2,4,6-trinitrotoluene TNB

ammonium picrate Explosive D	2,4,6-trinitrophenylmethylnitramine tetryl or nitramine

Figure 1. Chemical structures of some explosive materials

Many high explosives contain nitroaromatics, such as TNT. Often mixtures of high explosives are used. For example, Tetrytol is a mixture of tetryl and TNT used in demolition blocks and cast shaped charges. Because of its low cost, TNT is widely used in high explosive formulations (e.g., amatol, ednatol, pentolite, torpex, tritonal, picratol, cyclotol, ednatol, DBX, HBX-1, HBX-3, H-6, PTX-1, PTX-2, Minol2, and Compositions C-2, C-3, and B) [15]. Many ammonium nitrate based industrial explosives also contain DNT or TNT in the mixture [16]. This wide use of nitroaromatics makes them an important target class for explosives detection.

Organic polymers and optical fibres [17] have been previously used to detect vapours of explosive analytes [18,19]. The transduction methods include absorption, fluorescence, conductivity, etc [16]. Such simple techniques are promising, because they can be incorporated into inexpensive and portable microelectronic devices. For example, a chemically selective silicone polymer layer on a SAW (surface acoustic wave) device has been shown to provide efficient detection for the nitroaromatic compounds [20]. The fluorescence of pentiptycene conjugated polymers [21,22] and substituted polyacetylenes, such as PTMSDPA, are also highly sensitive to nitroaromatic molecules [23].

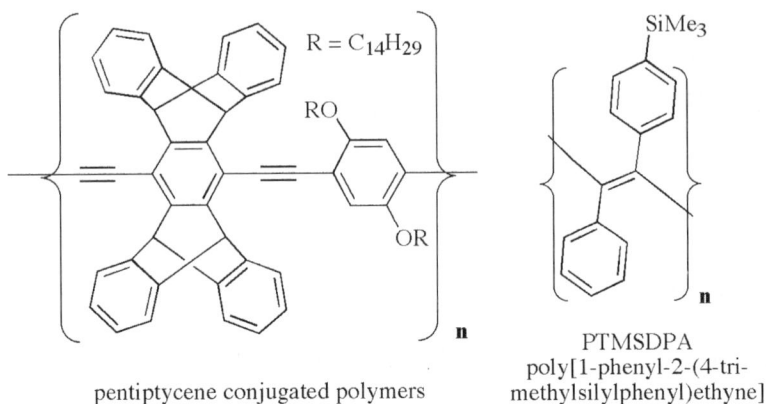

Figure 2. Chemical structure of Pentiptycene and PTMSDPA.

The bulky pentiptycene moiety provides a porous packing arrangement in the solid state, which prevents interchain π stacking and self-quenching of luminescence. These conjugated polymers show quenching of their blue luminescence when exposed to electron-acceptor organics, such as TNT or quinone. The long polymer chain lengths and exciton delocalization along the chain provide the exceptional sensitivity for detecting TNT vapour at sub part per trillion (ppt) levels. Nomadics Inc. is commercializing this remarkable technology. The pentiptycene-conjugated polymer has also been applied to construct a semi-selective sensor array for the tip of an optical fibre bundle [24,25]. This sensor primarily detects the DNT emanating from a buried landmine with a short (200 ms) sampling time. Luminescence quenching of the PTMSP polymer, which also exhibits a high degree of permeability (fractional free volume of 0.26), also can be used to attain low detection limits (part per billion, ppb) of nitroaromatic analytes [23]. Formation of a charge transfer complex between the nitroaromatic analyte (4-nitrotoluene, 1,4-dinitrobenzene, 2,6-dinitrotoluene, and 1,3-dinitrobenzene) and the electron-rich polyacetylene chain was postulated as important in the fluorescence quenching process. The potential for developing electronic noses for the detection of explosives, including those based on fluorescent polymers has been reviewed recently [26].

Luminescent inorganic polymer sensors had received little attention as explosive sensors. The observation of electron transfer quenching of luminescence in porous silicon nanostructures [27] suggested that polysilicon organometallic species might be especially promising for sensor applications. Several luminescent polysilicon based structures are known.

Polysilanes are air-stable Si-Si backbone polymers that exhibit efficient emission in the UV spectral region, high hole mobility, and high nonlinear optical susceptibility [28]. These properties arise from delocalization of σ electrons along the Si-Si chain. Polysilanes have been employed as fluorescent materials for radiation detection, as electroluminescent materials for display devices, and as photorefractive materials for holographic data storage [28].

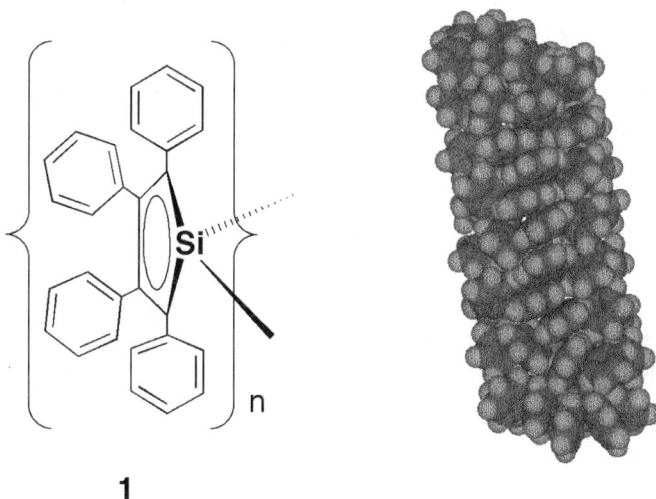

1

Figure 3. Chemical structure of polysilane.

Polysiloles, such as poly(2,3,4,5-tetraphenyl-1-silacyclopenta-2,4-diene), **1**, and have also received attention [28] because of their possible application as electron-transporting materials in display technologies, e.g., organic light-emitting diodes (LEDs) [29-31]. For molecular recognition, **1** displays a stabilizing shell of organic groups surrounding a central Si-Si backbone, as shown above. The backbone provides the correct electronic structure to allow migration of the excited state energy along the polymer chain. The organic sheath not only provides chemical protection needed to make the materials kinetically stable in air or water, but σ*-π* conjugation with the silole ring provides an efficient pathway for electron transfer quenching by analytes that can penetrate the hydrophobic exterior. Alkyl polysilanes are known to adopt 7/3 and other types of helical structures as a result of interactions between pendant organic substituents [32,33]. The other conformation observed is a rigid planar zig-zag configuration for the Si-Si backbone [33,34]. The latter structure results from an anti conformation along the Si-Si backbone, where the Si-Si-Si-Si torsion angle is 180°. While the structure of polysiloles is not known with certainty, the silole dimer exhibits a H-Si-Si-H torsoional angle near 90° [35]. This suggests a helical structure for **1**, which should be shape selective for intercalative binding of planar aromatic compounds, such as nitroaromatics.

A characteristic electronic feature of polysiloles **1**, and the corresponding polygermole, **2**, is a low reduction potential due to a low-lying LUMO arising from σ^*-π^* conjugation between the σ^* orbitals of the polysilane or polygermane backbone and the π^* orbital of the butadiene moiety of the five-membered ring [29-31,36]. This conjugation lowers the energy of the π^* orbital and shifts optical absorption spectra of polysiloles into the near UV (~400 nm) and their emission spectra into the visible (~520 nm) spectral regions [37]. Photoluminescence quantum yields near unity [38,39] have been measured, which makes them attractive candidates for fluorescent chemosensors.

2.   POLYMER SYNTHESIS AND LUMINESCENCE

Polymetalloles, polygermoles and various copolymers have been synthesized in 30-40% yield by a simple Wurtz-type polycondensation of the corresponding metallole dichloride, as shown in Scheme I [37,40,41].

Scheme I

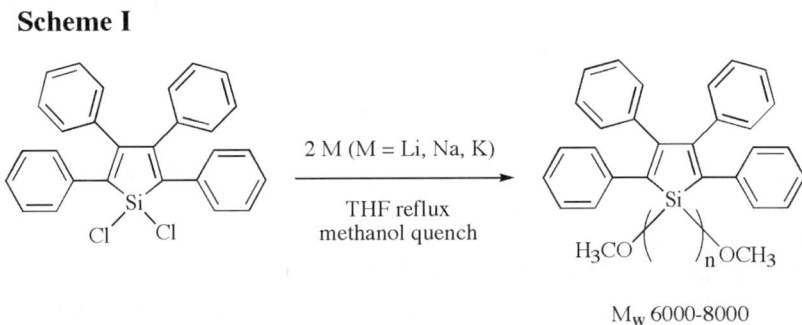

Figure 4. Wurtz-type polycondensation of the corresponding metallole dichloride.

Since a metallacyclopentadiene has a low lying LUMO compared with cyclopentadiene or thiophene, this unit is an excellent candidate for electron transporting materials.

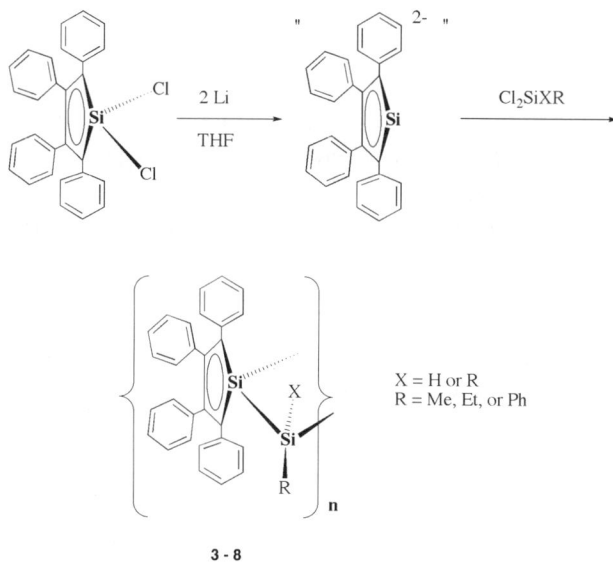

Figure 5. Metallole dichloride reactions.

Tuning of the HOMO-LUMO band gap and electroluminescent colours can be achieved by changing the R group on the 2,3,4, and 5 positions of the five-membered ring. Copolymers containing silole units (Figure 5), such as **3-8**, have also been prepared by this procedure [41]. Catalytic dehydrocoupling of silole and germole dihydrides has been developed recently as a route to polysiloles [35]. Previously, catalytic dehydrocoupling had been used to form polymers from monosubstituted silanes; however, disubstituted silanes generally only produce dimers or short oligomers [42]. However, the use of vigorous reflux conditions, and addition of an alkene as a hydrogen acceptor results in good yields (>60%) of poly(tetraphenyl)silole (Scheme II) with chain lengths of up to 20 monomeric units [35].

Scheme II

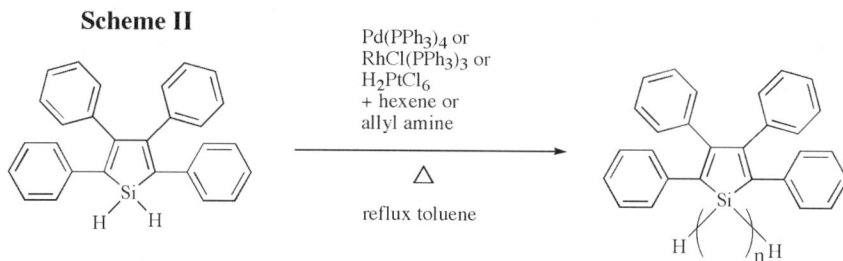

Figure 6. Creation of poly(tetraphenyl)silole.

The lowest energy UV-vis absorption and fluorescence spectral data for the poly(tetraphenyl)metalloles and tetraphenylmetallole-silane copolymers arise from the π-π* transition in the metallole ring [43]. The absorption occurs at a wavelength of about 370 nm for **1**, which is about 90 nm red-shifted relative to that of the monomeric dihydride. These red shifts are attributed to an increasing main chain length [44] and partial conjugation of the phenyl groups to the silole ring. One visible emission band ($\lambda_{max} \sim 513$ nm) is observed for **1** when excited at 340 nm. The bandwidth of the emission spectrum in solution is slightly larger than in the solid state; however, there is no shift in the maximum of the emission wavelength. This suggests that the polysilole exhibits neither π-stacking of polymer chains nor excimer formation, as often occurs for conjugated organic polymers.

3.   FLUORESCENCE QUENCHING STUDIES WITH NITROAROMATIC ANALYTES

The nitroaromatic detection method involves measurement of the quenching of photoluminescence of the polysiloles, such as **1**. Fluorescence spectra of a toluene solution of the polymers were obtained upon successive addition of aliquots of picric acid (purchased from Aldrich and recrystallized from ethanol solution before use), TNT (prepared from DNT [45] and recrystallized twice from methanol), DNT, and nitrobenzene. Photoluminescence quenching of the 12 different polysiloles, polygermole, and copolymers with polysilanes were measured in toluene solutions with picric acid, TNT, DNT, and nitrobenzene analytes. The purity of the TNT sample (prepared by nitration of 2,4-DNT) was found to be important to obtain reproducible results. When the quenching experiment was undertaken without recrystallization of TNT, higher (up to 10-fold) quenching percentages are obtained. Presumably, impurities with enhanced quenching efficiencies are present in crude TNT preparations. Therefore successive recrystallizations of analyte from methanol are necessary until a minimum in quenching ability is observed.

The Stern-Volmer equation was used to quantify the differences in quenching efficiency for various analytes in solution phase studies [46]. In this equation, I_o is the initial fluorescence intensity without analyte, I is the fluorescence intensity with added analyte of concentration [A], and Ksv is the Stern-Volmer constant.

$$(I_o/I) - 1 = K\text{sv [A]} \tag{1}$$

The graph below shows the Stern-Volmer plots for quenching of the luminescence of polysilole **1**, by TNT in toluene solution. A linear Stern-Volmer relationship is observed for the corresponding polygermole, **2**, and the copolymers **3-8**, but the Stern-Volmer plot for quenching by picric acid exhibits an exponential dependence when its concentration is higher than a value of 10^{-4} M [41].

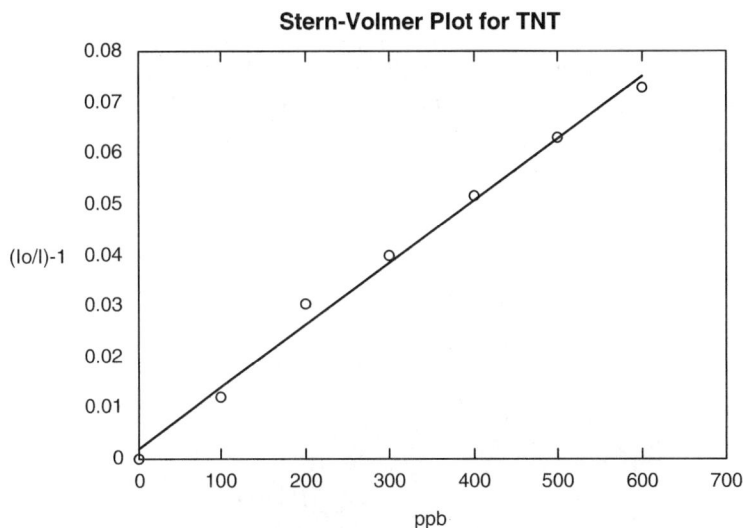

Figure 7. Stern-Volmer plot for quenching of the luminescence of polysilole **1** by TNT in toluene solution.

Photoluminescence quenching may arise from either a static process, by the quenching of a bound complex, or a dynamic process, by collisionally quenching the excited state [47,48]. For the former mechanistic case, *K*sv represents an association constant for analyte binding to the polymer chain in **1**. Thus, the collision rate between the analyte and polymer is not rate limiting for static quenching and the fluorescence lifetime is invariant with the concentration of analyte. With dynamic quenching, the fluorescence lifetime should diminish as quencher is added. The 'mean' lifetime (τ) measured for **1** is 0.7 ns. Luminescence decays were not single-exponential. Three lifetimes were needed to provide an acceptable fit over the first few nanoseconds and were weighted by their relative amplititudes in the average. Since the polymers span a size distribution, this behavior is not surprising. The fluorescence lifetimes as a function of TNT concentration were also measured and found to be constant. This suggests that the static quenching process is dominant for **1** and related polymers. When an electron acceptor molecule such as TNT is present, electron transfer quenching occurs from the excited state of the metallole polymer to the LUMO of the nitromatic analyte. The observed dependence of *K*sv (TNT > DNT > nitrobenzene) on analyte reduction potential suggests that for the static quenching mechanism, the polymer-quencher complex luminescence intensity depends on the electron acceptor ability of the quencher. An alternative explanation would be that the formation constant (*K*sv) of the polymer-quencher complex is

48

dominated by a charge-transfer interaction between polymer and quencher and that the formation constant increases with increased quencher electron acceptor ability.

For chemosensor 'electronic nose' applications, it is useful to have sensors with varied responses. The 12 different luminescent polymers (**9**-12 are germole copolymers) examined [41] exhibit a different ratio of the photoluminescence quenching for picric acid, TNT, DNT, and nitrobenzene and a different response with the same analyte, as shown in Figure 8.

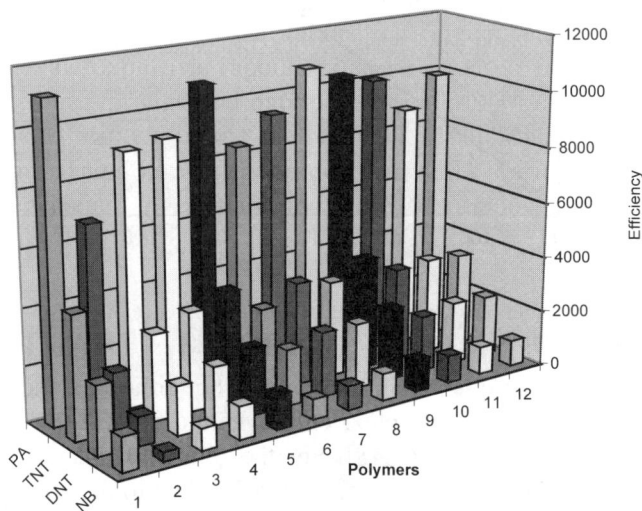

Figure 8. Response of twelve different luminescent polymers to four compounds.

An important aspect of the helical polysiloles **1** is its insensitivity to common interferents. Control experiments using both solutions and thin films of metallole copolymers (deposited on glass substrates) with oxygenated air displayed no change in the photoluminescence spectrum. Similarly, exposure of metallole copolymers both as solutions and thin films to organic solvents such as toluene, THF, and methanol or the aqueous inorganic acids H_2SO_4 and HF produced no significant decrease in photoluminescence intensity. The selectivity of luminescence quenching of polysilole **1** with TNT vs. benzoquinone is also greater than that of the pentiptycene conjugated polymer [21,22]. The Ksv value of 4.34×10^3 M^{-1} for polysilole **1** with TNT is 644 % greater than that for benzoquinone (Ksv = 674 M^{-1}), but the pentiptycene polymer exhibits only slightly better quenching efficiency for TNT (Ksv = 1.17×10^3 M^{-1}) (*ca.* 117 %) compared to that for benzoquinone. This result [41] indicates that polysilole **1** exhibits less response to interferences and greater selectivity to nitroaromatic compounds compared to the pentiptycene-derived polymer. Thin films of **1** can also be used to detect 50 ppb of TNT in sea water over a period of 60 seconds. The ability of **1** to bind TNT by intercalation into its putative helical structure may account for this ability.

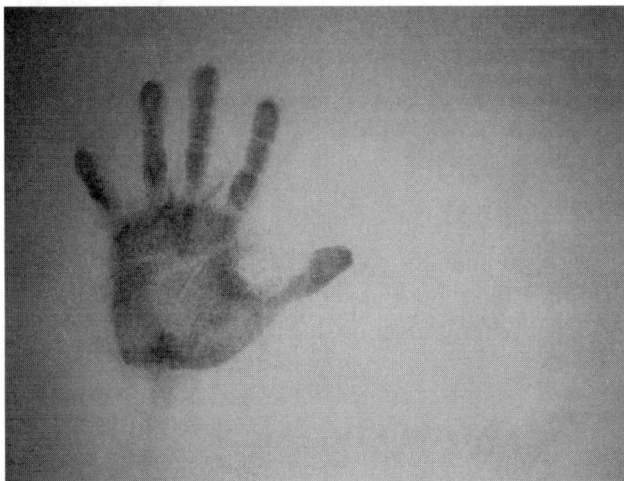

Figure 9. Imaged hand showing traces of TNT.

The solubility of **1** in organic solvents allows coating of surfaces by airbrushing or spin-coating to fabricate thin-film solid-state sensors. The high sensitivity of the material for detecting traces of TNT [49] is shown In Figure 9 where a hand contaminated with invisible traces of TNT was contacted with a paper surface on the left and the right side was contacted with the clean hand as a control. The image was developed by airbrush painting with an mM toluene solution of **1** (about 1 mg of polymer deposited on the 61 cm × 61 cm area) and illuminating with a black light after solvent evapourated. Traces of TNT quench emission of the thin fluorescent polymer coating, causing them to be imaged as dark regions against a bright background.

4. POLYSILOLE AND SILOLE NANOPARTICLE SENSORS

Addition of water to solutions of tetraphenylsilole monomer in organic solvents (e.g. ethanol) produces colloidal suspensions of highly luminescent particles [50]. This suggested to us that suspensions of such materials might be useful as sensors. The high surface area and surface charge of colloids should offer improved aqueous sensing, as well as providing differing selectivity for redox sensing of ionic species. Addition of 9 volumes of water to one volume of a 10^{-4} M THF solution of **1** produced highly luminescent particles, which are submicron in size by fluorescence microscopy. Atomic force microscope studies show the particles are about 100 nm in diameter [51]. They also exhibit enhanced sensitivity and altered analyte selectivity. For example, thin films of **1** do not readily detect carcinogenic CrO_4^{2-} unless the concentration of chromate approaches 0.5 M. This can be attributed to poor binding of the ionic chromate to the surface of the hydrophobic polysilole helix. Use of nanoparticles of **1** in aqueous suspension allows detection of chromate at 100 ppb, which is the EPA action level for the maximum amount of allowed chromium in drinking water. These nanoparticles also respond to arsenate and TNT analytes. The use of luminescent organometallic nanoparticles offers an unexplored area for future research. Unlike hard materials, such as CdSe and Si, organometallic nanoparticles are readily functionalized.

ACKNOWLEDGEMENTS

Support of our research by the U. S. National Science Foundation and the Environmental Protection Agency is gratefully acknowledged.

REFERENCES

1 A.M. Rouhi, Chem. Eng. News 75 (1997) 14.
2 U.S. Customs Service, Report 0000-0148, Washington, DC, 2000, pp. 123.
3 S.-A. Barshick, J. Foren. Sci. 43 (1998) 284.
4 K.D. Smith, B.R. McCord, W.A. McCrehan, K. Mount and W.F. Rowe, J. Foren. Sci. 44 (1999) 789.
5 A.W. Czarnik, Nature 394 (1998) 417.
6 K. Hakansson, R.V. Coorey, R.A. Zubarev, V.L. Talrose and P. Hakansson, J. Mass Spectrom. 35 (2000) 337.
7 J.M. Sylvia, J.A. Janni, J.D. Klein and K.M. Spencer, Anal. Chem. 72 (2000) 5834.
8 V.P. Anferov, G.V. Mozjoukhine and R. Fisher, Rev. Sci. Instr. 71 (2000) 1656.
9 R.D. Luggar, M.J. Farquharson, J.A. Horrocks and R.J. Lacey, J. X-ray Spectrom. 27 (1998) 87.
10 M. Krausa and K. Schorb, J. Electroanal. Chem. 461 (1999) 10.
11 L.C. Shriver-Lake, B.L. Donner and F.S. Ligler, Environ. Sci. Technol. 31 (1997) 837.
12 J. Lu, Z. Zhang, Anal. Chim. Acta 318 (1996) 175.
13 M. Dock, M. Fisher and C. Cumming, in Fifth International Symposium of Mine Warfare Association, Monterey, California, 2002, pp. 1.
14 Approaches for the remediation of federal facility sites contaminated with explosive or radioactive wastes, U. S. Environmental Protection Agency, Washington, D.C., 1993.
15 J. Köhler, R. Meyer and A. Homburg, Explosives, 5th ed., Wiley-VCH, Weinheim, 2001.

16 J. Yinon, Forensic and Environmental Detection of Explosives, John Wiley & Sons, Chichester, 1999.

17 K.J. Albert, M.L. Myrick, S.B. Brown, D.L. James, F.P. Milanovich and D.R. Walt, Environ. Sci. Technol. 35 (2001) 3193.

18 D.T. McQuade, A.E. Pullen and T.M. Swager, Chem. Rev. 100 (2000) 2537.

19 K.J. Albert, N.S. Lewis, C.L. Schauer, G.A. Sotzing, S.E. Stitzel, T.P. Vaid and D.R. Walt, Chem. Rev. 100 (2000) 2595.

20 R.A. McGill, T.E. Mlsna and R. Mowery, in IEEE International Frequency Control Symposium, 1998, pp. 630.

21 J.-S. Yang and T.M. Swager, J. Am. Chem. Soc. 120 (1998) 5321.

22 J.-S. Yang and T.M. Swager, J. Am. Chem. Soc. 120 (1998) 11864.

23 Y. Liu, R. Mills, J. Boncella and K. Schanze, Langmuir 17 (2001) 7452.

24 K. Albert, M. Myrick, S. Brown, D. James, F. Milanovich and D.R. Walt, Environ. Sci. Technol. 35 (2001) 3193.

25 K. Albert and D.R. Walt, Anal. Chem. 72 (2000) 1947.

26 J. Yinon, Anal. Chem. 75 (2003) 98A.

27 S. Content, W.C. Trogler and M.J. Sailor, Chem. Eur. J. 6 (2000) 2205.

28 R. West, H. Sohn, U. Bankwitz, J. Calabrese, Y. Apelog and T. Mueller, J. Am. Chem. Soc. 117 (1995) 11608.

29 K. Tamao, M. Uchida, T. Izumizawa, K. Furukawa and S. Yamaguchi, J. Am. Chem. Soc. 118 (1996) 11974.

30 K. Tamao and S. Yamaguchi, Pure Appl. Chem. 68 (1996) 139.

31 S. Yamaguchi and K. Tamao, Bull. Chem. Soc. Jpn. 69 (1996) 2327.

32 S. Furukawa, K. Takeuchi and M. Shimana, J. Phys. Condens. Matter 6 (1994) 11007.

33 W. Chunwachirasiri, I. Kanaglekar, M. Winokur, J. Koe and R. West, Macromolecules 34 (2001) 6719.

34 B. Farmer, J. Rabolt and R. Miller, Macromolecules 20 (1987) 1167.

35 S. J. Toal, S. Urbas, H. Sohn and W.C. Trogler, Organometallics 2004, to be submitted.

36 Y. Yamaguchi, Synthetic Met. 82 (1996) 149.

37 H. Sohn, R.R. Huddleston, D.R. Powell and R. West, J. Am. Chem. Soc. 121 (1999) 2935.

38 T. Sanji, T. Sakai, C. Kabuto and H. Sakurai, J. Am. Chem. Soc. 120 (1998) 4552.

39 S. Yamaguchi, T. Endo, M. Uchida, T. Izumizawa, K. Furukawa and K. Tamao, Chem. Eur. J. 6 (2000) 1683.

40 S. Yamaguchi and K. Tamao, in The Chemistry of Organic Silicon Compounds, Vol. 3 (Eds.: Z. Rappoport, Y. Apeloig), John Wiley & Sons, LTD, Chichester, 2001, pp. 641.

41 H. Sohn, M.J. Sailor, D. Magde and W.C. Trogler, J. Am. Chem. Soc. 125 (2003) 3821.

42 L. Rosenberg and D. Kobus, J. Organomet. Chem. 685 (2003) 107.

43 Y. Xu, T. Fujino, H. Naito, T. Dohmaru, K. Oka, H. Sohn and R. West, Jpn. J. Appl. Phys. 38 (1999) 6915.

44 K. Kanno, M. Ichinohe, C. Kabuto and M. Kira, Chem. Lett. 99 (1998).

45 J.W.H. Dennis, D.H. Rosenblatt, W.G. Blucher and C.L. Coon, J. Chem. Eng. Data 120 (1975) 202.

46 N.J. Turro, Modern Molecular Photochemistry, University Science Books, Sausalito, California, 1991.

47 K.A. Connors, Binding Constants: The Measurement of Molecular Complex Stability, Wiley-Interscience, New York, 1987.

48 J.R. Lakowicz, Principles of Fluorescence Spectroscopy, Plenum Press, New York, 1986.

49 H. Sohn, R.M. Calhoun, M.J. Sailor and W.C. Trogler, Angew. Chemie, Intl. Ed. 2001, 40, 2104.

52

50 J. Luo, Z. Xie, J. Lam, C. Lin, C. Haiying, C. Qui, H. Kwok, X. Zhan, Y. Liu, D. Zhu and B. Tang, Chem. Commun. 174 (2001).
51 S.J. Toal and W.C. Trogler, ACS Symp.Ser., Nanotechnology in the Environment 2004, in press.

Chapter 4

AMPLIFYING FLUORESCENT POLYMER ARRAYS FOR CHEMICAL DETECTION OF EXPLOSIVES

Colin Cumming, Mark Fisher and John Sikes
Nomadics, Inc., Stillwater, Oklahoma, USA

Abstract: Amplifying fluorescent polymers developed by collaborators at the Massachusetts Institute of Technology (MIT) have been integrated into a handheld sensor platform capable of detecting single femtogram masses of vapour-phase nitroaromatic explosives in real-time. This sensor, known as Fido, was originally developed under the Defense Advanced Research Projects Agency (DARPA) Dog's Nose program. During field tests against buried landmines, the sensor has demonstrated the ability to detect trace levels of vapours of nitroaromatic explosives emanating from landmines. The sensor utilizes amplification of fluorescence quenching as a transduction mechanism for detection of nitroaromatic explosives and other closely related compounds. Earlier Fido prototypes utilized a single band of amplifying polymer deployed inside a capillary waveguide to form the sensing element of the detector. A new prototype has been developed that incorporates multiple, discrete bands of different amplifying polymers deployed in a linear array inside the capillary. Vapour phase samples are introduced into the sensor as a sharp pulse via a gated inlet. As the vapour pulse is swept through the capillary by flow of a carrier gas (air), the pulse of analyte encounters the bands of polymer sequentially. If the sample contains nitroaromatic explosives, the bands of polymer will respond with a reduction in emission intensity proportional to the mass of analyte in the sample.

Key words: Fido, landmine, explosives, amplifying fluorescent polymer (AFP)

J.W. Gardner and J.Yinon (eds.),
Electronic Noses & Sensors for the Detection of Explosives, 53-69.
© 2004 *Kluwer Academic Publishers. Printed in the Netherlands.*

54

1.   INTRODUCTION

Regardless of the detection methodology utilized, explosives detection equipment must have adequate sensitivity to detect threat quantities of explosives while maintaining low false alarm rates. When trace chemical vapour detection is utilized for detection of explosives, the level of sensitivity required to achieve high probabilities of detection depends on a number of factors. One of these factors is the equilibrium vapour pressure of the explosive. The equilibrium vapour pressure of the explosive at a given temperature establishes an upper limit to the concentration of explosives vapour available for detection. Environmental factors and the physical properties of the explosive often reduce the actual vapour phase concentrations of explosive near an explosive device to levels lower than the equilibrium concentration by orders of magnitude. Only under ideal conditions will the explosive vapour approach equilibrium concentration levels. In order to sense trace levels of vapour emanating from the bulk explosive contained in an explosive device, detectors with high sensitivity are required. Chemical vapour detectors are commercially available that can routinely detect explosives with moderately high equilibrium vapour pressures. By contrast, explosives with low equilibrium vapour pressures generate lower concentrations of vapours, which in turn makes direct vapour-phase detection of these explosives impossible for all but the most sensitive detectors.

One method that has been utilized to enhance the performance of explosive vapour detectors is to incorporate preconcentrators into the detection system. Preconcentrators draw large volumes of air through a sorbent medium that strongly sorbs molecules of explosives. In some cases preconcentrators are also designed to trap microcrystalline particles of explosive. After capturing the explosives from a large volume of sample, the preconcentrator is typically heated, releasing the trapped molecules of explosive into a smaller volume of a carrier gas. This effectively raises the concentration of explosives in the sample delivered to the detector, increasing the probability that the explosive will be detected. While effective for certain explosives detection applications, there are valid reasons to avoid the use of preconcentrators. Because time is required to sample large volumes of air, preconcentrators can increase the cycle time of the sensor significantly, reducing sample throughput. Preconcentrators also increase the complexity of the sampling system. In addition to increasing the concentration of explosives delivered to the sensor, preconcentrators can also increase the concentration of chemical interferents delivered to the sensor, which can increase the false alarm rate of the sensor.

In recent years, the development of ultra-sensitive explosives detectors that can detect very low levels of explosive vapour in near real-time without the use of sample pre-concentrators has been a focal point of sensor development. Unfortunately, as the sensitivity of a detector increases, a corresponding increase in the sensor false alarm rate is almost always observed. A good example of the tradeoff between sensitivity and selectivity can be taken from the use of metal detectors to detect landmines. Landmines with metal cases are relatively simple to detect with sophisticated metal detectors. However, many modern mine cases are constructed of plastic, with the only metal content being contained in fuse components. Metal detectors with extremely high sensitivity are required in order to detect these low-metal content mines. Compounding the problem is the fact that mines are almost always deployed in the vicinity of battlefields, and metallic objects are common to battlefield environments. Objects such as shrapnel from previously exploded munitions, spent ammunition cartridges, unexploded ordnance, and other metallic battlefield debris are commonly present in the soil at sites of conflict. Hence, with metal detectors it is difficult to distinguish low metal content mines from other metallic objects in the soil. In some mined areas, thousands of clutter objects are located for each mine detected. In these areas, demining with metal detectors proceeds at a very slow pace because each alarm must be treated as though a mine generated it, even though there is a high probability that the signal is a false alarm.

As previously stated, the Fido detector was originally developed under the DARPA Dog's nose program [1-3]. One goal of this program was to develop chemical vapour sensors capable of detecting ultra-trace levels of explosives emanating from landmines. The intent was to develop electronic sensor systems that could mimic the ability of canines to detect landmines. Landmine detection using chemical vapour sensors is difficult for a variety of reasons [3]. First, the concentration of landmine chemical signature compounds in the environment near a buried landmine is typically very low [4, 5]. The concentration of TNT in the air over a buried landmine is often six or more orders of magnitude lower than the equilibrium vapour concentration of TNT, the most commonly utilized explosive in landmines. Hence, chemical vapour sensors must be extremely sensitive in order to detect mines. The Fido sensor is capable of detecting low femtogram (10^{-15} gram) masses of vapour phase TNT, and to our knowledge is the only vapour sensing equipment that has demonstrated the ability to detect mines under field conditions with probabilities of detection comparable to that of canines.

Sensors for landmine detection must also be very selective. The soil in which mines are deployed is a complex mixture composed of hundreds of

56

chemical constituents. In addition, battlefield environments are often contaminated with chemicals that are not encountered in a 'clean' environment. Because the sample space is so complex, when a vapour sample is presented to the sensor there are likely hundreds of chemical vapours present in sample. False alarms are likely if the sensor is not extremely selective. This problem is not unique to landmine detection. For example, screening luggage for explosives is very difficult because of the wide range of chemical substances present in luggage and its contents. In fact, every explosives detection application will present a different subset of chemical constituents that must be discriminated from target analytes. Hence, an explosives sensor that is universally applicable to a range of explosives detection tasks must be extremely selective.

The selectivity of previous Fido sensor prototypes for landmine chemical signature constituents has proven to be excellent. However, improvements in chemical selectivity for target analytes translates into better false alarm performance, so methods for reducing sensor false alarms have been a focal point of sensor research efforts. Previous sensor prototypes have achieved selectivity primarily through the chemical and physical properties of the amplifying fluorescent polymer (AFP) materials used as sensory materials in the system. Described here are changes to the sensor hardware that should further enhance the selectivity of the sensor.

2. SELECTIVITY OF AMPLIFYING FLUORESCENT POLYMER MATERIALS

Fido utilizes novel fluorescent polymers developed by collaborators at the Massachusetts Institute of Technology (USA). These polymers were specifically engineered to detect TNT [6-9], the explosive found in more than 85% of landmines now deployed [10].

Conventional fluorescence detection normally measures an increase or decrease (i.e., quenching) in fluorescence intensity that occurs when a single molecule of analyte interacts with a single fluorophore. The upper right frame of Figure 1 illustrates a transduction mechanism of this type. In AFP materials, binding of a single TNT molecule quenches the fluorescence of many polymer repeat units, thereby amplifying the effect of a single TNT binding event (refer to the lower right frame of Figure 1). When thin films of these polymers absorb a photon of light, excited state electrons (i.e., 'excitons') are able to migrate efficiently along the conjugated polymer backbone and between adjacent polymer chains.

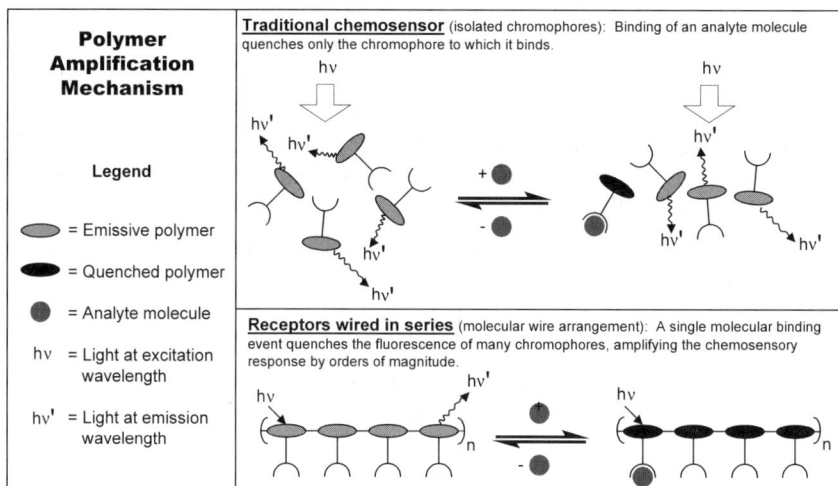

Figure 1. Description of the polymer quenching amplification mechanism.

During its excited state lifetime, the exciton propagates by a random walk through a finite volume of the polymer film. If an electron-deficient (i.e., electron accepting) molecule, such as TNT, binds to the polymer film, a low-energy 'trap' is formed. If the exciton migrates to the site of the bound electron-deficient molecule before transitioning back to the ground state, the exciton will be trapped (a non-radiative process), and no fluorescence will be observed from the excitation event. Since the exciton samples many potential analyte binding sites during its excited state lifetime, the probability that the exciton will sample an occupied 'receptor' site and be quenched is greatly increased.

In practice, there is evidence that these polymers result in an amplification of quenching response of between 100 and 1,000 fold as compared to conventional (monomeric) quenching mechanisms. The resulting amplification is a key factor in achieving the exceptional sensitivity of the detector, which has been demonstrated in laboratory tests to have a minimum detection limit for TNT of approximately 1 femtogram (1×10^{-15} grams).

The magnitude of fluorescence quenching (FQ) of the polymer per unit time is influenced by several factors. As described in the previous paragraph, the mechanism for fluorescence quenching is electron transfer from the electronically excited polymer to the quencher (i.e., through oxidative quenching). For electron transfer to occur spontaneously, the overall free energy change (ΔG^{0}) for this process must be negative. Hence, in order to quench the polymer an analyte must have a standard reduction potential large enough to result in a negative free energy change for the electron transfer process to occur. In addition, before electron transfer can occur an analyte molecule must first bind to the polymer film. Hence, the binding constant (K_b) for quenchers to the film is also important. Binding of electron-poor target analytes to electron-rich receptor sites in the polymer is enhanced by engineering the polymer so that the receptor sites are electrostatic mirror images of the target analyte. This increases the value of K_b for target analytes relative to potential interferents, improving selectivity. Figure 2 illustrates the electrostatic complementarity of a receptor site in the structure of an AFP used in the sensor. Steric constraints also affect analyte binding. Because of the chemical structure of AFP, films of these materials are porous. Small molecules such as TNT fit into cavities in the films, while larger molecules are excluded, limiting their ability to bind with and quench the polymer film. Finally, FQ is proportional to the concentration C of quencher in the sample. In equation form, FQ can be approximated as

$$FQ \propto C \exp(-\Delta G^0)^2 K_b \tag{1}$$

Hence, analytes that bind strongly to the film and have the appropriate redox potential could quench the fluorescence of the polymer. Analytes that bind poorly or that have weakly favorable reduction potentials will measurably quench the polymer only when present in high concentrations.

Blue=positive charge
Red=negative charge

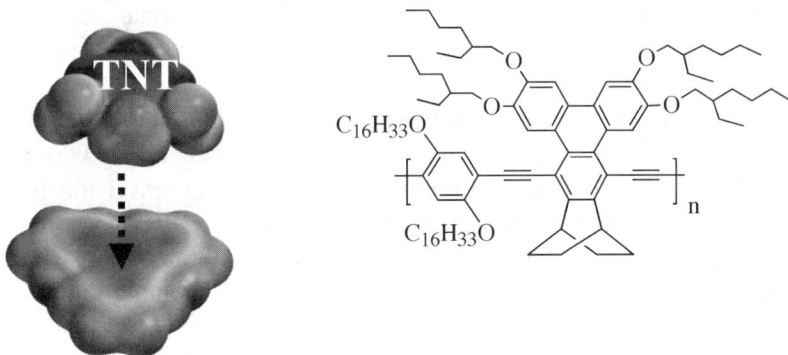

Figure 2. The chemical structure of an AFP and the electrostatic complementarity of an AFP receptor site to TNT.

In practice, except for nitroaromatic compounds, few compounds that quench the polymer have been encountered in actual field samples. As a part of field testing exercises, soil samples are routinely collected from minefields and are analyzed in our laboratory using Fido sensors interfaced to gas chromatographs. These samples have been collected on multiple occasions from test minefields and from sites potentially contaminated by UXO. A variety of soil types have been analyzed, ranging from desert sand to clay-loam soils with high organic content. Solvent extracts of these soils are injected into the chromatograph, and after separation on the column of the chromatograph the effluent from the column is introduced into the Fido sensor. In this way, the soil extract (which may contain hundreds of chemical constituents) is separated into its individual components, which are then introduced separately into the sensor.

This enables the response of the sensor to each constituent of the sample to be measured separately from the other constituents of the sample. By injecting reference standards containing target nitroaromatic compounds, it is possible to determine the time (retention time) at which a given target analyte elutes from the column and enters the sensor. By comparing retention times for target analytes in reference standards to peaks in unknown samples, it is relatively straightforward to determine if the sensor is responding to a target analyte or an interferent.

A chromatogram collected with Fido of a soil extract for a soil near an anti-tank landmine is shown in Figure 3. The response of Fido to a reference standard mix containing four target nitroaromatic compounds (100 picograms each) is also shown for reference. The four compounds in the reference standard mix (1,3-dinitrobenzene [1,3-DNB], 2,6-dinitrotoluene [2,6-DNT], 2,4-dinitrotoluene [2,4-DNT], and 2,4,6-trinitrotoluene [2,4,6-TNT]) are commonly found in the environment near landmines. Other nitroaromatic compounds have been identified in landmine chemical signatures, but these four are most commonly found in the vapour signature of TNT-containing landmines [5].

Responses due to a few unknown compounds are also seen in the chromatogram, but the responses are weak. As a point of reference, the chromatogram for this sample obtained using an electron capture detector (ECD), a detector commonly used for analysis of explosives, contained over 100 peaks. This data-set illustrates the excellent selectivity and sensitivity of the AFP for nitroaromatics.

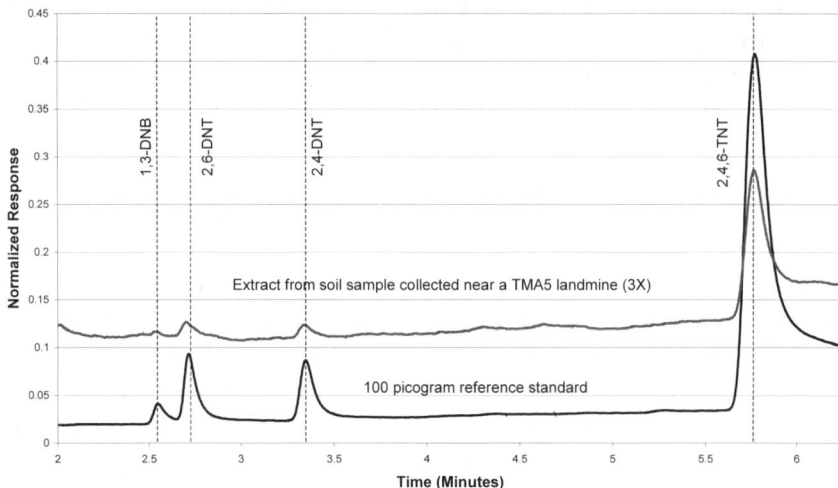

Figure 3. Chromatogram of a landmine soil extract compared to a reference standard mix, using Fido as the GC detector.

3. DESCRIPTION OF EARLY FIDO SENSOR PROTOTYPES

Binding of molecules of TNT or any one of a suite of related nitroaromatic compounds commonly associated with the presence of TNT (for example, 2,4-dinitrotoluene), or photochemical or microbial degradation products of TNT (1,3-dinitrobenzene or amino-dinitrotoluenes) results in a dramatic reduction in the emission intensity of films of AFP. The reduction in emission intensity is proportional to the mass of quencher adsorbed by the films and is measured by the sensor system. A schematic of an early Fido prototype is shown in Figure 4.

A blue light emitting diode (LED) or laser diode serves as the excitation source. Light from the source is focused at normal incidence onto two glass substrates coated with thin films of the polymer. The glass substrates act as planar waveguides for light emitted by the polymer and define the sensor sample chamber. The light exiting the edge of the substrate passes through an interference filter that passes light emitted by the polymer but blocks a significant fraction of stray light from the excitation source. The intensity of the fluorescence is then detected by a small photomultiplier tube (PMT) or photodiode.

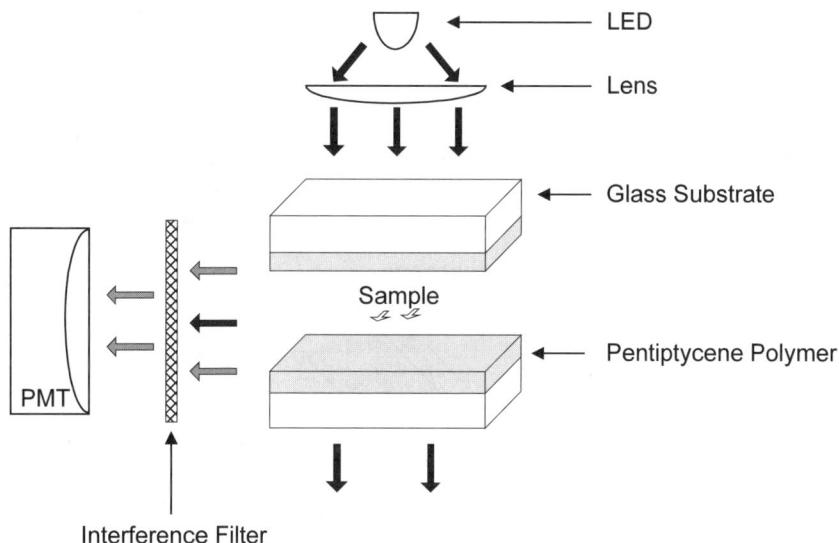

Figure 4. Early Fido sensor architecture based on AFP-coated planar waveguides.

A sampling cycle begins by establishing a baseline fluorescence reading in clean air (i.e. air free of nitroaromatics). Air that may contain target analytes is then drawn through the sampling chamber by a small pump. If the air contains nitroaromatic compounds, then the intensity of the fluorescence registered by the PMT will decrease proportionally to the concentration of target analytes in the sample. If a significant decrease in fluorescence is detected, then the operator is alerted to the possible presence of explosive. The response of the sensor to nitroaromatics is almost instantaneous upon sample introduction, enabling near real-time analysis of samples. Because binding of analytes to the films is reversible, the same polymer film can be exposed repeatedly to samples. A flow of clean air over the films will desorb analyte from the films, returning the fluorescence intensity to near the initial baseline reading.

Figure 5 is an illustration of the sensor architecture from which the Fido serial array sensor was derived. This sensor architecture replaced the planar waveguide geometry of the initial sensor prototypes. A borosilicate glass capillary coated with AFP on its inner surface forms the sensing volume of the sensor. The AFP film is excited by a blue LED positioned at normal incidence to the axis of the capillary. The capillary serves as a waveguide for the AFP emission. After passing through an optical filter that blocks stray light from the excitation source, the intensity of the AFP emission exiting from one end of the capillary is focused onto and measured by a photodetector. A small pump pulls samples through the bore of the capillary.

Operation of the sensor is identical to that of the planar waveguide design described above.

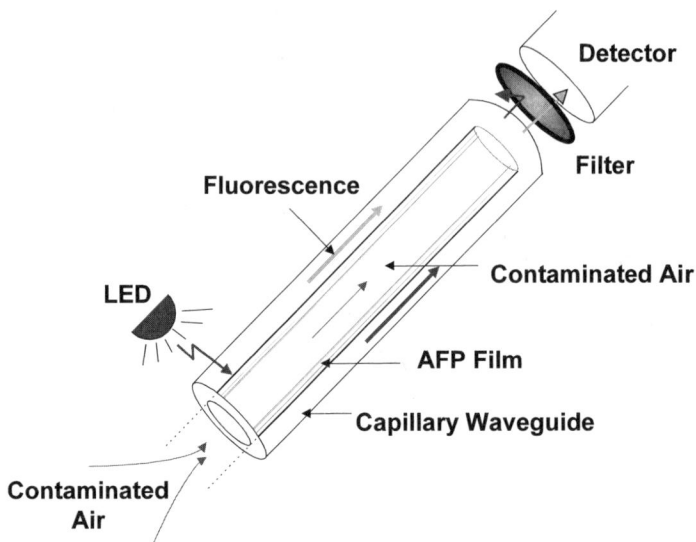

Figure 5. Sensor architecture based on AFP-coated capillary waveguide.

4. SELECTIVITY ENHANCEMENT OF FIDO SENSORS BY IMPLEMENTATION OF SERIAL AFP ARRAYS

Figure 6 is a schematic of a simple serial AFP array sensor. In this sensor prototype, samples are drawn into the sensor from the ambient through a gated inlet that enables introduction of pulses of sample of controlled duration. The samples are drawn through a glass capillary tube that has discrete bands of AFP coated inside the bore of the capillary. In this example, two bands of polymer are deployed. A separate laser diode excitation source is used to excite each polymer band. The lasers are modulated at different frequencies, enabling the emission of each band of polymer to be measured with a single photomultiplier tube (PMT). As with the earlier non-array capillary sensor prototype, the capillary is utilized as a waveguide, directing the light emitted from each polymer band onto the photodetector. A digital signal processor (DSP) is used to control the output of the laser diodes, and to process the signal from the photodetector. Earlier Fido sensor prototypes are very similar in design, with the exception that only a single polymer band and laser excitation source are utilized.

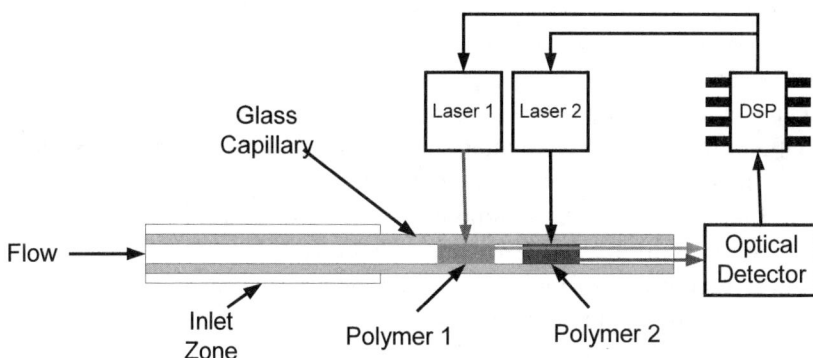

Figure 6. Schematic of serial array AFP-based sensor.

Pulses of analyte are introduced through a gated inlet, shown in Figure 7. Flow of sample into the sensor (F_a) can be calculated using the relationship

$$F_a = (F_s - F_b)$$ (2)

where F_s is the flow of air through the capillary array and F_b is the flow of clean carrier gas into the inlet delivered via a sidearm attached to the inlet. Prior to sample introduction (baseline mode), a pump supplies a flow of clean carrier gas (F_b) to the sensor inlet at a flow slightly greater than the flow through the capillary (F_s). A second pump and flow controller maintain the sample flow rate through the capillary. This results in a small net flow of air (less than 1 mL/minute) out of the inlet when the sensor is in baseline mode. Only clean carrier gas is introduced into the sensor during baseline mode. When the sensor is switched to sampling mode, F_b is reduced to a flow slightly greater than zero to prevent formation of a dead zone in the flow through the sidearm. During sampling mode, F_b is much less than F_s, resulting in a net flow of sample into the sensor at a sampling rate given by Equation 2.

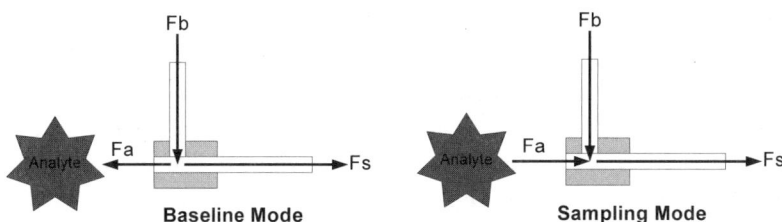

Figure 7. Schematic of gated sensor inlet.

Selectivity of the array sensor is improved relative to the earlier single-band prototypes via what is in essence a crude chromatographic separation that occurs as the analyte pulse traverses the length of the capillary. When the gated inlet is opened to allow introduction of a pulse of sample, the sample immediately encounters the first band of polymer. Because the strength of interaction of molecules of sample constituents with the AFP vary from compound to compound, some constituents of the sample bind more strongly to the AFP film than others. Binding of target analytes to the films is reversible, so molecules of analyte are swept through the capillary by the flow of carrier gas as they are repeatedly sorbed and then desorbed from the AFP film. Molecules that are weakly bound are rapidly swept past the bands of polymer by the flow of carrier gas, while those that are more strongly retained take longer to clear the sensor. If a given sample constituent quenches the polymer, the onset of response of the first polymer band to the quencher is very rapid after the sample gate is opened. After some period of delay, the second band begins to respond to the quencher. At the same time, the response of the first band begins to diminish as analyte is swept from the first zone to the second. Eventually, the response of both bands return to the baseline value as the quencher is swept from the sensor.

The serial arrangement of polymer bands inside the capillary provides additional selectivity for target analytes in the following way. The difference in time between the response maxima on channels 1and 2 is characteristic of a given quencher. Figure 8 shows the response of the array to a volatile chemical interferent (a non-explosive related compound), while Figure 9 illustrates the response of the sensor to 2,4-dinitrotoluene (2,4-DNT). The interferent is weakly retained by AFPs, so the difference in time between response maxima on the two channels is small. However, for 2,4-DNT the delay between response maxima is 1.07 seconds. Also shown in Figure 9 is the difference in response between channels 1 and 2. The differential response for a given analyte is representative of a given quencher, as can be seen from Figure 10 which is a comparison of the differential response for three quenchers. TNT and 2,4-DNT can easily be distinguished from the differential response of the potential interferent. The differential response enables straightforward differentiation of TNT from 2,4-DNT. Likewise, the chemical interferent has a different time response, enabling the interferents to be differentiated from the response of either target analyte. The peak-to-peak difference in time of maximum response for TNT, 2,4-DNT, and the interferent is 1.90, 1.07, and 0.38 seconds respectively. In addition, the differential band shapes for each compound are different, providing another indication of the identity of the quencher. This data was collected using the same type of AFP coating on each sensor channel.

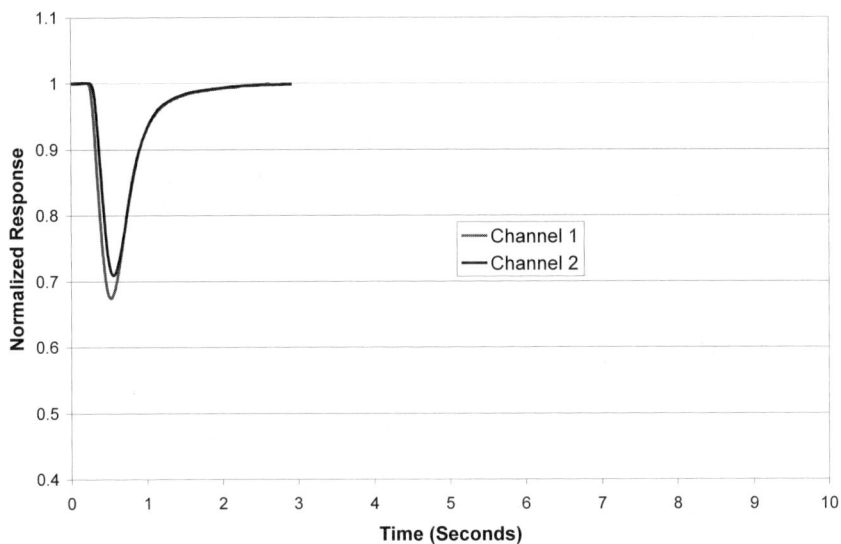

Figure 8. Array sensor response to a volatile chemical interferent.

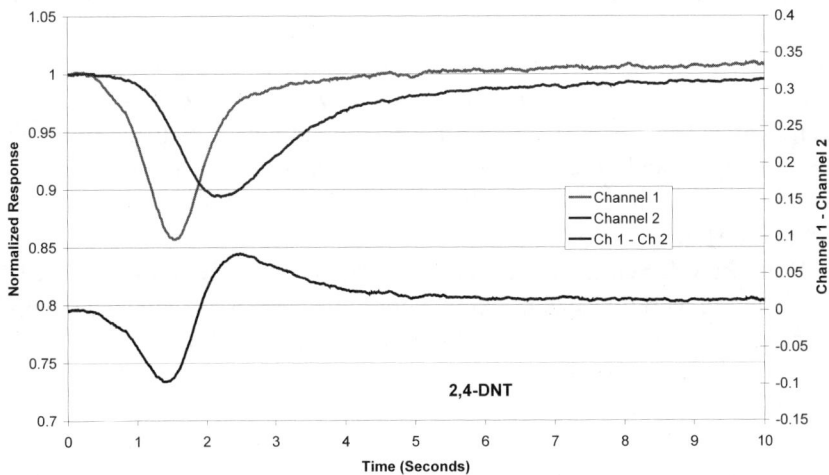

Figure 9. Array sensor response to 2,4-DNT.

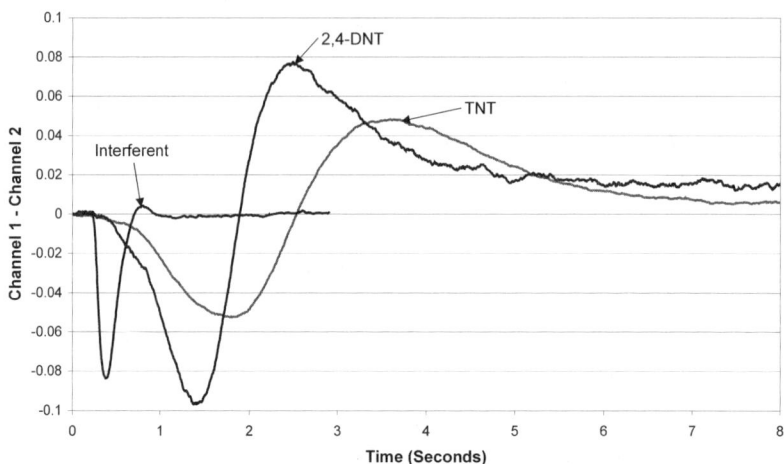

Figure 10. Array sensor differential response for TNT, 2,4-DNT, and an interferent.

The ratiometric response (i.e., the ratio of the quenching responses recorded by each sensor channel) can also be used to improve selectivity. Figure 11 illustrates the response of two different AFPs to TNT and 2,4-DNT. The ratio of the percent quench (Polymer A/B) was calculated for each polymer. The ratio of responses (A:B) for TNT was 2.1, while the ratio for 2,4-DNT was 0.5. The differences in the ratiometric responses for compounds on different polymers could provide a secondary means of discriminating interferents from target analytes. Over 20 different AFPs have been synthesized for use in the sensor. The responses of these polymers to target analytes and potential interferents vary, making it possible to prepare polymer arrays that differ widely in intensity of response to target analytes and to interferents.

Both the differential and ratiometric responses of the sensor can be tuned by changing certain sensor parameters. The sensor provides independent temperature control for each polymer zone, enabling tuning of the differential time response between polymer zones. In addition, the magnitude of the sensor response is temperature sensitive, so by varying the temperature of the polymer zones the ratiometric responses can be optimized for a given analyte or analytes. The carrier flow rate can also be adjusted to fine-tune the selectivity of the sensor.

In summary, the serial AFP array approach improves rejection of chemical interferents without sacrificing sensitivity. Because of the intrinsic selectivity of the polymers, data collected to date suggests that it may not be necessary to construct elaborate arrays with many orthogonal sensor elements to resolve interferents from target analytes. Finally, this discrimination is achievable while keeping sample analysis times short (less than 10 seconds).

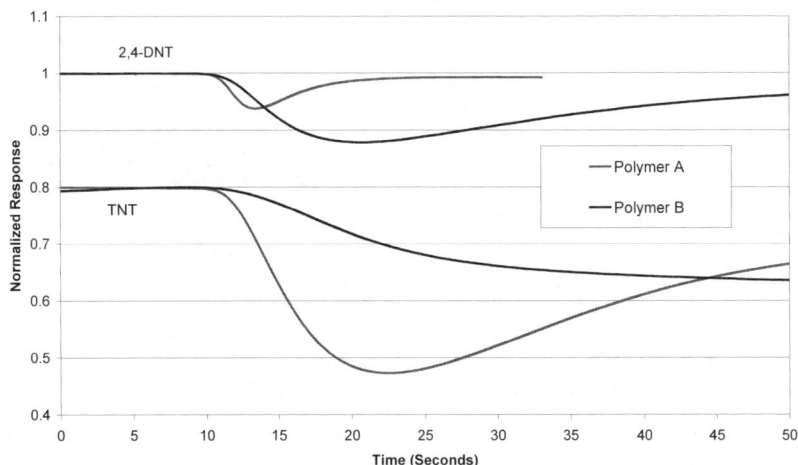

Figure 11. A comparison of the response of two different amplifying fluorescent polymers to 2,4-DNT and TNT.

5. CONCLUSIONS AND DISCUSSION OF FUTURE WORK

Preliminary testing of the Fido AFP array sensor suggests that the system can be used to improve selectivity relative to earlier single-channel Fido sensor prototypes. With a simple, two-element serial array sensor, improvement in the ability to discriminate target landmine chemical signature compounds from some potential chemical interferents has been demonstrated without any loss in sensitivity of the sensor. The sensor is small, lightweight, will operate for up to eight hours on a single battery charge, and can be produced at relatively low cost compared to similar sensor systems. Figure 12 illustrates the current handheld array sensor prototype.

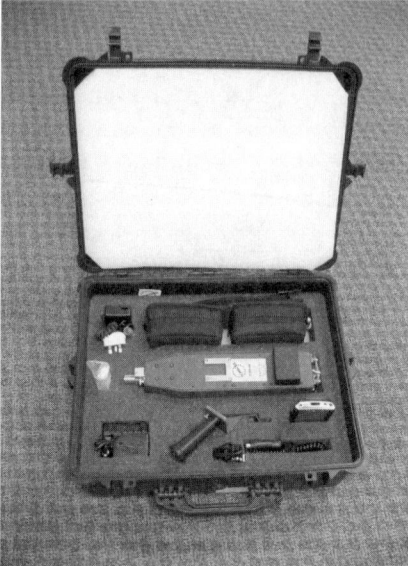

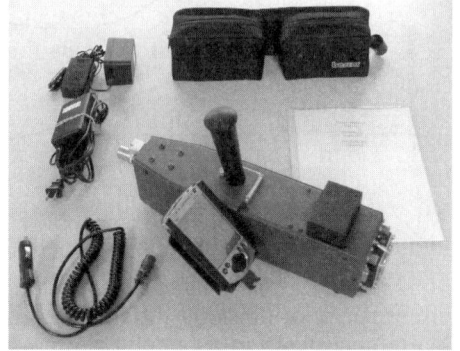

Figure 12. Handheld Fido AFP array sensor prototype.

Methods for improving the time resolution capabilities of the sensor array have been identified, and an improved prototype is nearing completion. The inclusion of additional sensor array elements has been considered as a mean to further improve selectivity. However, this will increase the complexity of the sensor and may not improve selectivity enough to justify the added complexity. An evaluation of the benefits and risks of utilizing more array elements in the sensor is underway. The two-channel array prototype promises significant improvement in the false alarm performance of the sensor, improving on the already excellent selectivity of previous generations of Fido sensors.

REFERENCES

1 M. laGrone, C. Cumming, M. Fisher, M. Fox, S. Jacob, D. Reust, M. Rockley and E. Towers, Detection of landmines by amplified fluorescence quenching of polymer films: A man portable chemical sniffer for detection of ultratrace concentrations of explosives emanating from landmines, Proc. SPIE, Detection and Remediation Technologies for Mines and Minelike Targets V, 4038 (2000) 553-562.
2 M. laGrone, C. Cumming, M. Fisher, D. Reust and R. Taylor, Landmine detection by chemical signature: detection of vapors of nitroaromatic compounds by fluorescence quenching of novel polymer materials, Proc. SPIE, Detection and Remediation Technologies for Mines and Minelike Targets IV, 3710 (1999) 409-420.

3   C. Cumming, C. Aker, M. Fisher, M. Fox, M. laGrone, D. Reust, M. Rockley, T. Swager, E. Towers and V. Williams, Using novel fluorescent polymers as sensory materials for above-ground sensing of chemical signature compounds emanating from buried landmines, IEEE Trans. on Geoscience and Remote Sensing, 39 (2001) 1119-1128.

4   T.F. Jenkins, M.E. Walsh, P.H. Miyares, J.A. Kopczynski, T.A. Ranney, V. George, J.C. Pennington and T.E. Berry, Analysis of explosives-related signature chemicals in soil samples collected near buried landmines, ERDC Technical Report 00-5 (2000).

5   V. George, T.F. Jenkins, D.C. Leggett, J.H. Cragin, J. Phelan, J. Oxley and J. Pennington, Progress on determining the vapor signature of a buried landmine, Proc. SPIE, Detection and Remediation Technologies for Mines and Minelike Targets IV, 3710 (1999) 258-269.

6   J.S. Yang and T.M. Swager, Porous shape persistent fluorescent polymer films: an approach to TNT sensory materials, J. Am. Chem. Soc. 120 (1998) 5321-5322.

7   J.S. Yang and T.M. Swager, Fluorescent porous polymer films as TNT chemosensors: electronic and structural Effects, J. Am. Chem. Soc. 120 (1998) 11864-11873.

8   V. Williams and T.M. Swager, Iptycene-containing poly(aryleneethynylene)s, Macromolecules 33 (2000) 4069-4073.

9   Q. Zhou and T. Swager, Methodology for enhancing the sensitivity of fluorescent chemosensors: energy migration in conjugated polymers, J. Am. Chem Soc. 117 (1995) 7017-7018.

10 Jane's Mines and Mine Clearance, Colin King ed., 1999-2000 (Fourth Edition).

Chapter 5

FAST DETECTION OF EXPLOSIVES VAPOURS AND PARTICLES BY CHEMILUMINESCENCE TECHNIQUE

Dao Hinh Nguyen, Shirley Locquiao, Phuong Huynh, Qiaoling Zhong, Wen He, David Christensen, Lin Zhang, and Bill Bilkhu
Scintrex Trace Corporation, 300 Parkdale Ave, Ottawa, Ontario, CANADA

Abstract: A chemiluminescence (CL) apparatus has been devised for the detection of nitro-containing explosives based on its reaction with luminol. This reaction is highly selective for NO_2 under certain conditions, so that other nitrogen-containing compounds such as ammonia, organic nitrite, organic nitrate, NO and hydrocarbons do not interfere. The detection system uses a gas-liquid reaction leading to light emission with a maximum at a wavelength of 425 nm. The reaction cell is packaged in a semi-permeable container, which allows NO_2 to react with the luminol and at the same time prevents leakage from the cell. The apparatus is highly sensitive and responds quickly to the fast release of NO_2 emanating from the pulsed decomposition of explosives vapours and particles. The simplicity of the technique allows for easy combination with other technologies and multi-detection platforms, such as CL-GC/IMS are possible.   In some instances, explosives can be identified by their decomposition rate without prior chromatographic separation. The use of a disposable chemiluminescence cartridge permits the sensor to be packaged into a small portable unit that can be easily deployed at airport or other security checkpoints

Key words: Chemiluminescence, luminol, explosives.

71

J.W. Gardner and J.Yinon (eds.),
Electronic Noses & Sensors for the Detection of Explosives, 71-80.
© 2004 *Kluwer Academic Publishers. Printed in the Netherlands.*

1.   INTRODUCTION

A thermochemiluminescence (TCL) apparatus has been devised for the detection of nitro-containing explosives based on its reaction with luminol. This chemiluminescence reaction is highly selective to NO_2 under certain conditions, so that other nitrogen-containing compounds such as ammonia, organic nitrite, organic nitrate, NO and hydrocarbons do not interfere. A different method [1] for detection of NO_2 is to first convert it to nitric oxide (NO), which then subsequently reacts with O_3 to emit light by chemiluminescence as well. Inaccuracies of that method result from the fact that small amount of NO originating from reduction of NO_2 is compared with high quantity of NO in the air.

The above mentioned disadvantage is avoided by the direct measurement of NO_2, which chemiluminescence reaction is described below:

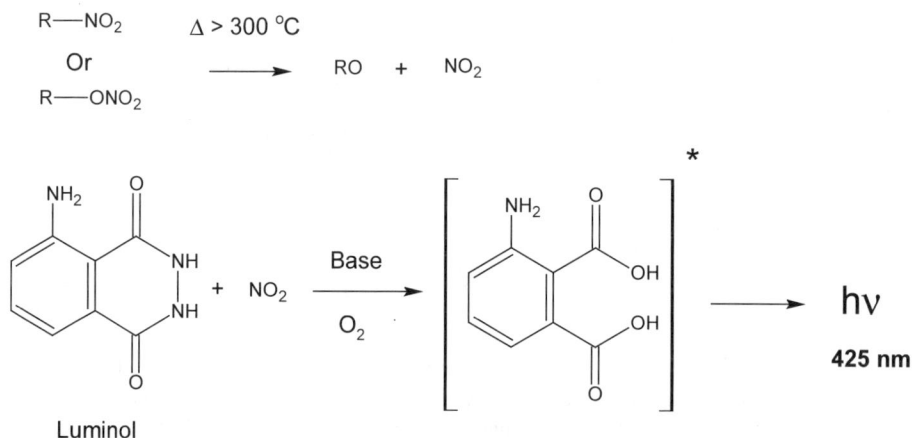

R—NO$_2$ $\Delta > 300\ °C$
Or $\longrightarrow$ RO   +   NO$_2$
R—ONO$_2$

Luminol + NO$_2$ $\xrightarrow{\text{Base}, O_2}$ [...]* $\longrightarrow$ hν **425 nm**

Other common techniques for the determination of NO_2 are laser absorption spectroscopy, fluorescence or photo-acoustic spectrometry, or also differential optical absorption spectroscopy. Instrumentation of those methods is mostly complicated, and expensive to operate and maintain.

In the present technique, organo-nitrate compounds are first trapped into a collection tube that contains a proprietary sampling material. The desorbed material is then catalytically converted to NO_2 upon exposure to pulse heated platinum wire. Chemiluminescence is then produced by the reaction of NO_2 with luminol. The detection system uses a gas-liquid reaction leading to light emission with a maximum at 425 nm. This wavelength corresponds to the maximum sensitivity for most photomultiplier tubes. The biggest problem with this method for direct measurement of NO_2 has been interferences from other soluble oxidants, particularly peroxy acetyl nitrates (PAN, R-C=OO-O-NO_2) [2]. Moreover, ozone and sulfur dioxide have been found to positively interfere with this reaction whereas carbon dioxide negatively. Munemorl et al. [3] discovered that the ozone and sulfur dioxide interference is removed by addition of sodium sulfite (Na_2SO_3) to the luminol solution. Whereas the negative contribution from carbon dioxide can be rectified by adjusting the alkalinity of the solution.

The solution is packaged into a scintillation vial equipped with a hydrophobic membrane, which allows NO_2 to react with the luminol and at the same time prevent from any possible leakage from the cell. The operator easily replaces the cartridge after several weeks of operation.

The unit is operated in a pulsed mode which increases the momentary concentration of NO_2, thereby improving its sensitivity. Figure 1 shows the typical response to various concentration of NO_2 in a continuous mode.

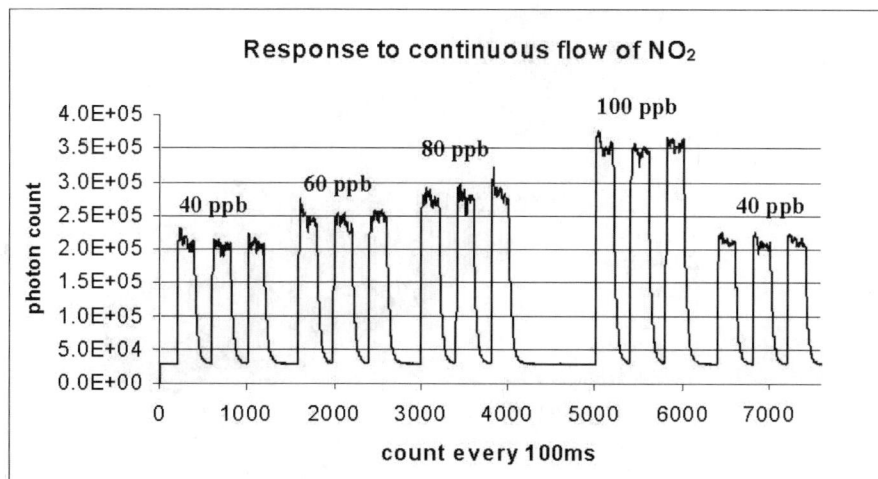

Figure 1. Typical response to various concentration of NO_2 in a continuous mode.

The simplicity of the technique makes it ideal to use it as a detector for explosives or in combination with other complementary techniques, such as GC/IMS.

2.   E-3500

The apparatus is highly sensitive and responds quickly to the fast release of NO_2 emanating from the pulsed decomposition of explosive vapours and particles.

To collect vapours the user samples in and around an area where explosives are suspected. To collect particles, the user wipes the surface of suspect articles with gloves and transfers the sample onto a metal sample screen. The sample screen (Figure 2) is then inserted into the E-3500 for analysis. The analysis is usually completed within 10 seconds and the results displayed on the LCD (liquid crystal display). An alarm is indicated by a red light, audible signal, and a value displayed on the LCD.

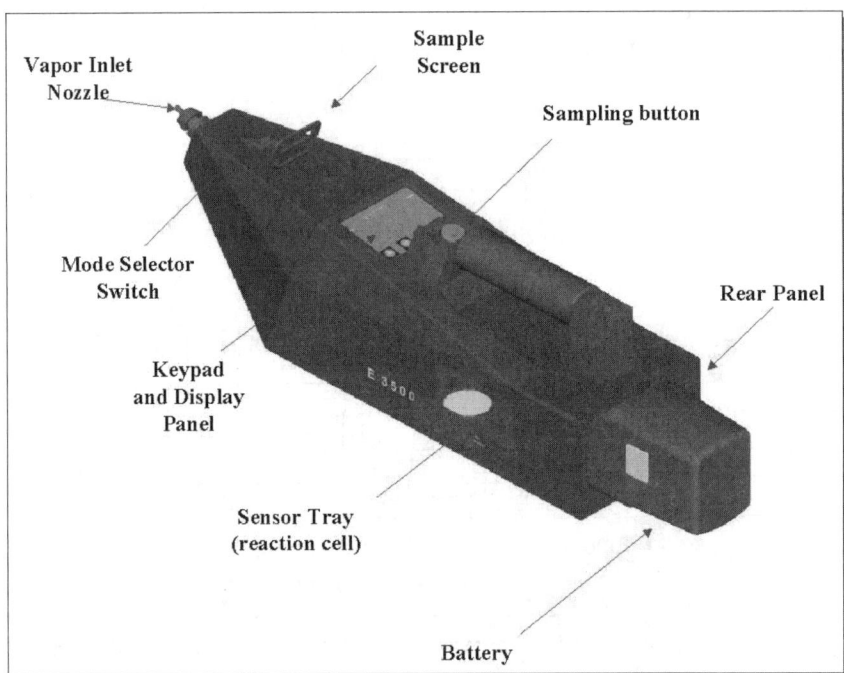

Figure 2. Schematic of E-3500 detector.

Because the technique relies on thermal decomposition of nitro-containing explosives, and no prior chromatographic separation is performed, no chemical information of the studied material is obtained. Currently, the unit will alarm on compounds such as RDX, PETN, TNT, urea nitrate, ammonium nitrate, nitroglycerine, EGDN and DMNB.

Figures 3 and 4 show, respectively, results from particle and vapour sampling.

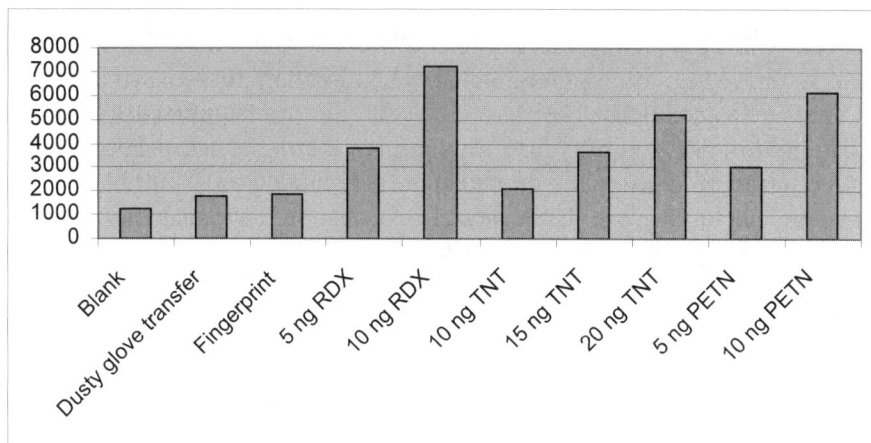

Figure 3. Results from particle sampling.

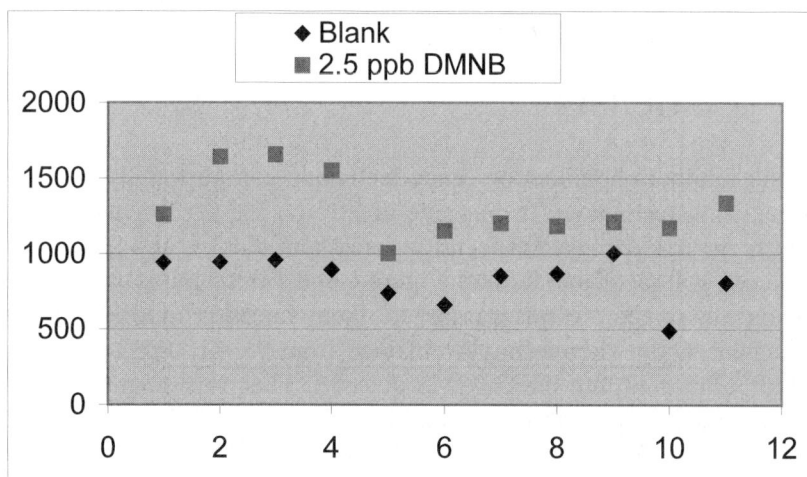

Figure 4. Results from vapour sampling. Samples are: blank: 832 (+/- 151); 2.5 ppb DMNB: 1306 (+/- 215).

The unit can also detect triacetone-triperoxide (TATP), which does not contain any nitro functionality. However, when thermally decomposed TATP seems to release oxidants capable of inducing luminol chemiluminescence as shown in Figure 5.

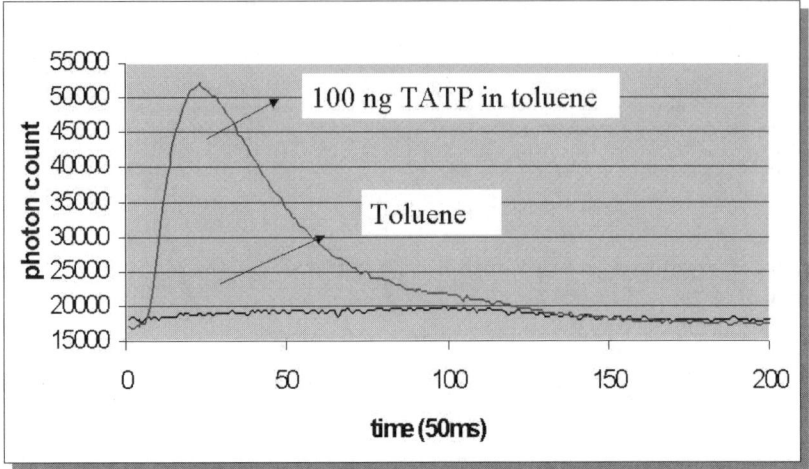

Figure 5. Response of detector to 100 ng of TATP in toluene.

3. APPLICATIONS

3.1 CL-GC/IMS

The luminol chemiluminescence technique can be rapidly incorporated into existing technology to improve greatly on the performance of current security devices deployed at security checkpoint. For instance, when coupled to the GC-IMS detector (E 5000, Figure 6 and Table 1), it provides a rugged method to screen samples prior to their injection into the analytical instrument. Only when a 'hit' is obtained from the CL detector, the sample would be injected into the GC-IMS detector. This process greatly enhances the lifetime of the chromatographic column and also keeps it from contamination. By combining both techniques, false alarm level rate and instrument downtime is considerably reduced.

Table 1. Results from CL and GC-IMS for different compounds.

	Blank	1 ng RDX	1 ng PETN	1 ng TNT	1 ng NG
CL	2.28 (0.45)	2.41 (0.07)	3.8 (0.22)	2.17 (0.04)	2.49 (0.18)
GC-IMS		1157 (433)	10133 (244)	11920 (309)	1848 (15)

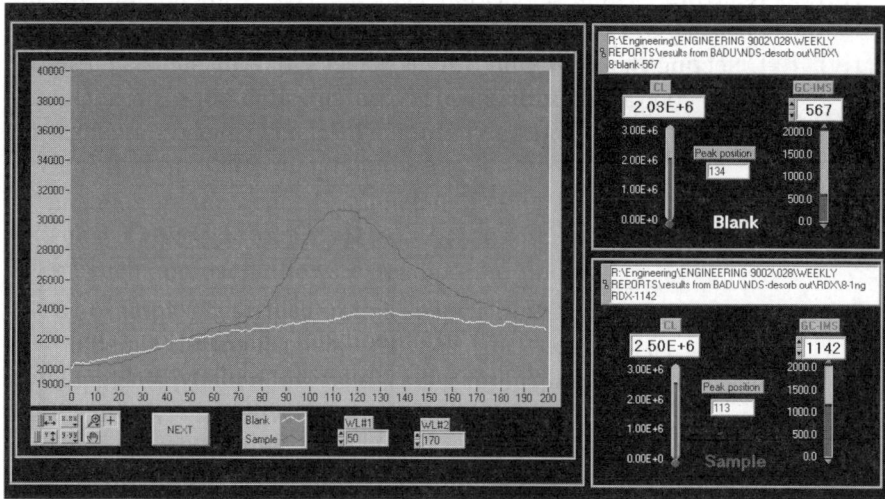

Figure 6. Front panel of the E 5000 detector.

3.2 Laser-CL

The use of the luminol reaction enables a compact scanning system to detect the presence of a wide variety of contraband substances in an accurate and reliable manner. The system rapidly and accurately discriminates among different substances and provides quantitative indication of the amount and location of a critical substance. It is especially well suited for use in applications which require high throughput and accuracy, such as security screening associated with airline and other forms of public transportation.

Advantageously, the system provides in some aspects for automated screening. It can be configured to automatically scan substantially the entire exterior surface of luggage and other hand-carried personal items, as well as cargo, without the need for hand wiping or sampling by an operator or other physical contact. Vagaries of human performance are virtually eliminated, and detection efficacy is improved. The system's greater speed, accuracy, reliability, and flexibility, as well as its lower cost, and expanded range of detectable substances overcome problems associated with commercial scanning systems.

Generally stated, the method comprises the steps of:

(i)   Producing NO_2 by decomposition of at least a portion of the contraband substance

(ii)   Transferring the NO_2 to a reaction cell, a portion of which contains an aqueous, alkaline luminol solution.

(iii)   Reacting, within a reaction cell, the NO_2 with luminol in the presence of O_2 to produce light by chemiluminescence.

(iv)   Detecting the light with a light detector to indicate the presence of the contraband substance (Figure 7).

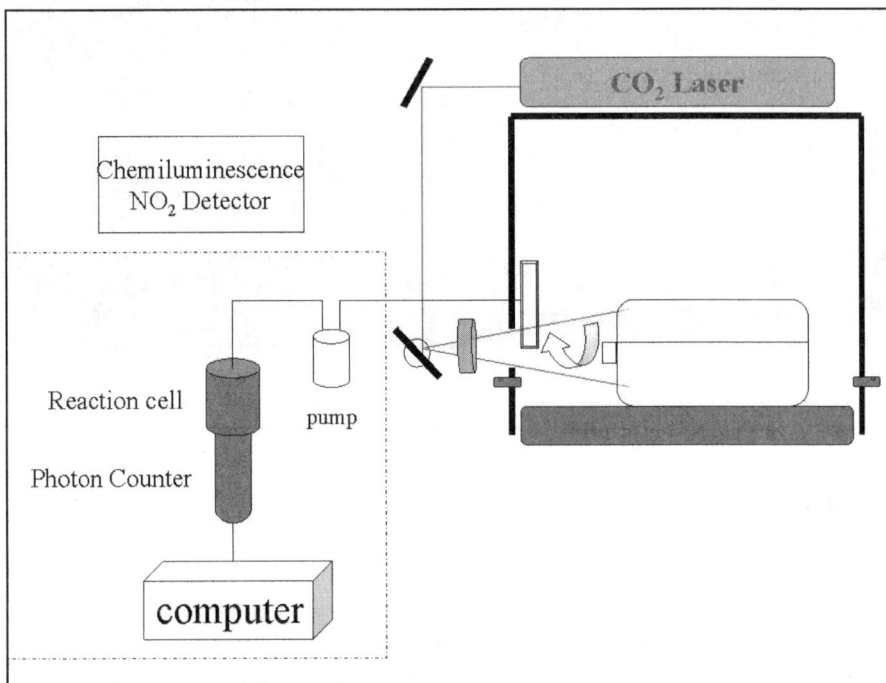

Figure 7. Arrangement of system with CL NO_2 detector.

Partial decomposition of explosives [4], upon laser irradiation of the surface, yields a considerable amount of NO_2 which can be detected in near real time by the chemiluminescence sensor. It has been found that this process can be done by either using a continuous wave laser (Figure 8) or a pulsed laser (Figure 9). In the last case, more NO_2 is evolved upon interrogation of the surface.

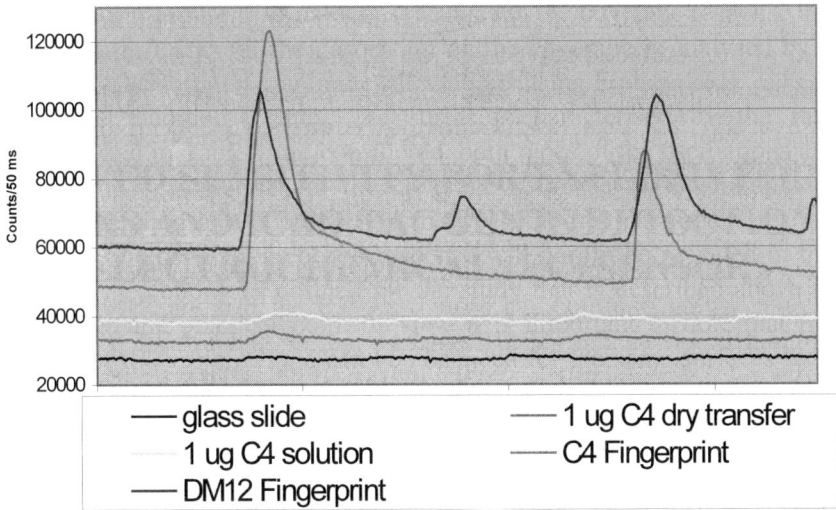

Figure 8. Results with continuous wave laser.

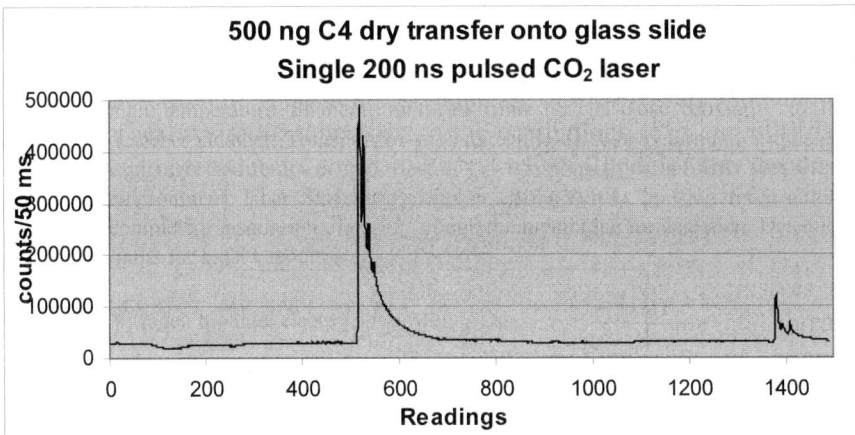

Figure 9. Results with pulsed laser.

REFERENCES

1 D.P. Rounbehler, S.J. McDonald, D.P. Lieb and D.H. Fine, Analysis of explosives using high speed gas chromatography with chemiluminescence detection, Proc. First Int. Symp. Explos. Detect. Technol., Khan, S.M. Ed., FAA Atlantic City, NJ, November 13-15, 1991 pp. 703-713

2 J.S. Gaffney, R.M. Bornick, Y.-H. Chen and N.A. Marley, Capillary gas chromatographic analysis of nitrogen dioxide and PANs with luminol chemiluminescent detection, Atmospheric Environment 32 (1998)1145-1154.

3 Y. Maeda, K. Aoki and M. Munemorl, Chemiluminescence method for the determination of nitrogen dioxide, Anal. Chem. 53 (1986) 307-311.

4 C. Capellos, S. Lee, S. Busulu and L.A. Gam, Infrared laser multiphoton decomposition of 1,3,5-trinitrohexahydro-s-triazine (RDX), Advances in Chemical Reaction Dynamics (1986) 395-404.

Chapter 6

OPTICAL MICROSENSOR ARRAYS FOR EXPLOSIVES DETECTION

David Walt and Tamar Sternfeld
Department of Chemistry, Tufts University, Medford, Massachusetts, USA.

Abstract: The electronic nose described in this paper uses a cross-reactive sensor array based on fluorescence sensors. The sensors are fabricated by attaching solvatochromic dyes to different microspheres. The microspheres are then placed into wells chemically etched on the distal end of an optical fibre bundle. The system uses an olfactometer to deliver a pulse of analyte vapour to the sensors. An optical imaging system is employed to monitor fluorescence intensity over time. We use a heterogeneous array that contains different types of sensors, which allows us to classify a large number of analytes and complex odours. The array is formed by randomly distributing microspheres on the end of the fibre array. The position of each microsphere is determined by using a method that compares different sensor responses to their responses to known analytes. The electronic nose has been used to detect explosives and explosive-like vapours at low levels, and was able to detect nitro aromatic compound (NAC) vapour concentrations as low as 5 ppb. In addition, by fabricating a model of a nasal cavity and placing identical sensors at different positions, we demonstrated how the flow environment affects sensor response. By using the information from multiple sensors placed in different spatial positions in the complex flow environment, we demonstrated it is possible to obtain better discrimination between analytes.

Key words: Electronic noses, odour detection, explosive vapours, cross-reactive array.

J.W. Gardner and J.Yinon (eds.),
Electronic Noses & Sensors for the Detection of Explosives, 81-92.

1.   INTRODUCTION

'Electronic noses' are vapour detection systems that mimic key principles of biological olfaction [1]. The functioning principles of biological, olfactory systems do not rely upon selective interactions with specific analytes, but rather on cross-reactive receptors [2]. The receptors respond to many odours, generating unique response patterns, which serve as 'fingerprints' for each odour.

Using the principles of biological olfaction, electronic nose systems contain arrays of different types of cross-reactive vapour-sensitive sensors. While it is difficult to discriminate analytes entirely by their responses to a single type of sensor, using an array of sensors yields response patterns that can readily distinguish many different vapours. Ideally, the response mechanisms of the sensors are highly varied and encompass both physical and chemical phenomena [1].

2.   OPTICAL ELECTRONIC NOSE

Our electronic nose measures fluorescence intensity responses from an array of polymer sensors over time during exposure to different analyte vapours. The sensors contain solvatochromic dyes that undergo intensity and wavelength shifts depending on changes in micro-environmental polarity. During exposure of the array to vapour-phase analytes, each of the sensor types in the array produces a temporal fluorescence response profile that depends on the sensor-analyte interaction. The combination of different responses from the sensor array to an analyte creates a pattern that characterizes the specific analyte and, therefore, enables identification. Such an array can recognize a variety of volatile organic vapours, and moreover, it can discriminate complex odours that contain multiple components, such as different types of coffees or perfumes. In these complex odours, the electronic nose recognizes the response pattern of the mixture, eliminating the necessity to quantify every single component.

The sensor array system is comprised of several important elements. The sensor substrate is an optical imaging fibre bundle, which is a high-density array of micrometer scale optical fibres [3]. The structure of each single fibre comprising this bundle contains a cylindrical 'core' material surrounded by an outer 'cladding' (Figure 1). The refractive index of the core is slightly higher than the refractive index of the cladding, and as a result, light is transmitted via total internal reflection through the core over long distances. The imaging fibre bundle generally contains thousands to tens of thousands of closely packed individual optical fibres (Figure 2), with each fibre retaining its own isolated optical path, enabling it to transmit a signal from one end to the other. Each individual optical fibre acts as a single pixel, and the assembly of many individual pixels into a fused coherent bundle enables images to be transmitted through the entire length of the bundle (Figure 2) [3c].

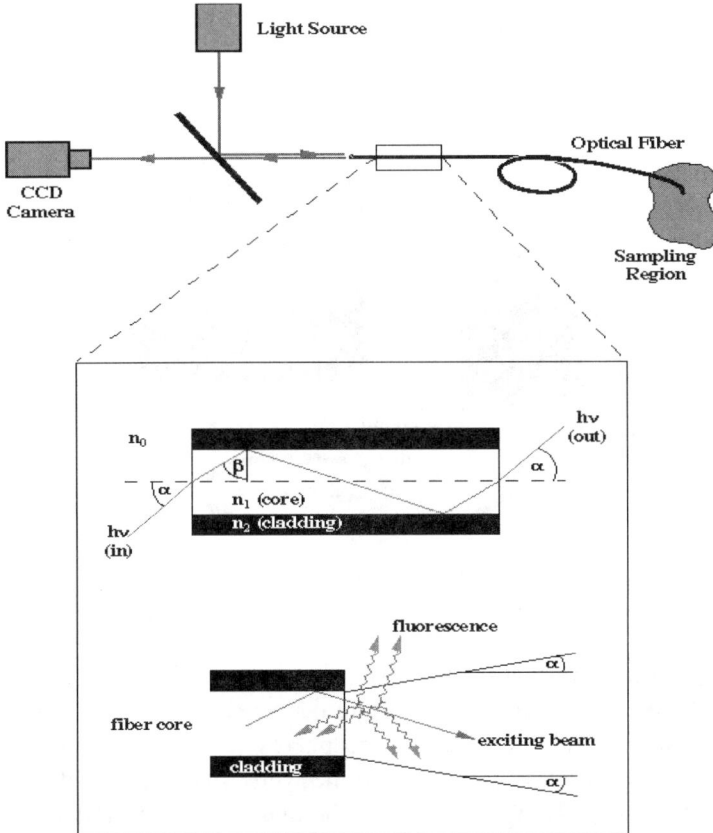

Figure 1. Schematic of the optical fibre system. Excitation light is launched into the fibre. Due to the refractive index differences between the fibre core and cladding materials, the light is internally reflected and travels through the fibre with minimal loss (see inset). The emitted light is carried back from the fluorescent sensor located on the tip of the fibre to a CCD camera detector. Reprinted with permission from Science, 2000, 287, 451-452. Copyright 2000 AAAS.

The fibre bundles used for the electronic nose platform are polished and chemically etched with hydrofluoric acid to create an ordered array of micrometer sized wells on the tip of the fibre (Figure 3) [4]. The etching process takes advantage of the difference in reactivity of the core and the

84

cladding materials. All the wells in the array have a depth that is controlled by the etch reaction time. The base of each well in the array is the distal end of an optical fibre which carries light from the well to the fibre's proximal end attached to a detection system. Thus, the optical fibre well array enables each microwell to be addressed and serves as a useful platform for the sensors on the electronic nose.

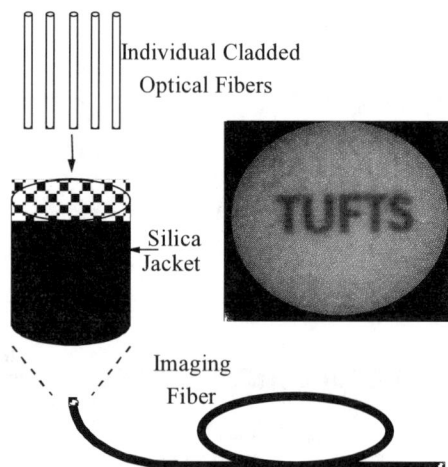

Figure 2. Coherent imaging bundle. The fibre optic imaging bundles are made by fusing many single core fibres [3c].

The sensors are fabricated by adsorbing or covalently attaching solvatochromic dyes to various microspheres, including differentially modified silica and polymer microspheres (also called beads). The microspheres are slightly smaller than the wells such that the diameter of a microbead sensor is complementary to the diameter of an individual well. A single bead fits inside a well and remains immobilized as a result of complementary binding forces between the glass surface and the functional groups on the microsphere. With tens of thousands of wells in each array, thousands of replicate beads of each sensor type are represented in the array (Figure 3) [3a]. By using a simple procedure for filling the array, 96% of the wells are occupied by the microbeads (Figure 3b) [5], resulting in a high-density array.

Using the microspheres as a 'substrate' for solvatochromic dyes provides many advantages. The silica microspheres have a high surface area and a high affinity for the solvatochromic dye molecule, which ensures adsorption on the surface. These microsphere properties help improve their ability to classify analyte vapours. The loading of dye molecules on microsphere surfaces is adjusted to achieve an optimal concentration, and the dense surface functionality maximizes vapour–dye interaction. As a result, the fluorescent signal-to-noise (S/N) ratio is enhanced. Moreover, by using microspheres that have diverse chemical and material characteristics, such as surface functionality or bead composition, the number of possible different types of sensors is increased. Hence, by adsorbing the same dye to different types of microspheres, multiple sensor types can be generated. Each sensor type interacts differently with different vapours, resulting in a unique fluorescence response profile (to be discussed in detail later) [6]. The solvatochromic dye Nile Red has been used for most of our experiments. This preference is based on the large shifts in the dye's emission wavelength maximum, and its relatively high photostability [7].

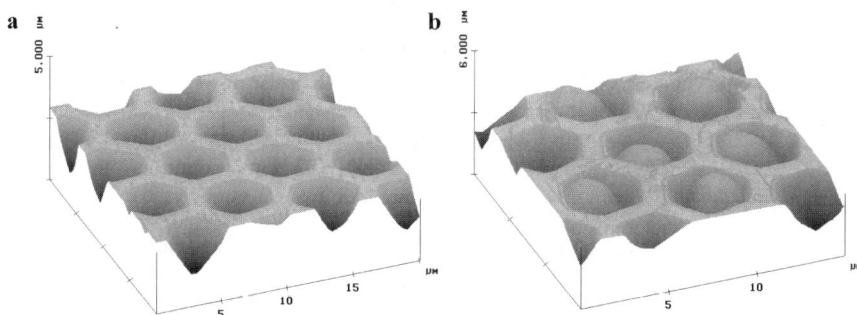

Figure 3. Atomic force microscopy image of an etched fibre bundle (a) before and (b) after microspheres were distributed into the array. Reprinted with permission from ref. [5]. Copyright 1998 American Chemical Society.

The silica microspheres provide some diversity but not enough for many complex discrimination tasks. To introduce more sensor variety, hollow polymeric microspheres have been fabricated [8]. The preparation of these hollow microspheres involves coating silica microspheres by living radical polymerization, using the surface as the initiation site. Once the polymer layer forms on the silica microbead surface, the silica core is removed by chemical etching. These hollow spheres can be derivatized with the dye of interest. The main advantage of these polymer microspheres is the variety of monomers that can be employed in their fabrication to produce sensors with many different surface functionalities and polymer compositions

A fluorescence measurement is performed by directly delivering a vacuum controlled pulse of the analyte vapour diluted with air to the distal end of the optical fibre containing the sensors (Figure 4). The optical instrument includes a fluorescence microscope and a charge coupled device (CCD) camera. The excitation light is launched into the fibre, and the fluorescent light, emitted from the sensors, is carried back along the fibre to the camera (Figure 1). The fluorescence intensity response for each sensor is plotted versus time, which creates a pattern that characterizes the sample.

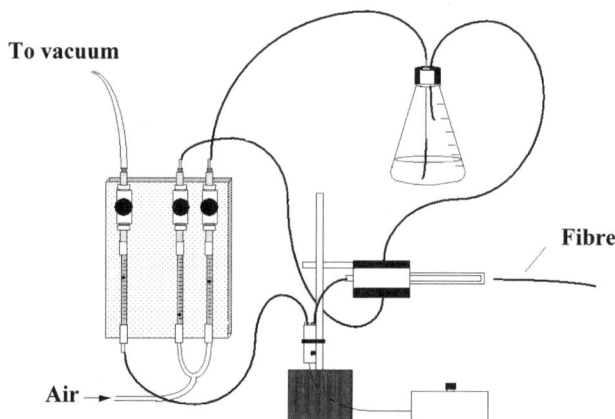

Figure 4. A schematic diagram of the olfactometer system.

In analogy to the biological olfactory system that contains a high number of different sensors, the electronic nose can be assembled from many different sensor materials, all of them located on a single fibre bundle to form a heterogeneous array. In such cross-reactive arrays, complex vapour responses are simultaneously collected including changes in intensity, wavelength, and spectral shape.

The sensors in a heterogeneous array are prepared from different types of microspheres, as well as different dyes. A mixture of sensors is randomly distributed onto the fibre tip, which means that the occupied positions of each type of sensor are different from array to array. Therefore, the first step after array fabrication involves the positional registration of each individual sensor in the array, i.e. decoding of the array [9]. Since each sensor type exhibits a unique response to a given vapour, the decoding is done by exposing the array to known analytes and matching the acquired fluorescence responses to the responses collected separately for each sensor type (Figure 5). We define this method as a 'self-encoded bead array' where each bead's analytical signal can also be used as its encoding signal.

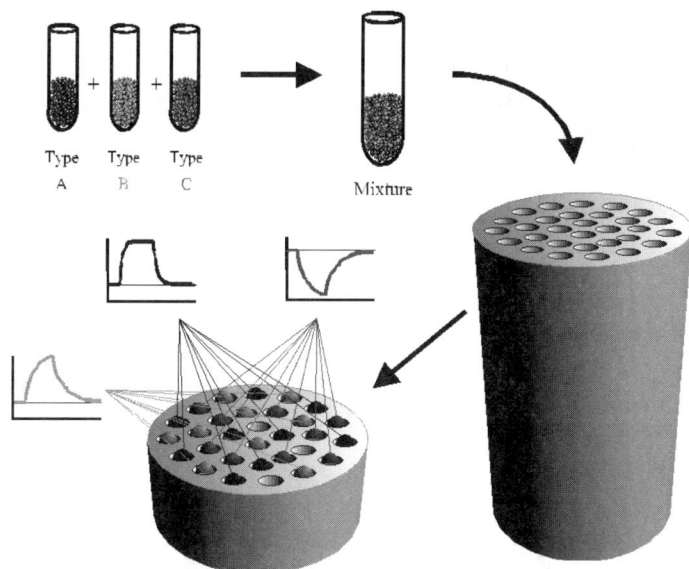

Figure 5. Schematic depiction of a self-encoded bead array. A mixture of three sensor types fills the fibre tip wells randomly. The sensors are identified by their characteristic responses to a test vapour pulse. Reprinted with permission from ref [9b]. Copyright 1999 American Chemical Society.

Many advantages arise from a heterogeneous array including the miniaturization of the array, the high density of different sensor types, shorter detection times due to the small sensor sizes, low materials cost, and the ease of preparation.

A critical issue that arises in the development of all electronic noses is array-to-array reproducibility. If arrays cannot be made reproducibly, new training is required for each new array. These microsphere beads show good sensor-to-sensor reproducibility and stability over time [10]. Moreover, these sensors are made in large quantities, which eliminate the issue of batch-to-batch reproducibility. Hence, the main question is whether a sensor response pattern is reproducible from array to array. In one study, arrays of four different microsphere types were exposed to ethanol and 4-nitrotoluene on three different days over a six-month period. The first array was used as the 'training array', while the other two were used as 'testing arrays'. Although the responses of each microsphere sensor type to the analytes varied from array to array, the classification was always high (over 93%). These results demonstrate the ability of the electronic nose to 'remember' the odour of different analytes and to recognize them over extended periods of time. Therefore, we are able to create an 'odour memory' library that can be maintained from one array to another over extended time periods [10].

88

Another approach to analyze the multi-sensor array responses is the ensemble approach [11], which does not require decoding and therefore, speeds up the identification process. The analysis in this approach uses combined response profiles of the entire undecoded array as a single response (Figure 6).

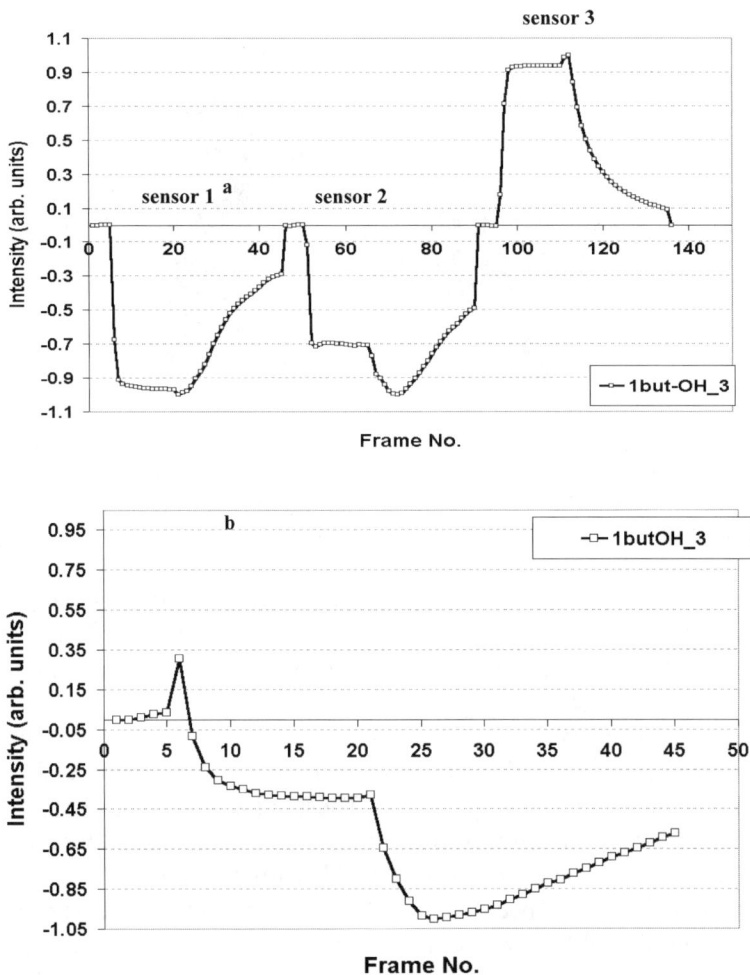

Figure 6. Response profiles from three different sensor types. (a) Responses of each of the individual sensor types (decoded array). (b) The collective response (undecoded array). Reprinted with permission from ref [11]. Copyright 2003 American Chemical Society.

Generally speaking, the higher the number of sensor types in the array, the more likely the array will yield a unique fluorescence response profile for each odour, leading to better classification of odours. For example, a classification of 97% correct was achieved by using a heterogeneous array that contained 18 bead types. In this experiment, 20 different odours, including pure analytes as well as mixtures, were compared using a total of 100 observations (five exposures per odour) [11].

3. RESPONSE TO EXPLOSIVE VAPOURS

An important demonstrated application of this artificial nose system is the high-speed detection of low levels of explosives and explosive-like vapours. Several sensors, based on Nile Red attached to silica microspheres, show high sensitivity to nitroaromatic compounds (NAC) within a mixture [12]. Different fluorescence response profiles were observed for several NAC's, such as 1,3,5-trinitrotoluene (TNT) and 1,3-dinitrobenzene (DNB), despite their similar structures. These responses were monitored at low concentrations of the NAC vapours (ca. 5 ppb) and at short vapour exposure times (less than 200 ms). The detection of such low concentrations is possible since we are able to measure simultaneously the responses of a high number of individual sensors and sum their responses, thereby increasing the S/N ratio. Theoretically this ratio increases by $n^{1/2}$, where n is the number of sensors analyzed. Therefore, by combining the responses of a large number of individual beads (1000 beads), the noise is essentially removed and low concentrations of analytes can be detected (Figure 7).

High-speed detection is a necessity for many artificial nose applications. In one study [12a], it was shown that even at short exposure times (< 1 sec), the nose could identify different vapours and the responses were reproducible. Figure 8 demonstrates this quality, when three high-speed exposures (0.38 s exposure time) produced reproducible response profiles.

In addition, a linear dependence was found between the concentration of DNB and its fluorescence response profile. All these characteristics demonstrate that this sensor array is suitable for use in detecting explosive vapours.

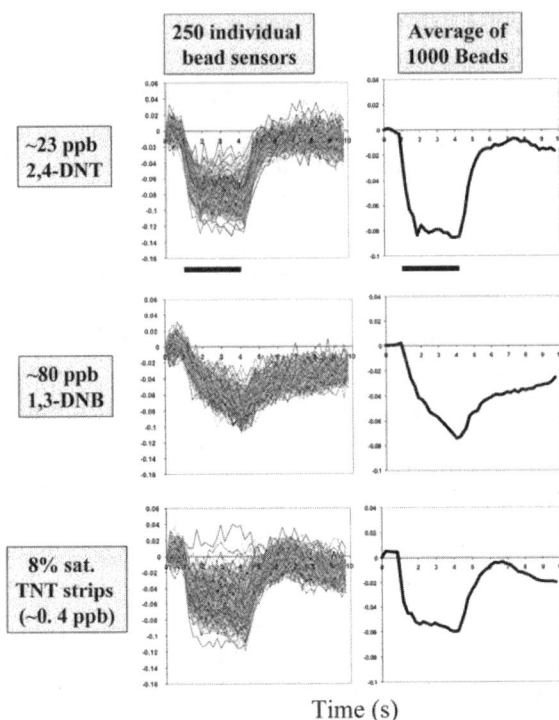

Figure 7. Simultaneously monitoring vapour signatures of 1000 sensors for 2,4-DNT, 1,3-DNB, and TNT vapour strips at 8% saturated vapour levels. The (noisy) responses for 250 individual sensors are compared to the averaged response profile for 1000 individual sensors. Reprinted with permission from ref [12]. Copyright 2000 American Chemical Society.

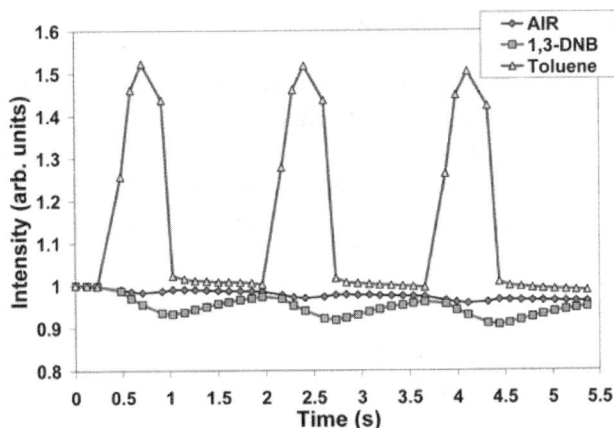

Figure 8. Seventy-six sensor beads (Jupiter C4/Nile Red) monitored to show that the average responses for three consecutive 0.38 s exposures of 50% saturated vapour levels result in reproducible and high-speed response profiles. The sensors are positioned on the distal tip of an optical imaging fibre and relative analyte concentrations are 0.5 and 18700 ppm for 1,3-DNB and toluene, respectively. Reprinted with permission from ref [12a.] Copyright 2000 American Chemical Society.

The sensor response depends not only on the polarity of the analyte, but also on the flow environment. In the mammalian olfactory system, the nasal cavity structure plays an extremely important role in odour discrimination [13]. Identical olfactory neurons are located in different places in the cavity, and therefore occupy different positions in the flow path. By using a nasal cavity model, we investigated the influence of the dynamic flow on the sensors' response [14]. The responses from identical fibre optic sensors located at different positions in a nasal cavity model using different exposure conditions were measured (Figure 9). These identical sensors produced different fluorescence response profiles to the analytes depending on their positions. As a result of the sensors' arrangement, the classification of odours improved. When rums, vodkas, and perfumes were analyzed, the percent classification rates increased by 8±5%, 8±6% and 7±3%, respectively. Moreover, in some cases using the nasal cavity model made it possible to distinguish between analytes that could not be distinguished by using the same sensor in a single position.

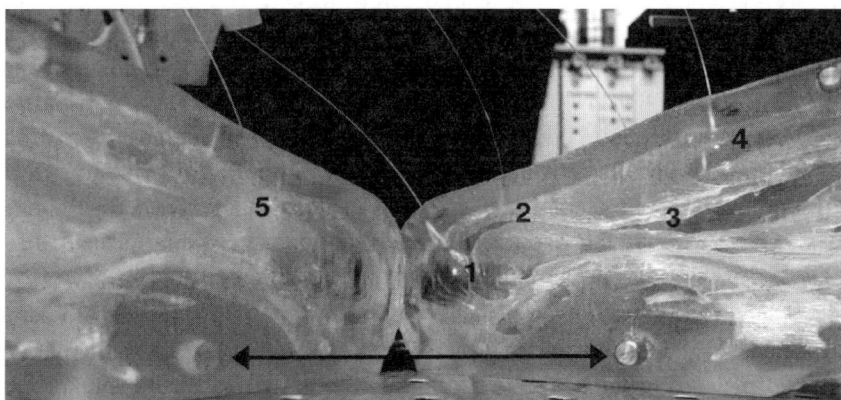

Figure 9. A plastic model of a nasal cavity showing the different positions (numbers 1-5) of the sensors. Reprinted with permission from ref [9]. Copyright 2003 American Chemical Society.

This research demonstrated the significance of the flow environment on the sensor responses. By placing the same sensors in different flow environments, more information could be obtained. We have been able to mimic some of the properties of the olfactory system by exploiting some of the biological system's design principles. To improve the vapour classification ability, a cross-reactive fluorescence sensor array was designed and implemented along with a flow environment modeling the nasal cavity structure. The system showed high sensitivity, and an ability to identify explosive vapours at very low concentrations.

ACKNOWLEDGMENTS

This work was funded through generous support of DARPA and the Office of Naval Research.

REFERENCES

1 K.J. Albert, N.S. Lewis, C.L. Schauer, G.A Sotzing, S.E. Stitzel, T.P. Vaid and D.R. Walt, Chem. Rev. 100 (2000) 2595-2626.
2 S. Firestein, Nature, 413 (2001) 211-218.
3 (a) D.R. Walt, Science 287 (2000) 451-452. (b) D.R. Walt, Acc. Chem. Res. 31(1998) 267-268. (c) J. Epstein and D.R. Walt, Chem. Soc. Rev. 32 (2003) 203-214.
4 P. Pantano and D.R. Walt, Chem. Mater. 8 (1996) 2832-2835.
5 K.L. Michael, L.C. Taylor, S.L. Schultz and D.R. Walt, Anal. Chem. 70 (1998) 1242-1248.
6 (a) K.J. Albert, D.S. Gill, T.C. Pearce and D.R. Walt, Anal. Chem. 73 (2001) 2501-2508. (b) S.R. Johnson, J.M. Sutter, H.L. Engelhardt, P.C. Jurs, J. White, J.S. Kauer, T.A. Dickinson and D.R. Walt, Anal. Chem. 69 (1997) 4641-4648. (c) J. White, J.S. Kauer, T.A. Dickinson and D.R. Walt, Anal. Chem. 68 (1996) 2191-2202.
7 (a) E. Vauthey, Chem. Phys. Lett. 216 (1993) 530-536. (b) J.F. Deye and T.A. Berger, Anal. Chem. 63 (1990) 615-622.
8 T.K. Mandal, M.S. Fleming and D.R. Walt, Chem. Mater. 12 (2000) 3481-3487.
9 (a) T.A. Dickinson, J.S. White, J.S. Kauer and D.R. Walt, Nature 382 (1996) 697-700. (b) T.A. Dickinson, K.L. Michael, J.S. Kauer and D.R. Walt, Anal. Chem. 71 (1999) 2192-2198.
10 E. Stitzel, L.J. Cowen, K.J. Albert and D.R. Walt, Anal. Chem. 73 (2001) 5266-5271.
11 K. Albert and D.R. Walt, Anal.Chem. (2003) 4161-4167
12 (a) K.J. Albert and D.R. Walt, Anal. Chem. 72 (2000) 1947-1955. (b) K.J. Albert, M.L. Myrick, S.B. Brown, D.L. James, F.P. Milanovich and D.R. Walt, Environ. Sci. Technol. 35 (2001) 3193-3200.
13 (a) K. Keyhani, P.W. Scherer and M.M. Mozell, J. Theor. Biol. 186 (1997) 279-301. (b) J.S. Kauer and J. White, Annu. Rev. Neurosci. 24 (2001) 963-979.
14 S.E. Stitzel, D.R. Stein and D. R. Walt, J. Am. Chem. Soc. 125 (2003) 3684-3685.

Chapter 7

METAL OXIDE SEMICONDUCTOR SENSORS FOR DETECTION OF TOXIC AND EXPLOSIVE GASES

Dzmitry Kotsikau and Maria Ivanovskaya
Research Institute for Physical Chemical Problems of the Belararusian State University, Leningradskaya 14, Minsk, BELARUS.

Abstract: The influence of chemical composition and structural peculiarities of Fe_2O_3-SnO_2 and Fe_2O_3-In_2O_3 nano-scale composites on their gas-sensitive properties has been studied. The correlation between the structural factors and the functional characteristics of the sensitive layers (long-term parameter stability, threshold sensitivity, selectivity) when detecting some explosive and toxic gases (CH_4, CO, C_2H_5OH, NO_2, O_3) has been established. The active layers in form of thin and thick films were prepared by liquid phase deposition of hydroxide sols stabilised with various additives (Sol-Gel Technology). It was shown that the indicated preparative approach allows achieving the required gas-sensitive features of the metaloxide semiconductor sensors by fine adjusting of an active layer structure. Advanced materials demonstrating optimal sensitive characteristics regarding the studied gases have been proposed.

Key words: Explosive gases, toxic gases, gas sensor, selectivity, Fe_2O_3, In_2O_3, SnO_2

1. INTRODUCTION

The detection of the certain gases is drastically important not for technological applications solely, but for anti-terrorist actions. Thus, reducing odourless and colourless gases, when mixed with air or oxidising gases give explosive agents. Besides, explosions commonly lead to the

93

J.W. Gardner and J.Yinon (eds.),
Electronic Noses & Sensors for the Detection of Explosives, 93-115.
© 2004 *Kluwer Academic Publishers. Printed in the Netherlands.*

evolution of other toxic and flammable gases, often odourless and colourless, which have to be detected as well.

In previous papers, we have shown the possibility of selective detection of a series of gases by using heterojunction oxide composites based on In_2O_3, MoO_3, NiO, ZnO as well as by oxides doped with noble metals like Pd, Pt, Au. High gas-sensitive performance of the mentioned systems has been established. For details see reviews [1-3]

The present paper is mainly devoted to the binary oxide composites based on Fe_2O_3, which appear to be promising materials for gas sensing applications.

The existing literature reports that the sensitivity of ceramic sensors based on Fe_2O_3 layers to reducing gases is rather low [4-6]. However, doping Fe_2O_3 with quadrivalent metal ions (Sn, Ti, Zr) as well as the modification of this material with SO_4^{2-} ions can significantly enhance the gas-sensitive properties of the corresponding sensors towards ethanol and hydrocarbons [6]. In particular, the addition of Fe_2O_3 to SnO_2 thick films leads to an increased response to ethanol [7]. There are also some papers concerning the effect of Fe_2O_3 additives on the properties of In_2O_3 based sensors; for example, the sputtering of Fe_2O_3 layer over In_2O_3 thin film increases its sensitivity to O_3 and reduces the optimal operating temperature [8].

A considerable improvement of In_2O_3 thin film sensors with respect to O_3 by their modifying with γ-Fe_2O_3 is reported by Gutman et al. [9].

We have recently reported in details our studies on the particular Fe_2O_3-In_2O_3 [1, 10-13] and Fe_2O_3-SnO_2 [1, 11, 12, 14] systems, their gas-sensitive performance and peculiar structural features. We now summarize the results placing emphasis on aspects for the detection of some toxic and explosive gases. An attempt to establish the correlation between the structure of the materials and their functional performance has been made.

2. EXPERIMENTAL

The sensitive elements based on Fe_2O_3-In_2O_3 and Fe_2O_3-SnO_2 (Fe:Me 9:1; 1:1; 1:9) nanocomposites were formed from the stabilised sols of the corresponding metal hydroxides, which were prepared by the sol-gel technique. The procedure of a sol preparation used in this study consisted of the following steps:

- Forced hydrolysis of inorganic metal salt solution ($FeCl_2$, $Fe(NO_3)_3$, $In(NO_3)_3$) with a basic agent (NH_3)
- Precipitation of a metal hydroxide followed by its separation
- Formation of a sol through the peptization of the deposit with a peptizing agent or as a result of self-peptization

Both α-Fe_2O_3-In_2O_3 and α-Fe_2O_3-SnO_2 composites were prepared by either mixing the preliminary obtained α-Fe_2O_3 and SnO_2 sols (procedure I) or by combined precipitation of $Fe(OH)_2$ and $In(OH)_3$ hydroxides followed by their oxidation with oxygen (procedure II). To perform oxidation a flow of air was passed through the corresponding suspension during 5-6 h at 30°C. The composites based on metastable γ-modification of Fe_2O_3 (γ-Fe_2O_3-In_2O_3 and γ-Fe_2O_3-SnO_2) were obtained by mixing of individual sols of γ-Fe_2O_3 and $In(OH)_3$ in the required proportions.

In order to fabricate the thin/thick film sensors, the sols/powders were deposited onto polycrystalline Al_2O_3 substrates supplied with a Pt interdigital electrode structure and a Pt meander heater. In the case of In_2O_3-containing samples, an In_2O_3 sub-layer was preliminary deposited onto the substrate in order to form the hetero-junction structure and provide suitable sensor conductance. The indicated samples denoted as Fe_2O_3-In_2O_3/In_2O_3. Single-layer sensors consisting only of Fe_2O_3, In_2O_3 and SnO_2 were also studied in parallel for comparison. The samples were dried at 25°C and annealed at 300°C for 10 h in air. The active elements were mounted on cases and were put inside a chamber for the DC electrical measurements in presence of fixed gas concentrations and RH levels.

The gas-sensitive properties of thin and thick film sensors (Figure 1) based on both Fe_2O_3-In_2O_3 and Fe_2O_3-SnO_2 (Fe:In/Sn 9:1; 1:1; 1:9) with respect to C_2H_5OH (100-500 ppm), CH_4 (50 ppm), CO (50 ppm) ozone (200 ppb) and NO_2 (0.5-5 ppm) were investigated. Besides, electrical conductivity of some layers in air was also estimated.

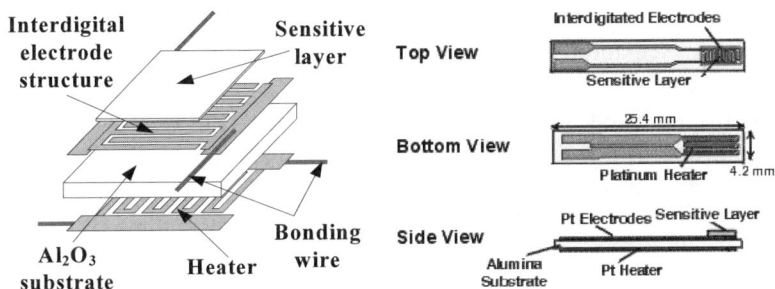

Figure. 1. Construction of thin (left) and thick (right) sensitive elements.

The sensor response S was calculated as dG/G_{air} at C_2H_5OH, CH_4, CO detection and as dG/G_{gas} at NO_2 and O_3 detection, where G is electrical conductance of a sensitive layer.

The structure of the simple oxides and nanocomposites was characterised by means of X-ray diffraction (XRD), Transmission Electron Microscopy

(TEM), Infra-Red Spectroscopy (IR), Electron Paramagnetic Resonance (EPR) and Mössbauer spectroscopy.

3   RESULTS AND DISCUSSION

3.1   Peculiar features of the sol-gel technology products

The choice of the sol-gel technology was driven by its advantages when compared to other preparation approaches, such as magnetron sputtering, thermal decomposition of inorganic and organic precursors etc [1]. The preparative technique used provides the formation of materials in form of films, ceramics and powders. It allows easy variation of the chemical composition of the samples, adding modifying agents in different forms and at different steps of the synthesis. And the main advantage of this approach is the possibility to precisely control the material nanostructure in a wide range. The products of the sol-gel technology are characterized by a number of particular features, which are of drastic importance regarding the gas sensing phenomenon:

- Nanocrystallinity
- High defectiveness
- Stabilisation of metastable point defects (V_o, $M^{(n-1)+}$, $M^{(n+1)+}$)
- Formation of metastable phases (H-In_2O_3, γ-Fe_2O_3, SnO, H-MoO_3)
- Developed specific surface area
- Elevated activity in a course of adsorption processes
- Increased additive solubility as compared to the equilibrium values

Due to the used preparation technique, one can expect unique gas-sensitive properties of the layers. Their behaviour differs from that typical of sensors fabricated by using other routs. But the ultimate goal of this study is applying systems of complex structure, since the advantages of the sol-gel technique are the most prominent not for single oxides but for binary oxide systems. Their structural features provide a variety of gas-sensitive properties of such materials. Note, that the properties of the layers are rather influenced by the type of sensors used as well.

3.2 Fe_2O_3-In_2O_3 composites

It was found [10] that the γ-Fe_2O_3-In_2O_3(Fe:In 9:1)/In_2O_3 and α-Fe_2O_3-In_2O_3(9:1)/In_2O_3 sensors are characterised by high sensitivity to O_3 and NO_2 over a low temperature range (70-135°C), as shown in Figures 2a, b. These response values are greater than those ones typical of single-layer sensors based on In_2O_3 and Fe_2O_3.

The response values of the α-Fe_2O_3 and γ-Fe_2O_3 samples to O_3 and NO_2 at various operating temperatures are given in the Table 1. It is clearly seen from these data that the γ-Fe_2O_3-In_2O_3(Fe:In 9:1)/In_2O_3 sensor shows a high conductance variation in the O_3 atmosphere at 135°C while its response to NO_2 at the same temperature is negligible.

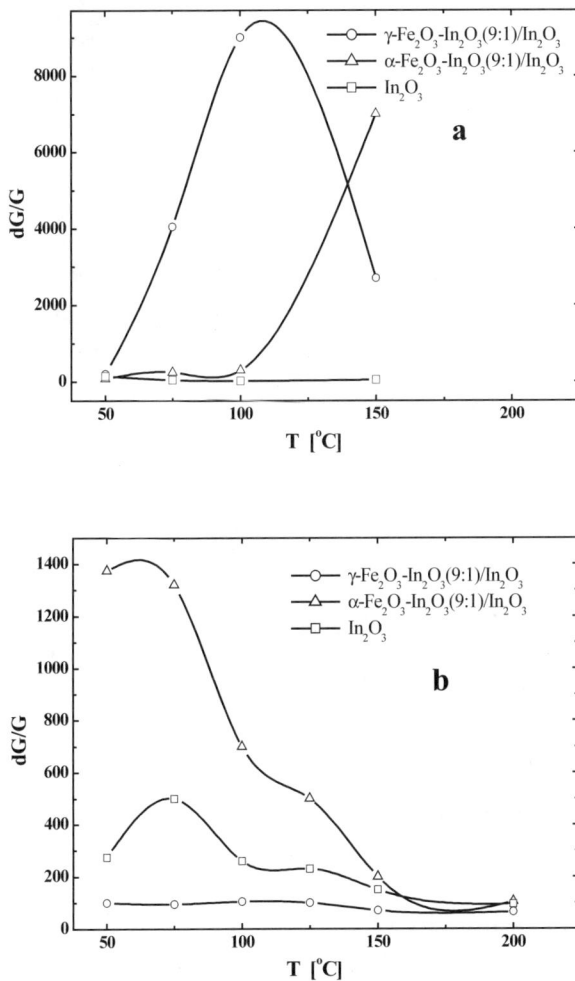

Figure 2. Temperature-dependent responses of In_2O_3 based sensors to (a) 200 ppb O_3 and (b) 5 ppm NO_2.

In contrast, the α-Fe$_2$O$_3$-In$_2$O$_3$(Fe:In 9:1)/In$_2$O$_3$ sample shows a good response to NO$_2$ in the temperature range 50-100°C together with a rather low one to O$_3$. These distinctions observing in the behaviour of both composites can be used for a selective analysis of O$_3$ and NO$_2$ in gas mixture. A pictorial diagram (Figure 3) depicts the dependence of the predominant sensitivity to oxidizing gases (NO$_2$ and O$_3$) of the Fe$_2$O$_3$-In$_2$O$_3$ composites based on either α-modification of iron oxide or γ-Fe$_2$O$_3$.

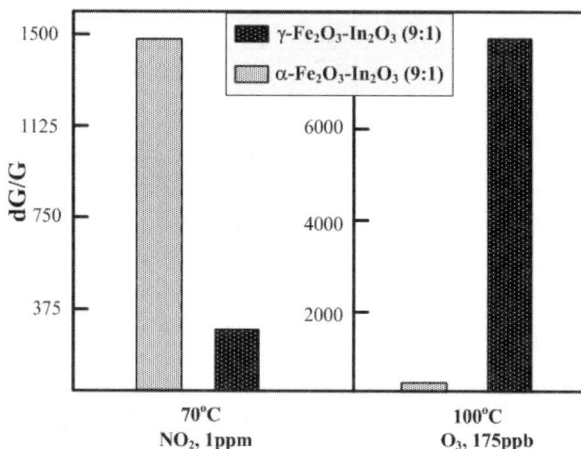

Figure 3. Selectivity of the sensors based on Fe$_2$O$_3$-In$_2$O$_3$ of different phase composition to oxidizing gases – NO$_2$ and O$_3$.

Table 1. The comparison of response values of the sensors based on both α-Fe$_2$O$_3$ and γ-Fe$_2$O$_3$ to O$_3$ and NO$_2$

Detected gas	C_{gas}, ppb	T, °C	Response, dG/G α-Fe$_2$O$_3$-In$_2$O$_3$(9:1)/In$_2$O$_3$	γ-Fe$_2$O$_3$-In$_2$O$_3$(9:1)/In$_2$O$_3$
O$_3$	100	100	65	130
	100	135	450	8670
NO$_2$	500	100	65	15
	500	135	75	10
	5000	100	600	90
	5000	135	440	50

Generally, the sensitivity of the In$_2$O$_3$ and SnO$_2$ films to O$_3$ is lower in comparison with their sensitivity towards NO$_2$ [15,16]. In contrast, the Fe$_2$O$_3$-In$_2$O$_3$ layers are characterised by a higher sensitivity to O$_3$ than to NO$_2$. Moreover, the indicated compositions show better NO$_2$ detection performances than the previously investigated thin film sensors based on In$_2$O$_3$-NiO [17] and In$_2$O$_3$-MoO$_3$ [15]. Figure 4 shows the dependence of the response values to NO_2 on the operating temperature for thin film sensors with different composition of sensitive layers.

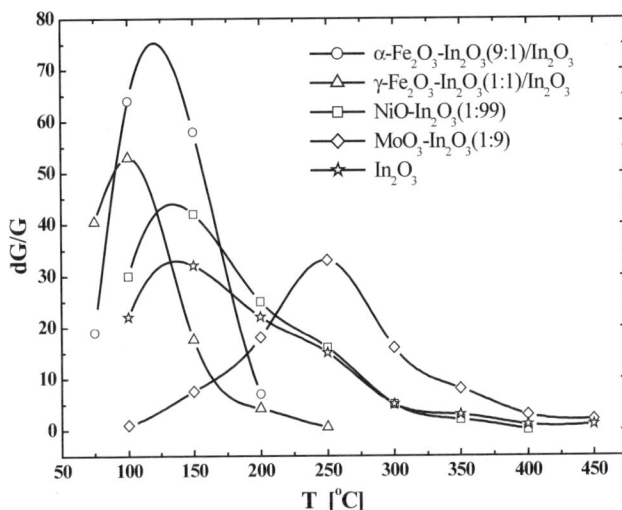

Figure 4. Comparison of the sensitivity of the In_2O_3 based sensors doping with oxides of different metals to 1 ppm NO_2.

As it is seen from these curves, the Fe_2O_3-In_2O_3 sensors have not only the greatest signals, but they can operate properly at relatively low temperatures. The Fe_2O_3-In_2O_3/In_2O_3 and Fe_2O_3/In_2O_3 sensors possess poor responses to low concentration of CO (50 ppm), as it is shown in Figure 5; they are also almost insensitive to both CH_4 and NH_3.

It is important to note that all double-layer sensors are much more sensitive towards alcohol (C_2H_5OH, CH_3OH) vapours than the single-layer In_2O_3 and Fe_2O_3 samples [13]; the maximum response is showed by the γ-Fe_2O_3/In_2O_3 composite. One should also point out that the Fe_2O_3-containing films are insensitive to O_3 and NO_2 over the temperature range of the most efficient ethanol detection (250-400°C). At the same time, their sensitivity regarding ethanol is negligible at 50-150°C when O_3 and NO_2 interaction with an oxide surface has the maximum value. An increasing of In_2O_3 content within the Fe_2O_3-In_2O_3 composite up to 50 % (mol) leads to the growth of the sensor responses to both NO_2 and ethanol.

According to the responses to various gases, the sensing layers can be placed as follows [2]:

O_3: γ-Fe_2O_3–In_2O_3/In_2O_3>γ-Fe_2O_3/In_2O_3>α-Fe_2O_3–In_2O_3/In_2O_3>In_2O_3

NO_2: α-Fe_2O_3–In_2O_3/In_2O_3>γ-Fe_2O_3/In_2O_3>In_2O_3>γ-Fe_2O_3–In_2O_3/In_2O_3

CO: γ-Fe_2O_3/In_2O_3>γ-Fe_2O_3–In_2O_3/In_2O_3≥α-Fe_2O_3–In_2O_3/In_2O_3>In_2O_3

Alcohol: γ-Fe_2O_3/In_2O_3>α-Fe_2O_3–In_2O_3/In_2O_3>γ-Fe_2O_3–In_2O_3/In_2O_3>In_2O_3

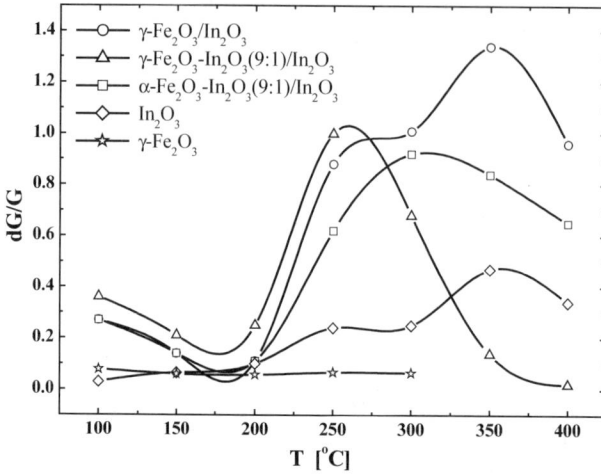

Figure 5. Temperature-dependent responses of the layers of different composition to 50 ppm CO.

Both TEM and XRD data give evidence that all the studied films appear to be nanosized systems. According to the XRD analysis of the samples, the $\alpha\text{-Fe}_2O_3\text{-In}_2O_3$ (Fe:In 9:1) composite consists of $\alpha\text{-Fe}_2O_3$ phase with increased parameters of unit cell. An increase of the cell parameters is caused by the substitution of part of Fe(III) ions with In(III) ones. Besides, the X-ray reflexes assigned to $\alpha\text{-Fe}_2O_3$ phase are strongly broadened; this fact can be explained both by the nano-dimension of particles and high defectiveness of the crystalline structure.

It was also assumed that $\alpha\text{-Fe}_2O_3$ phase, obtained through the oxidation of $\gamma\text{-Fe}_2O_3$ phase is quite different from $\alpha\text{-Fe}_2O_3$ phase prepared by thermal dehydration of α-modification of iron (III) hydroxide. The irregularity of Fe(III) state within the $\alpha\text{-Fe}_2O_3\text{-In}_2O_3$ (Fe:In 9:1) can be observed from the Mössbauer pattern recorded from the indicated sample.

Regarding to the magnetic properties the oxides prepared by the Sol-Gel Technology differ from the corresponding standard sample (Figures 6a, b).

We can distinguish three types of Fe(III) ions with discriminate parameters in Mössbauer spectrum of $\alpha\text{-Fe}_2O_3\text{-In}_2O_3$ composite (Table 2).

1. About 78 % of total amount of Fe(III) ions is characterised by the magnetic parameters and coordination environment, which is typical of Fe(III) ions within amorphous or poorly crystallised $\alpha\text{-Fe}_2O_3$ phase.
2. About 15 % of Fe(III) has a cubic coordination environment. This type of coordination can be assigned to isolated Fe(III) ions in octahedral environment of oxygen, which is characteristic of cubic In_2O_3 modification. Moreover, $\gamma\text{-Fe}_2O_3$ can possess cubic structure

as well. Under the sample heating at 150-200°C γ-Fe_2O_3 phase remains stable within the α-Fe_2O_3-In_2O_3 (Fe:In 9:1) sample obtained from Fe(II) precursor, which was used in this study. Annealing the composite at temperatures over 250°C leads to the transformation of γ-Fe_2O_3 phase doped with In(III) ions into α-Fe_2O_3, whereas individual γ-Fe_2O_3 oxide remains stable regarding γ-Fe_2O_3 → α-Fe_2O_3 phase transformation up to 485°C. Moreover, γ-Fe_2O_3 phase within the γ-Fe_2O_3-In_2O_3 composite is stable at temperatures up to 700°C depending on the component ratio. Thus, in the case of the γ-Fe_2O_3-In_2O_3 (Fe:In 9:1) sample, γ-Fe_2O_3 → α-Fe_2O_3 phase transformation occurs at about 500°C.

3. A minor part of Fe(III) ions (7 %) can be only assigned to γ-FeOOH structure.

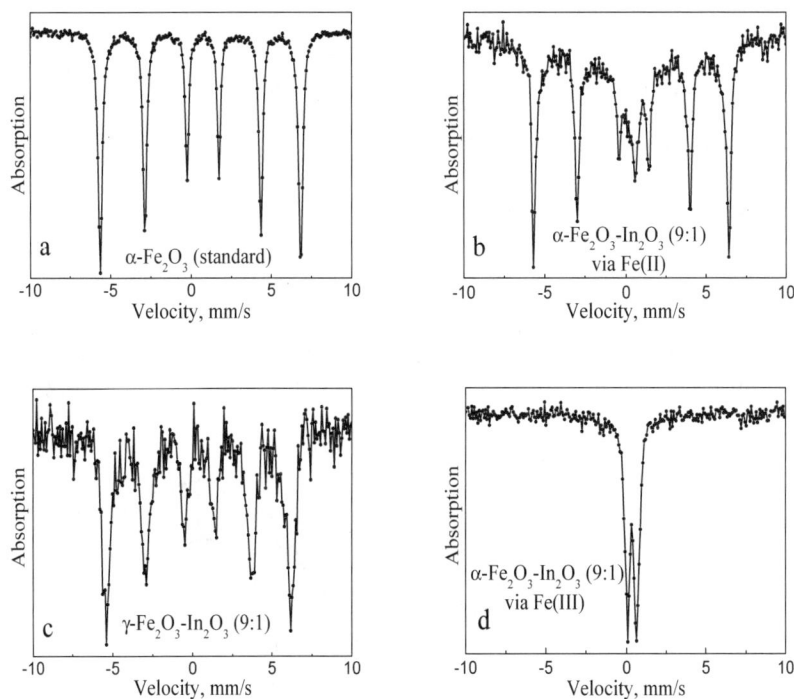

Figure 6. ^{57}Fe Mössbauer spectra recorded from Fe-containing species at 298 K: a – α-Fe_2O_3 standard, b – α-Fe_2O_3–In_2O_3 (Fe:In 9:1), via Fe(II), c – γ-Fe_2O_3–In_2O_3 (Fe:In 9:1), d – α-Fe_2O_3–In_2O_3 (Fe:In 9:1), via Fe(III).

The sample based on the γ-Fe_2O_3-In_2O_3 (Fe:In 9:1) consists of γ-Fe_2O_3; a small amount of C-In_2O_3 phase is also present. In this case, the grain size is greater than in the case of the occurrence of α-Fe_2O_3 with the same composition. Mössbauer pattern of the γ-Fe_2O_3-In_2O_3 system differs from that one recorded from the standard sample by broadening and asymmetric shape of the resonance peaks (Figure 6c). A paramagnetic doublet is typical of the α-Fe_2O_3-In_2O_3 composites prepared via Fe(III) precursor. The particles have the size of a single magnetic domain, thus demonstrating superparamagnetism phenomenon. (Figure 6d)

Table 2. Parameters of ^{57}Fe Mössbauer spectra recorded from Fe-containing samples at 298 K

Sample	δ, mm s^{-1}		Δ, mm s^{-1}	B, T
α-Fe_2O_3-In_2O_3 (Fe:In 9:1) (300°C)	0.38	78 %	0.08	50.7
	0.53	15 %	0	0
	0.22	7 %	0.69	0
γ-Fe_2O_3-In_2O_3 (Fe:In 9:1) (300°C)	0.33		0.02	48.6
γ-Fe_2O_3 (sol) (300°C)	0.34		-0.03	49.1
γ-FeOOH (300°C)	0.33		0.78	0
α-Fe_2O_3 (amorphous) (300°C)	0.39		0.09	50.7
γ-Fe_2O_3 (standard sample)	0.34		-0.05	49.6
α-Fe_2O_3 (standard sample)	0.47		0.24	51.8
	0.38		0.12	51.5

EPR data give the evidence that the mentioned composites contain Fe–O–Fe associates, $(FeO)_x$ clusters or micro-inclusions of Fe_2O_3 amorphous phase. Fe(III) ions are characterised by cubic symmetry of coordination environment. The very areas of poorly crystallised (or amorphous) phase are active in gas adsorption that follows from the increasing of the corresponding EPR signal intensity under exposure of the Fe_2O_3-In_2O_3 layers in NO_2 ambient (Figure 7). Isolated Fe(III) ions are not participating in this process.

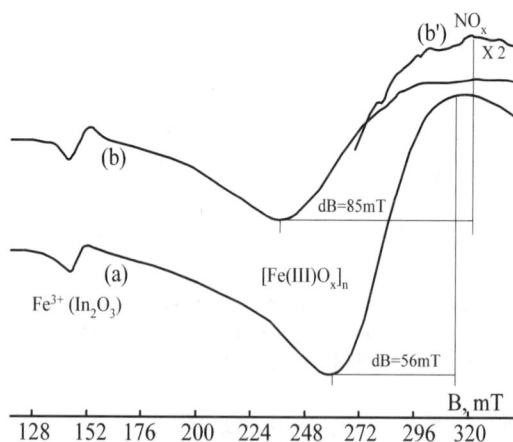

Figure 7. ESR spectra of the α-Fe_2O_3–In_2O_3 (Fe:In 1:1) composite, annealed at 800°C: a – initial, b – treated in NO_2 at 120°C, 0.5 h.

In the case when the species is only consisted of amorphous phase, suitable value of sensing layer conductivity is not reaching. The presence of amorphous phase is essential in quantity, sufficient to provide chemisorption of detected gas, but not making difficult the charge transfer. The later should be provided by other highly conductive phases (C-In_2O_3).

The distinctions in shape and parameters observed in the Mössbauer spectra recorded from the γ-Fe_2O_3-In_2O_3 and γ-Fe_2O_3 samples probably result from the following factors: γ-Fe_2O_3 cubic lattice distortion, irregularity of Fe(III) octahedral environment or Fe–O bond ionicity shift in the presence of In(III) ions within γ-Fe_2O_3 crystal lattice.

3.2.1 Regularities of detection of some gases by the layers based on Fe_2O_3-In_2O_3 composites

On the base of the obtained results we made an attempt to find the correlation between the gas-sensitive behaviour of the Fe_2O_3-In_2O_3/In_2O_3 and Fe_2O_3/In_2O_3 active layers and their structural features.

i) Nitrogen dioxide

In order to obtain an advanced sensor for NO_2 detection, it is necessary to use materials which are characterised by high dispersion and defectiveness [18]. It is well known that doping of In_2O_3 with Ni(II) and Mo(VI) ions results in an increased sensitivity to NO_2 of the In_2O_3 based sensor [15, 17]. Additions of these ions cause the formation of strongly defective In_2O_3 structure and favour the decreasing of oxide grains.

Similar changes were observe for α-Fe_2O_3 oxide doped with In(III) ions. In the case of the α-Fe_2O_3-In_2O_3 (Fe:In 9:1) sample together with α-$In_xFe_{2-x}O_3$ solid solution, it is possible to distinguish other structural elements based on Fe_2O_3. We have yet not succeeded in identification of the supposed additional phases using XRD analysis, but the presence of several types of Fe(III) ions was confirmed by Mössbauer spectroscopy. An increased Fe–O bond length and the distortion of octahedral environment of Fe(III) ions favours the effective adsorption of NO_2, whereas the presence of two types of ions (Fe(III) and In(III)) within the solid solution of α-$In_xFe_{2-x}O_3$ facilitates to a certain extent the desorption of oxygen in comparison with the simple oxides.

Thus, the high sensitivity of the α-Fe_2O_3-In_2O_3 films to NO_2 at low temperatures can be explained by high system dispersion and the presence of Fe(III) ions in irregular coordination environment, which is evoked by doping Fe_2O_3 phase with In(III) ions.

The growth of the response value of the γ-Fe_2O_3-In_2O_3/In_2O_3 sensor with the increasing of In_2O_3 content from 10 to 50 % within the composite can be connected with grain size decreasing and the formation of highly defective sample with a developed specific surface.

ii) Ozone

The most important requirement for the efficient detection of ozone at low temperatures (70-100°C) is suitable catalytic activity of an oxide in reaction of ozone decomposition: $O_3 \rightarrow O_2 + O$.

Iron oxide is known to be an active catalyst in this process; the main factors, which influence the catalytic ability of Fe_2O_3 are oxidation state of a sample and its dispersion [19]. The oxidation level of a sample is closely related to its activity. Thus, α-Fe_2O_3 possesses better catalytic properties than γ-Fe_2O_3.

With regards to O_3, α-Fe_2O_3 phase demonstrates considerably higher sensitivity in comparison with γ-Fe_2O_3 species [9]. However, the origin of different behaviour of these systems is still unclear. In the case of the γ-Fe_2O_3-In_2O_3 composite, the presence of separate γ-Fe_2O_3 phase probably provides an elevated activity of this sample towards O_3 at low temperatures. In contrast, the α-Fe_2O_3-In_2O_3 sample consists of In(III)–α-Fe_2O_3 solid solution; catalytic ability of α-Fe_2O_3 phase in O_3 decomposition reaction is insignificant.

Since at low temperatures a limiting stage of the reaction is a removal of chemisorbed oxygen, the presence of the second phase (In_2O_3) in γ-Fe_2O_3 is capable to facilitate the desorption of oxygen from the oxide surface.

Detection of O_3 is going at Fe_2O_3-In_2O_3 through the adsorption of not molecular (O_2), but atomic (O) oxygen species. Thus, one can explain the observed differences in optimal detecting temperature of O_3 using the α-Fe_2O_3-In_2O_3 and γ-Fe_2O_3-In_2O_3 composites by change of detection mechanism as a result of operating temperature varying.

iii) Alcohols

The sensors based on hetero-junction oxide structures show considerable response in alcohol (ethanol, methanol) media. The hetero-junction between oxide and solid solution phases appear to be very active in a course of both adsorption and oxidation of alcohol.

It is shown in [4] that the presence of two types of centres possessing the discriminate reductive-oxidation and acid-base properties and participating in transformation processes of an alcohol molecule is an essential requirement to achieve high sensor response values when alcohol detection is mentioned. Alcohol detection is considered as a multi-step process involving both red-ox and acid-base interactions. Oxide phases within composites differ by oxygen-oxide surface bonding energy which can be the relative measure of oxide activity in oxidation reactions. The reactivity of oxides in acid-base reactions depends on the electronegativity of metal cation. The electronegativity is the measure of Lewis acid site activity. Thus, the centre of one type can mainly participate in adsorption-desorption processes of alcohol molecules, whereas a complete oxidation of intermediates is going effectively at the centre of another type.

An increased response of the γ-Fe_2O_3-In_2O_3(Fe:In 1:1)/In_2O_3 sample as compared with the γ-Fe_2O_3/In_2O_3 one can be explained by the presence of an higher contact interface between In_2O_3 and γ-Fe_2O_3 phases within the γ-Fe_2O_3-In_2O_3 composite.

3.3 Fe_2O_3-SnO_2 composites

The Fe_2O_3-SnO_2 composites of the same chemical composition (Fe:Sn ratio), prepared via Fe^{2+} and Fe^{3+} precursors differ considerably regarding their phase composition [1]. The sharpest distinctions are observed for the samples heated at temperatures below 500°C. It is important to note that all the oxide systems mentioned in this paper are characterised by high system dispersity and relatively low degree of crystallinity. The grain size does not exceed 6 nm (500°C).

The XRD reflections assigned to the α-Fe_2O_3-SnO_2 composites of the whole series of the applied component ratios (Fe:Sn 9:1, 1:1, 1:9) appear to be broad and low-intensive. This can be explained by very small size of SnO_2 grains (about 2 nm) and poor crystallinity of α-Fe_2O phase. The composites with Fe:Sn ratio 1:9 and 1:1 has the structure of Fe(III)–SnO_2 solid solution; in the case of the Fe_2O_3-SnO_2 (Fe:Sn 9:1), Sn(IV)–α-Fe_2O_3 solid solution is typical.

Fast removal of water and crystallisation of α-Fe_2O_3 occurs at temperatures above 500°C. The reflections of this phase become narrow. Noticeable SnO_2 particle agglomeration accompanying the crystallisation processes is only observed at 800°C. High-temperature treatment results in partial decomposition of the sample structure and isolation of simple oxide phases – SnO_2 and α-Fe_2O_3; the samples become heterogeneous.

As opposed to α-Fe_2O_3, γ-Fe_2O_3 phase obtained by using the mentioned synthesis technique is characterised by rather well crystallinity and absence of constitutional water after drying at 150°C. However, annealing the sample based on metastable γ-Fe_2O_3 phase at 400-500°C results in its transforming into α-Fe_2O_3 one. Thus, the XRD patterns of both the α-Fe_2O_3-SnO_2 and γ-Fe_2O_3-SnO_2 species (of the same component ratio) when heated at 800°C become similar. Besides, as have been established by Mössbauer spectroscopy, small areas of amorphous highly dispersive Fe_2O_3 phase are also preserved within the α-Fe_2O_3-SnO_2 and γ-Fe_2O_3-SnO_2 (Fe:Sn 9:1, 1:1) samples up to 400°C. EPR and Mössbauer studies give evidence that α-

Fe_2O_3 phase obtained through amorphous Fe_2O_3 (via Fe^{3+} precursor) and by the thermally stimulated transformation of γ-Fe_2O_3 into α-Fe_2O_3 (via Fe^{2+} precursor) are significantly different: the motive of γ-Fe_2O_3 structure (cubic symmetry) preserves within the latter species [20]. This phenomenon is also typical of highly dispersive α-modifications of iron oxides/hydroxides (trigonal symmetry). Note, that in the case of the γ-Fe_2O_3-SnO_2 composites grain size of both oxide phases is greater as compared to the α-Fe_2O_3-SnO_2. This effect is proper for all series of composite component ratios and annealing temperatures. Consequently, one can expect that the mentioned types of samples will demonstrate different gas-sensitive behaviour.

Electrical measurements in air

Electrical conductivity of the Fe_2O_3-SnO_2 thin film layers with Fe:Sn 1:9 and 1:1 is found to be considerably lower as compared to the Fe_2O_3-SnO_2 (Fe:Sn 9:1) composite and simple oxides in air (Figure 8). The latter one along with SnO_2 demonstrates minimum resistance values. The conductivity of the Fe_2O_3-SnO_2 (Fe:Sn 1:9, 1:1) films is equal at points 100 and 500°C; however, the trends of the curves are quite dissimilar within the indicated temperature region. In the case of Fe:Sn 1:9 ratio the conductivity begins to rise gradually starting from 200°C; the sample with equal content of the components demonstrates maximum conductivity change within a low-temperature region (100-200°C). The conductivity of SnO_2 is mainly determined by the presence of singly charged oxygen vacancies ($V_o^{\cdot}$) [21].

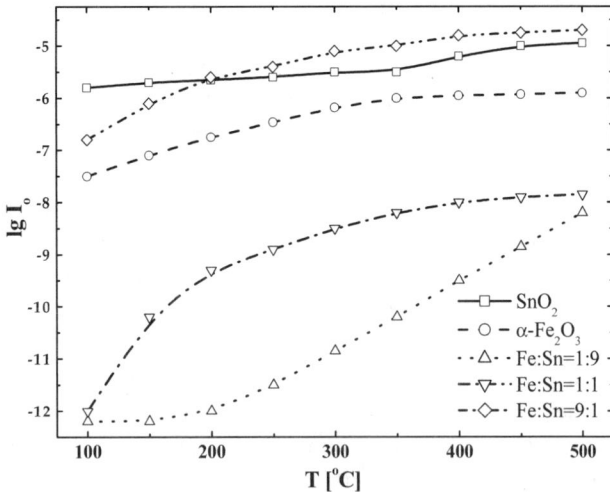

Figure 8. Temperature-dependent conductivity of thin film simple oxides and α-Fe_2O_3-SnO_2 composites in air, RH 30%

Fe^{3+} ions occupying the Sn(IV) positions within SnO_2 crystal lattice, are acting as electron acceptors that results in decrease of charge carrier amount. Meanwhile, Fe^{3+}, when bounded in $[Fe^{3+}-V_o^{\cdot}]$ associates, are not participating in the charge transfer. Temperature increasing up to 400-550°C is an important requirement to provide electron activation and sufficient resistance drop. As it was noted above, the Fe_2O_3-SnO_2 composites consist of highly dispersive iron and tin oxide phases. The indicated phases are characterised by an elevated activity and react readily under heating. Sn–OH–Fe and Sn–O–Fe bonding is possible at SnO_2/Fe_2O_3 phase interface that leads to decreasing of the contact resistance and increasing of the potential barrier transmissivity.

Surface conductivity mainly contributes the total conductivity of the thin oxide films at direct current measurements. The surface layer of amorphous Fe_2O_3 doping with Sn^{4+} ions provides an increased conductivity of the Fe_2O_3-SnO_2 (Sn:Fe 1:1) film as contrast to the Fe_2O_3-SnO_2 (Fe:Sn 1:9) one. Adding SnO_2 to Fe_2O_3 brings to increasing the free charge carrier concentration, and consequently, to heightening the conductivity of the Fe_2O_3-SnO_2 (Fe:Sn 9:1), having the structure of Sn(IV)–α-Fe_2O_3 solid solution.

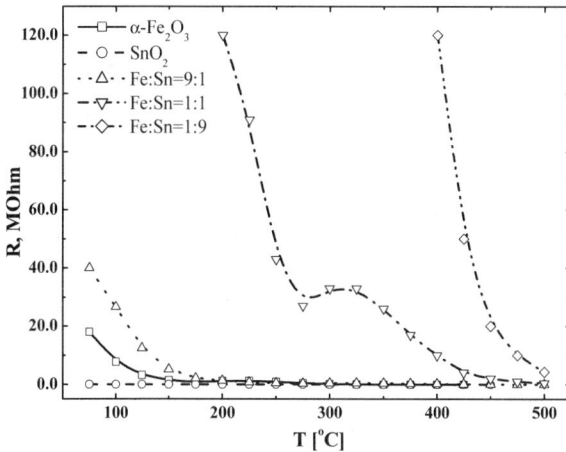

Figure 9. Temperature-dependent resistance of thick film simple oxides and Fe_2O_3-SnO_2 composites in air, RH 30%.

The thick film layers based on the Fe_2O_3-SnO_2 composites and simple oxides demonstrate the same regularities intrinsic to the corresponding thin films, but absolute conductivity values for the thick films appear to be significantly lower (Figure 9), probably because of poor contact between separate particles of the powder. Because of the indicated point we were not able to perform the functional characterisation of some samples at low

108

operating temperatures both in air and gas ambient. The sample with Fe:Sn ratio 1:9 appears to be of extremely high resistance in air.

Electrical measurements in gas ambient

The thin and thick Fe_2O_3-SnO_2 films of different structure and composition are characterised by dissimilar behaviour in gas ambient of different chemical nature: oxidising (NO_2) and reducing (C_2H_5OH) gases. The thin film sensors are found to be the most suitable for detection of NO_2; meanwhile, the thick films demonstrate good performance when detecting ethanol vapours. The α-Fe_2O_3-SnO_2 samples only show measurable response values to NO_2 among the thin film layers (Figure 10). The α-Fe_2O_3-SnO_2 (Fe:Sn 1:1) composite demonstrates maximum sensitivity at 100°C. Note, that NO_2 molecules proceed as electron donors within low-temperature region. It can be connected with the point that the conductivity of the Fe_2O_3-SnO_2 composites (up to 150°C) is determined by the presence of surface OH-groups, which are desorbing at 150-400°C. Besides, the response of the α-Fe_2O_3-SnO_2 sensors of all series of compositions to NO_2 reaches its maximum value at temperatures when the maximum layer conductivity drop in air is observed.

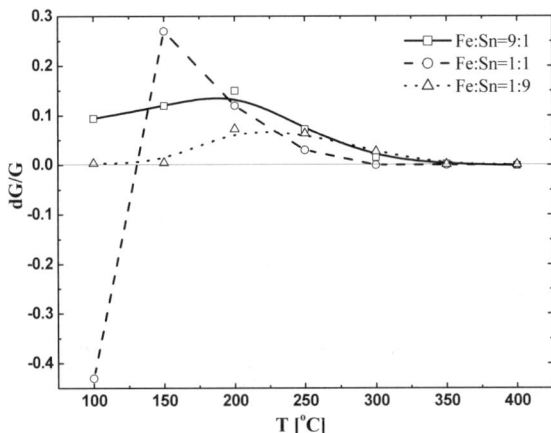

Figure 10. Temperature dependent response of Fe_2O_3-SnO_2 thin films to 1 ppm NO_2, RH 40 %.

In the case of thick films, the α-Fe_2O_3-SnO_2 and γ-Fe_2O_3-SnO_2 (Fe:Sn 9:1) composites show maximum sensitivity to humid C_2H_5OH vapours. As it was noted above, the layers were annealed at 550°C to provide measurable sensor conductivity. Thus prepared samples have the structure of Sn(IV)–α-Fe_2O_3 solid solution, γ-Fe_2O_3 is transformed into α-Fe_2O_3. However, certain insignificant differences are observed in the α-

Fe_2O_3-SnO_2 and γ-Fe_2O_3-SnO_2 behaviour that can be caused by certain distinctions in grain size and microstructure of the samples. Thus, the composites obtained via Fe^{3+} precursor (α-Fe_2O_3-SnO_2) are characterised by higher response and an increased optimal detecting temperature (Figure 11).

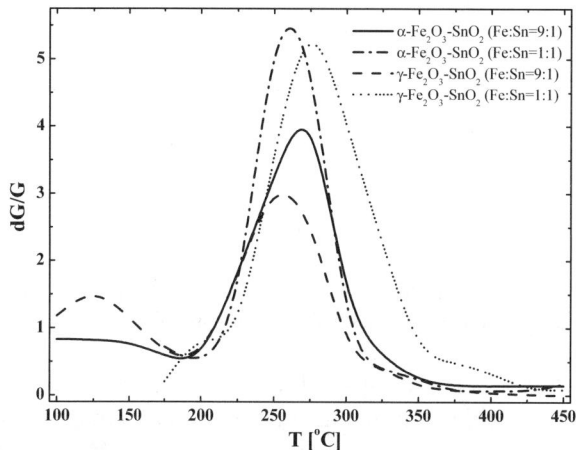

Figure 11. Temperature-dependent response of α-Fe_2O_3-SnO_2 and γ-Fe_2O_3-SnO_2 thick film sensors to 0.05‰ of C_2H_5OH.

Maximum response of the Fe_2O_3-SnO_2 (Fe:Sn 9:1, 1:1) layers to ethanol appears to be greater than that one shown by simple oxides (SnO_2, Fe_2O_3) (Figure 12). We failed to measure the Fe_2O_3-SnO_2 (Fe:Sn 1:9) sensor because of its extremely high resistance. Remarkably, the optimal operating temperature for the whole series of the thick film sensors lies in narrow temperature range (280-320°C).

There is a clear difference in the behaviour of the α-Fe_2O_3-SnO_2 and γ-Fe_2O_3-SnO_2 species in the mixtures containing different amounts of ethanol. The α-Fe_2O_3-SnO_2 layers demonstrate the performance proper for the detection of alcohol within the concentration range 0.05-1.0 ‰. In contrast, both the γ-Fe_2O_3-SnO_2 composites and simple oxides are characterised by unsuitably low ratio $\Delta R/\Delta C$ that makes the mentioned materials usable for the detecting devices determining the presence of ethanol vapours in air, but they could not be used for the accurate registration of the gas concentration.

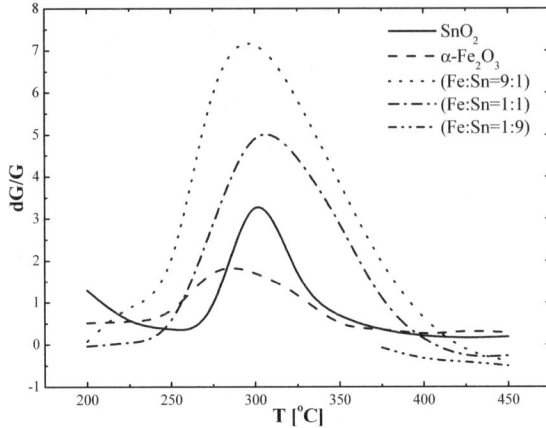

Figure 12. Temperature-dependent response of thick film sensors based on simple oxides and α-Fe$_2$O$_3$-SnO$_2$ composites to 0.05‰ of C$_2$H$_5$OH.

According to the EPR data available, the stabilisation of different types of Fe ions is possible within the Fe$_2$O$_3$-SnO$_2$ composites. Types of coordination of Fe ions within oxide matrix and the corresponding values of g-factors are listed in the Table 3.

Table 3. Main ESR signals of Fe(III) canters incorporated into oxide matrix

g-factor	Nature of centre	Remarks
8-10	Rhombic centres	Strong crystal field
6-8	Axial centres	Strong crystal field
4.3	Centres in octahedral symmetry with rhombic distortion	Strong crystal field
2-4	Centres in cubic symmetry with slight distortion (octahedral, tetrahedral)	Weak crystal field
2.0	Centres in cubic symmetry	Weak crystal field

All the mentioned paramagnetic centres were encountered in the studied composites. Their distribution depends strongly on the method of synthesis and mode of thermal treatment of the samples, Figure 13. Samples, annealed at 300°C only demonstrate a broad signal with g value about 2.06. It corresponds to amorphous Fe$_2$O$_3$. Annealing the species at 500°C leads to the appearing the weak signal with g 4.3 together with the broad one. It means that besides of Fe–O associates the presence of isolated Fe ions within SnO$_2$ matrix is evident. At higher temperatures the signal of isolated ions increases essentially; meanwhile amorphous phase disappears. Besides, there are some signals assigned to Fe ions in various environments in this spectrum.

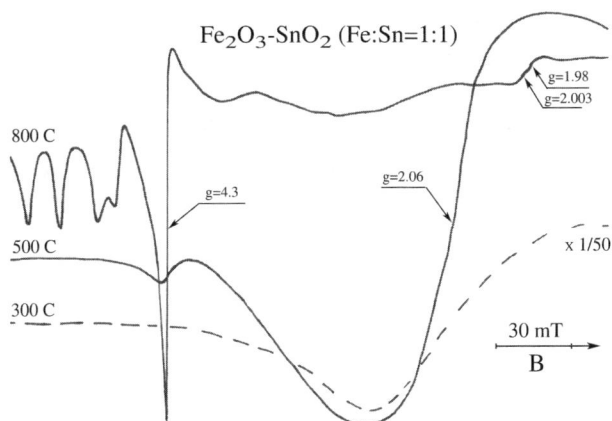

Figure 13. EPR spectra of the Fe_2O_3-SnO_2 (Fe:Sn=1:1) composite annealed at different temperatures (recorded at 300 K).

Recording spectra at lower temperatures as well as decreasing Fe content in the samples (Figure 14) allows us to distinguish a lot of small peaks in the range of strong and weak fields. The analysis of the spectra gives a set of signals, which are listed in the Table 4. Degree of cubic symmetry distortion (λ), g-factors and nature of the signals are included. Between the mentioned species there are various Fe ions, oxide phases and free electron captured by oxygen vacancy (F-centre).

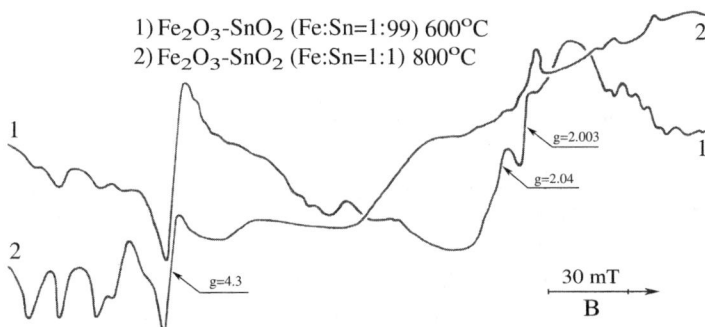

Figure 14. EPR spectra of the Fe_2O_3-SnO_2 (Fe:Sn=1:99 and 1:1) composites (recorded at 77 K).

Treating the Fe_2O_3-SnO_2 (Fe:Sn 1:1) powder with NO_2 (120°C, 10 min) leads to the increase of the signal **II** only; the signal **I** remains unchanged (Figure 15) [1]. This fact gives evidence that the amorphous Fe_2O_3 only participates in NO_2 adsorption; the isolated Fe(III) ions within SnO_2 lattice are inactive. This explains significant sensitivity of the Fe_2O_3-SnO_2 (Fe:Sn 1:1 and 9:1) films when detecting NO_2.

Table 4. EPR spectra of the Fe_2O_3-SnO_2 (Fe:Sn=1:99 and 1:1) composites (recorded at 77 K)

g_1	g_2	g_3	$\lambda=E/D$	Nature of signal
\multicolumn{3}{l}{g-tenzor}		
6.7	5.11	1.98	0.037	Fe(III) in interstitial positions
5.58	2.58	2.34	0.117	Fe(III) in Sn lattice points
4.40	4.25	4.08	0.314	Fe(III)$_{surface}$ in Sn lattice points
	4.3		0.33	[Fe(III)–V$_o$]$_{surface}$ (octahedral symmetry)
8.08	2.61	1.98	0.115	Fe(III) in Sn lattice points (octahedral symmetry)
7.3	3.68	1.4	0.117	Fe(III) in Sn lattice points (tetrahedral symmetry)
	3.0÷3.3			Fe_2O_3 (amorphous)
	2.02÷2.08			Fe^{3+}–O–Fe^{3+} superparamagnetic clusters
	2.02			γ-Fe_2O_3, Fe_3O_4-like noncrystalline structures

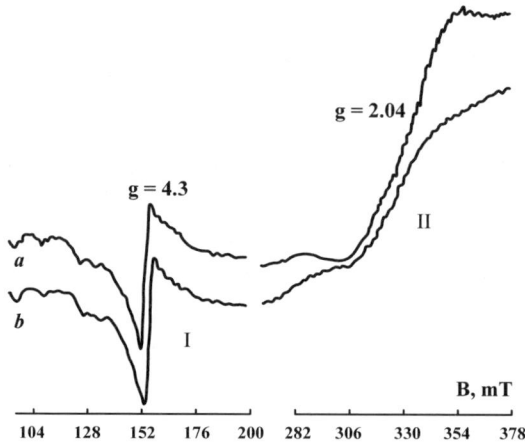

Figure 15. EPR spectrum of the Fe_2O_3-SnO_2 (Fe:Sn = 1:1) powder: a – initial, b – treated with NO_2 (10 min, 130°C).

Mössbauer spectra of the α-Fe_2O_3-SnO_2 and γ-Fe_2O_3-SnO_2 samples annealed at 300-500°C represent a broadened doublet, which is the evidence of superparamagnetic Fe_2O_3 particles (d ~ 3-4 nm) formation.

Thus, the samples with Fe:Sn = 9:1 and 1:1 contain areas of highly dispersive poorly crystallised Fe_2O_3 phase. The parameters of the Fe_2O_3-SnO_2 spectra are different from the parameters, which are typical of the spectra of both α-Fe_2O_3 and γ-Fe_2O_3 bulk phases. However, the values of isomeric shift (δ) and hyperfine magnetic field (B) for amorphous Fe_2O_3 are closer to the parameters of cubic γ-Fe_2O_3 rather than to trigonal α-Fe_2O_3 (Table 5) that is in agreement with the EPR data.

The coordination of Fe(III), which is specific of oxide/hydroxide iron compounds possessing cubic structure of unit cell (γ-FeOOH, γ-Fe$_2$O$_3$), preserves within the Fe$_2$O$_3$ amorphous phase. Considerable increase of the quadrupole splitting (Δ) values as compared to the simple oxides indicates that the crystal environment of the Fe(III) points within Fe$_2$O$_3$ matrix is strongly irregular in the presence of Sn(IV).

Table 5. Parameters of ^{57}Fe Mössbauer spectra of the Fe$_2$O$_3$ and Fe$_2$O$_3$-SnO$_2$ composite. Recorded at 300 K

Sample	δ, mm s^{-1}	Δ, mm s^{-1}	B, T
Fe$_2$O$_3$-SnO$_2$ (Fe:Sn 1:1), 300°C	0.35±0.02	0.72±0.02	50.8
Fe$_2$O$_3$-SnO$_2$ (Fe:Sn 1:1), 500°C	0.35±0.02	0.87±0.02	50.8
γ-Fe$_2$O$_3$ (bulk)	0.34±0.01	-0.053±0.020	49.6
α-Fe$_2$O$_3$ (bulk)	0.47±0.03	0.23±0.01	51.8
Fe$_2$O$_3$ (amorphous)	0.39±0.02	0.09±0.01	50.7

Thus, the isolated Fe(III) ions substituting Sn(IV) in the points of SnO$_2$ lattice decrease considerably electrical conductivity of SnO$_2$ films and their sensitivity to both oxidising (NO$_2$) and reducing (C$_2$H$_5$OH) gases. As it was noted above, the Fe$_2$O$_3$-SnO$_2$ nanocomposites, which have a structure of Sn(IV)–α-Fe$_2$O$_3$ solid solution, show high response to ethanol vapours due to the presence of two types of adsorption centres – Sn(IV) and Fe(III). The two centres are characterised by different activity in the course of both reduction-oxidation and acid-base reactions. The mechanism of ethanol molecule detecting at the surface of the Fe$_2$O$_3$-SnO$_2$ composite is oxidising dehydration, in which Sn(IV) operates like a catalyst [22]. Due to the low Fe–O bonding energy and increased basicity of Fe$_2$O$_3$ in comparison to SnO$_2$, Fe(III) centres promote further oxidation of intermediate products of ethanol molecule transformation [23]. Moreover, high dispersity of Sn(IV)–α-Fe$_2$O$_3$ solid solution provides efficient electron exchange between the cations: Fe(III) $\leftrightarrow$ Fe(II). All this produces a greater conductivity drop of the active layer and, consequently, improves the sensor performance.

4. CONCLUSIONS

Referring to the results of the functional and structural investigations, we can recommend a series of Fe$_2$O$_3$-In$_2$O$_3$ samples with different structural and phase state to be used as advanced materials for O$_3$, NO$_2$ and C$_2$H$_5$OH detection. The particular compositions, detected gases and optimal operating temperatures are listed in the Table 6. The first three sensors appear to be absolutely selective to O$_3$, NO$_2$ and C$_2$H$_5$OH at the indicated operating temperature. The fourth one can be used for NO$_2$ detection (at low temperatures) as well as for C_2H_5OH analysis (at higher temperatures). The SnO_2-based composites are characterised by lower response values as compared to In_2O_3-based sensors.

Table 6. The most promising Fe_2O_3-In_2O_3 and Fe_2O_3-SnO_2 gas-sensitive materials

Sample	T (°C)	Detected gas
γ-Fe_2O_3-In_2O_3 (Fe:In 9:1)	135	O_3
α-Fe_2O_3-In_2O_3 (Fe:In 9:1)	70-100	NO_2
γ-Fe_2O_3	250	C_2H_5OH
γ-Fe_2O_3-In_2O_3 (Fe:In 1:1)	70-100	NO_2
	300	C_2H_5OH
α-Fe_2O_3-SnO_2 (Fe:Sn 1:1)	150	NO_2
α-Fe_2O_3-SnO_2 (Fe:Sn 9:1)	320	C_2H_5OH

Note, that the elaborated sensors can not be used for the direct determination of explosive gases with low vapour pressure like TNT or RDX. However, there is a possibility to detect the indicated solids after certain refashion of the sensing elements and pre-concentration of the diluted vapours.

ACKNOWLEDGEMENTS

This work has been supported by the EC Programs: INTAS (project No 2000-0066), INCO-COPERNICUS (project GASMOH).

REFERENCES

1 M. Ivanovskaya, D. Kotsikau and D. Orlik, Chemical problems of development of new materials and technologies. Collection of papers, Issue 1, Minsk 2003, p151.
2 M. Ivanovskaya, D. Kotsikau and D. Orlik, Chemical problems of development of new materials and technologies. Collection of papers, Issue 2, Minsk 2003, p135.
3 M. Ivanovskaya, D. Kotsikau, D. Orlik *et al.*, Proc. EMRS 2003 Spring Meet., Strasbourg, France, June 10-13, 2003, N/PII.09.
4 Y. Nakatani and M. Matsuoka, Jap. J. Appl. Chem. (1982) L758.
5 T.G. Newor and St.P. Yordanov, Ceramic gas sensors: technique and application, Techmomic, Lancosh, 1996.
6 W. Chang and D. Lee, Thin Solid Films 200 (1991) 329.
7 O.K. Tan, W. Zhu, Q. Yan *et al.*, Sens. Actuators B 65 (2000) 361.
8 T. Takada, K. Suzuki and M. Nakane, Sens. Actuators B 13-14 (1993) 404.
9 F.H. Chibirova and E.E. Gutman, Russ. J. Phys. Chem. 74 (2000) 1555.
10 M. Ivanovskaya, D. Kotsikau, G. Faglia *et al.*, Sens. Actuators B 93 (2003) 422.
11 M. Ivanovskaya and D. Kotsikau, Proc. 9th Europ. Conf. Solid State Chem., Stuttgart, Germany, Sept. 3-6, 2003, p26.

12 M. Ivanovskaya, D. Kotsikau, G. Faglia *et al.*, Proc. 203rd Meeting of Electrochem. Soc., Paris, France, Apr. 27-May 2, 2003, No 2884.

13 M. Ivanovskaya, D. Kotsikau, G. Faglia *et al.*, Sens. Actuators B 96 (2003) 498.

14 D. Kotsikau, M. Ivanovskaya, G. Faglia *et al.*, Proc. EMRS 2003 Spring Meeting, Strasbourg, France, June 10-13, 2003, N/PI.11.

15 A. Gurlo, N. Bârsan, M. Ivanovskaya *et al.*, Sens. Actuators B 47 (1998) 92.

16 M. Ivanovskaya, A. Gurlo, P. Bogdanov *et al.*, Sens. Actuators B 77 (2001) 264.

17 M. Ivanovskaya, P. Bogdanov, G. Faglia *et al.*, Sens. Actuators B 68 (2000) 344.

18 P. Bogdanov, M. Ivanovskaya, F. Comini *et al.*, Sens. Actuators B 57 (1999) 153.

19 V.V. Lunin, M.P. Popovich and S.N. Tkachenko, The physical chemistry of ozone, Moscow State University, Moscow, 1998.

20 D.R. Orlik, M.I. Ivanovskaya, P.A. Bogdanov *et al.*, Sens. Actuators B 13-14 (1993) 605.

21 M.I. Ivanovskaya, V.V. Romanovskaya and G.A. Branitsky, Russ. J. Phys. Chem. 68 (1994) 232.

22 M.I. Ivanovskaya, P.A. Bogdanov and A.Ch. Gurlo, Russ. J. Inorg. Mater. 34 (1998) 329.

23 M.I. Ivanovskaya, E.V. Lutynskaya, L.S. Ivashkevich, Russ. J. Inorg. Chem. 43 (1998) 1716.

Chapter 8

DETECTION OF LANDMINES AND OTHER EXPLOSIVES WITH AN ULTRA-TRACE CHEMICAL DETECTOR

Mark Fisher and John Sikes
Nomadics, Inc., Stillwater, Oklahoma, USA

Abstract: The Nomadics Fido sensor was developed to meet the challenges of detecting modern plastic-cased landmines with low metal content. The search for these mines using metal detectors often results in an unmanageable number of false alarms due to detection of shrapnel, debris, and mineralized soil. Trained dogs have been used for many years as an effective means for finding landmines. However, dogs are expensive to purchase and maintain, are inconsistent, difficult to train, tire within a few hours, have problems in areas with many strong competing scents, are unacceptable for use in certain cultures, cannot work under all environmental conditions, and are prone to health problems. In tests to date, the Fido sensor has demonstrated canine-comparable performance.

Key words: Fido, landmine, explosives, MECHEM Explosives and Drug Detection System (MEDDS), Remote Explosive Scent Tracing (REST), dogs

1.   INTRODUCTION

Once a landmine is deployed, a complex process begins in which the environment near the mine becomes contaminated with explosives and explosive-related compounds (ERCs) derived from the charge contained in the mine. It has been known for decades that mine detection dogs can detect the chemical vapour signature of explosives emanating from landmines [1]. More recently, detection of landmines by vapour-phase sensing of key

J.W. Gardner and J.Yinon (eds.),
Electronic Noses & Sensors for the Detection of Explosives, 117-130.
© 2004 *Kluwer Academic Publishers. Printed in the Netherlands.*

chemical signature compounds using ultra-sensitive chemical sensors has been demonstrated. As part of the Defense Advanced Research Projects Agency's (DARPA) Dog's Nose Program, in Nomadics in 1998 first demonstrated chemical vapour detection of landmines using an electronic vapour sensor. This sensor, known as Fido (Figure 1), utilizes novel fluorescent polymers to detect ultra-trace concentrations of explosives (TNT) and other nitroaromatic compounds emanating from landmines. The sensor has recently been adapted to enable analysis of modified Remote Explosive Scent Tracing (REST) filters. Using the REST methodology, Nomadics and MECHEM Division of DENEL (Pty) Ltd participated in testing of the Nomadics Fido sensor and the MECHEM MEDDS system as a tool for minefield area reduction. This work, funded by the US Army Night Vision and Electronic Sensors Directorate (NVESD), enabled comparison of the Fido sensor with canines as a tool for minefield area reduction. While more testing is needed, the initial results were promising.

Figure 1. Nomadics' Fido sensor.

By performing laboratory analysis of soil samples collected near landmines, researchers have been able to learn more about landmine chemical signatures [2-6]. The results of studies published thus far suggest that the chemical contamination emanating from mines tends to be non-uniformly distributed and can be dispersed a significant distance from the mine. In general, the concentration of signature compounds decreases as the distance from the mine increases, but depending on a myriad of environmental factors may not fall to zero (or below detection limits of dogs or the Fido sensor) for a significant distance from the mine. While much has been learned in recent years regarding the release of explosives into the environment near landmines, more study is needed. Most of the information available in the literature is derived from data gathered on a limited number of mines, and at only a few test sites. While our field test results are largely in agreement with much of the data that has been published, more data of this type are needed before general conclusions should be drawn.

If the conclusion that the chemical signature of landmines is often non-uniformly dispersed and not localized directly over the mines, it would be logical to conclude that it would be difficult to pinpoint the exact location of the mine using trace chemical detection methods. From discussions with mine detection dog handlers, free-running mine detection dogs usually indicate within a metre to at most a few metres from a mine. Similar results have been obtained using the Fido sensor. However, the use of REST sampling methods appears, in many cases, to extend the range of detection to many meters from the mine position. This is because the REST sampling method can be used to concentrate very low levels of contamination that may occur many metres from a mine onto a filter prior to analysis by dogs or a chemical vapour sensor. The vapour-concentrating effect provided by REST sampling enables low-level landmine chemical signatures that may be present a substantial distance from a mine to be sampled, concentrated, and detected by dogs or a sensitive chemical sensor at a distance much farther away from the mine than may be possible by direct searching with a dog or sensor.

The REST method, while not particularly useful for determining the exact location of a mine, is possibly quite useful for isolating the location of a mine to within a well-defined area. In theory, this makes the method ideal for use as a minefield area reduction tool. Data supporting the viability of this method for area reduction tasks will be presented.

2. FIDO SENSOR PRINCIPLE OF OPERATION

To our knowledge, Fido was the first chemical vapour sensor to detect landmines under field conditions. In these blind field tests administered by DARPA, the sensor was able to detect buried TMA5 and PMA1A landmines with the fuses and detonators removed, with shipping plugs capping the detonator well. Canines were also tested at the site during these tests. The performance of Fido was comparable to that of the canines in this test [6].

The Fido sensor is an extremely sensitive detector for nitroaromatic explosives. The sensor has been described in detail elsewhere [7], and a serial array version of the sensor is described in another chapter of this book. Therefore, only a brief description will be presented here. The sensor detects TNT and other explosives that contain TNT such as Composition B. The sensor is approximately 1000 times more sensitive than most explosive detection systems currently used for passenger screening in airports. This extreme sensitivity is necessary to detect the explosives vapours released from landmines. The sensitivity of the sensor is achieved by using novel polymer materials developed by collaborators at the Massachusetts Institute of Technology (MIT) [8]. In the absence of TNT, the polymers fluoresce (emit visible light) when exposed to light of the correct wavelength. When molecules of TNT are present, the intensity (brightness) of the fluorescence is greatly reduced. The drop in fluorescence intensity is then detected by a sensitive photodetector. The sensor detects TNT, 2,4-DNT, amino-DNTs and other nitroaromatic compounds derived from TNT. In laboratory tests, the sensor has demonstrated lower limits of detection of 1 femtogram (i.e. 1 $\times 10^{-15}$ g) of TNT.

The sensor is small (handheld), weighs about 4.5 pounds (2 kg), and can run for approximately 8 hours on a battery charge. While not yet commercially available, it is projected that the sensor will eventually have a cost comparable to that of a metal detector.

3.   NOMADICS REST FILTER DESIGN

The REST method is derived from the MECHEM Explosive and Drug Detection System (MEDDS). Using this methodology, the scent of an area suspected of being mined is sampled and transported to a detector dog for analysis. Samples are collected by drawing large volumes of air and entrained soil particulates from a suspect area through an inexpensive and disposable filter designed to trap vapours of explosives. High-volume air pumps are used to draw air through the filters. After collecting a sample onto the filter, the filter is presented to highly trained dogs for analysis. These dogs are trained to detect traces of TNT that may have been collected on the filter during sampling of a mined area. African pouched rats are also being trained for this purpose. When a dog or rat indicates the presence of TNT on a filter, the area from which the sample was collected is regarded as contaminated, and then investigated using traditional methods.

If no explosive scent is found in a sample area, it is returned to productive use by the local community. Because most areas that are suspected of containing mines are actually free of mines, this method has the advantage of preventing unnecessary and costly demining efforts. Once proven as a minefield area reduction tool, the REST concept using an on-site vapour sensor will enable real-time analysis of samples, enabling rapid screening of large areas for contamination by mines. If successful, this will result in a dramatic reduction in demining costs, and will increase the rate at which areas can be declared free of mines. An overview of the methodology is shown in Figure 2.

Large areas are sampled with a high-volume pump mechanism.

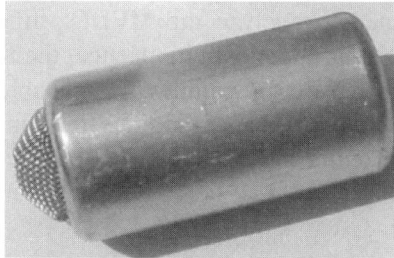

The pump concentrates any explosives present onto specially designed cartridges.

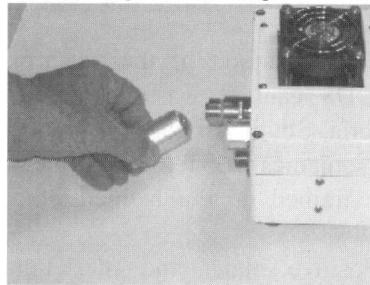

Cartridges are presented to a trained animal or instrument for analysis.

Figure 2. REST overview.

The MECHEM MEDDS filter design has proven to be very effective when used with canines. However, after initial attempts at analyzing the MEDDS filter with Fido, it was determined that the MECHEM filter design was not compatible with the sensor. After several presentations of MEDDS filters to the sensor, the sensor became essentially non-responsive to TNT. After further investigation it was determined that other researchers found it impossible to analyze the MEDDS filter using traditional laboratory methods. After considerable research, a material used in the MEDDS filter was identified as the agent reducing the sensitivity of Fido with repeated use. Unfortunately, this material is believed to be essential to the performance of the filter when the filters are used with dogs. Hence, no attempt was made to eliminate this material from the MEDDS filter.

Because of incompatibilities of the MEDDS filter with Fido, Nomadics designed a REST-type filter that was compatible with both Fido and the MEDDS dogs. The filter is the same basic geometry and size as the REST filter and can be used with traditional sampling equipment without modification of the pumps. The filter is constructed from a thin-walled metal tube packed with small, spherical beads coated with a thin film of a proprietary material. The beads are held in place within the tube by metal screens. Testing of this filter using the Fido sensor yielded promising results. In addition, after a limited amount of training on this filter, canines initially trained to analyze the MEDDS filter were able to analyze the Nomadics filter with good results. Hence, the filter is compatible for use with both the Fido sensor and canines. This enabled direct comparisons of the sensor and canine performance on the same sample. To our knowledge, this is the only filter currently available that has been proven compatible with sensors and dogs.

4. LABORATORY COMPARISON OF FIDO AND CANINES USING THE NOMADICS REST FILTER

A comparison of the performance of the Fido sensor to MECHEM MEDDS canines was performed at the MECHEM MEDDS facility in Pretoria, South Africa in February 2003. These tests were conducted using the Nomadics REST filter. At the time of testing, the MECHEM canines had been trained on the Nomadics filter for a period of approximately four months.

Positive, blank, and interferent samples were prepared using standard MECHEM methods. All samples were marked by sampling personnel in a manner that made it impossible for analysts to determine the composition of the sample during analysis. Nomadics personnel and dog handlers were not given any information on sample identity until analysis of samples was completed and results were submitted for scoring (i.e., the tests were conducted in a 'blind' fashion).

Samples were first analyzed by the canines, and were then analyzed by Fido. Samples were analyzed in two batches. Each batch contained positive, blank and interferent samples. Batch 1 contained a total of 25 samples, 4 of which were positive. Both Fido and the canines detected three of the four positives. The sensor and the dogs missed the same sample. All samples from the first batch were analyzed at room temperature. In the second batch of samples, there were three positive samples out of 24. Fido and the canines detected all three positive samples. Prior to presentation of samples in the second batch to Fido, the samples were heated slightly to enhance the vapour phase concentration of target analytes in the samples. As would be expected, responses to the positive samples that were heated were stronger than the room temperature samples. The performance of Fido and the canines against interferents was also identical. Of the 20 potential interferents included in the test, Fido and the canines responded to the same interferents, indicating on two of the twenty interferents.

The results of the laboratory comparison were promising. The performance of the sensor during this series of tests was comparable to that of the canines. One outcome of these tests was the notion that the Fido sensor could possibly be used as a canine training aid. For example, when positive samples are prepared there is currently no easy way to determine if the samples are actually positive. The sample that was missed by the canines and by Fido was prepared in exactly the same manner as the three samples that were detected, yet this sample was not detected. If the sample in question were used as a positive sample during training, but was actually blank, confusion of the dog could occur, reducing the effectiveness of the training session. In addition, a properly designed electronic sensor should exhibit reproducible and quantifiable levels of performance from day to day. The performance of canines can vary for a variety of reasons, and it can be difficult to determine when a dog is not performing at its best. The sensor could possibly be used to help verify the performance of canines. This is not to say that the performance of Fido is presently adequate to replace dogs in certain roles, but it may have a role in enhancing and complementing the performance of dogs.

5.   FIELD TEST RESULTS

From July 2001 to August 2003, Nomadics and MECHEM performed a series of trials in a test minefield near Sisak, Croatia (Figure 3). These tests were sponsored by the US Army Humanitarian Demining Program and coordinated by the Croatian Mine Action Centre. One purpose of the tests was to test the ability of both the Nomadics and MECHEM trace chemical vapour collection and analysis systems in detecting the presence of mined areas within a larger area clear of landmines.

Figure 3. Test minefield in Croatia.

The test field consisted of two segments. The first was a 40,000 square metre 'blind area' laid out in a grid pattern and containing 8 to 15 mines with locations, type, and burial depth unknown to the team. The second was a 'proximity area' which contained 3 each of 4 different mine types (12 mines total) at known positions separated by 30 metres. The purpose of this area was to determine how far explosive contamination could be detected from a mine.

Air samples were taken from the field prior to mine emplacement and analyzed by both the Fido sensor and trained canines. All samples collected were negative for explosives contamination, showing that the area was free of explosive contamination prior to emplacement of the mines.

Over the life of the project, five samplings were taken after burial of the mines, in environmental conditions ranging from very hot and dry to moderately cold and damp. In every sampling both systems detected the presence of explosive contamination. Even three days after burial of the mines, both systems detected the presence of mines in the blind test area. This was a surprise to the team, because it was expected that there would not have been time for explosives to leach from the mines to the soil surface. In general, there was an increase in contamination of the area with time, with more positive samples being obtained as the time the mines were in the ground increased.

In the proximity area, samples were taken along and 2 m to each side of 3, 7, and 11-m radii marked around each mine during each sampling event. Fido and the MECHEM canines routinely detected contamination up to 11 metres from the mine centres. Because of the layout of the test field (the mines were only 30 metres apart), it was impossible to determine if contamination spread past 11 metres from the mines. Results from the blind test area suggest that contamination spread more than 11 metres, but it was not possible to determine on average how far the contamination spread from a given mine location.

The goal of these tests was to determine whether both systems were effective in detecting areas that were mined. Based on the test results, it was determined that both systems could detect mined areas. In retrospect, the blind test area probably contained too many mines and did not contain a large area that was free of mines. Because of the large number of mines in the area, contamination of the test area was widespread. Hence, it was not possible to delineate a mined area from a non-mined area. It should be again noted that both systems found the area to be free of contamination prior to emplacement of the mines.

Certain results from the field tests were somewhat surprising. The location of positive samples as determined by Fido and the dogs was largely uncorrelated. One possible explanation for this is that the dogs were trained to detect TNT, while the Fido sensor detects TNT as well as other nitroaromatic compounds derived from TNT. Hence, Fido and the dogs may not have been detecting the same scent compounds in all samples. Another interesting finding was that a portion of the test area that was positive in one sampling was not necessarily positive in other samplings. This suggests that the contamination in a minefield is dynamic, changing along with changes in environmental conditions. Ultimately, it was concluded that the systems detected contamination of the test field with mines, but that there is still much to be learned about the spread of explosive contamination from mines.

6. OTHER SENSOR APPLICATIONS

Nomadics has also developed versions of the sensor for sensing of detonable quantities of TNT in firing range scrap and for detection of unexploded underwater ordnance.

6.1 Detection of explosives in containers

Tests conducted by Nomadics in conjunction with Sandia National Laboratories have important implications for detecting terrorist devices using the Fido sensor. The two organizations demonstrated how the Nomadics Fido sensor can be used to sense quickly and easily explosive residues in military range scrap using vapour only. The same method can be applied to applications for screening cargo and pallets of material.

The ability to detect detonable masses of high explosives, such as TNT, in scrap collected from firing ranges is of extreme interest to the US military. During normal use, firing ranges accumulate enormous piles of scrap materials that must eventually be disposed of. There is valid concern that within the piles of scrap metal are devices containing detonable masses of high explosives (HE) that could explode during disposal and recycling efforts. Dud munitions or pieces of munitions that have undergone low-order detonations can have large masses of unexploded HE still attached. These can explode with catastrophic consequences.

The team was funded by the Strategic Environmental Research and Development Program (SERDP) to determine if the Nomadics Fido sensor could address this problem by detecting detonable masses of HE within piles of scrap materials.

Laboratory tests were first performed by Sandia to quantify the sensitivity of the Fido sensor. The vapour-sensing threshold was determined to be 10 to 20 ppt for TNT and 150 to 200 ppt for DNT.

Field tests with Fido then demonstrated the proof of concept that energetic material residues can be identified with vapour sensing in enclosed scrap bins. During the tests, piles of scrap were collected from firing ranges and placed inside large, wooden shipping containers. The materials collected consisted of old tires, wooden items, shrapnel from exploded ordnance, scrap metal, plastic, and miscellaneous items such as parts from rocket motors. These items were carefully screened for the presence of detonable masses of HE and were determined to be free of detonable HE before inclusion in the test. The items were not screened for the presence of trace contamination by HE, and some contamination was likely, given the nature of the items. Figure 4 shows the contents of a typical box of scrap used for these tests.

Figure 4. Typical scrap utilized in screening tests.

Fido was then used to sample vapour from inside the boxes by inserting the inlet of Fido through an access port in the top of the box. On some boxes, a very small shift in baseline was observed when the sensor was inserted. This may have been due to very low concentrations of vapours of HE from surface contamination of scrap items. Once the background response was measured, the lids were removed from the boxes and a single item containing detonable masses of HE was placed in the box. The item was placed in the scrap pile at a point at least two feet away from the sampling port. The target items consisted of either live mortar rounds, a piece of shrapnel covered with TNT (from a munition that had undergone a low-order detonation), or shavings from a demolition block of TNT. Depending on the target item selected, the presence of the item in the pile of scrap could be easily detected almost immediately after the lid was placed back on the box.

Figure 5 compares the Fido response for background characterization (scrap only) to that for a box with shrapnel from a low-order munition added. The response to shrapnel is very strong and easily discernable from the background response. These were real-time responses obtained without sample preconcentration (i.e., high-volume sampling) of any type.

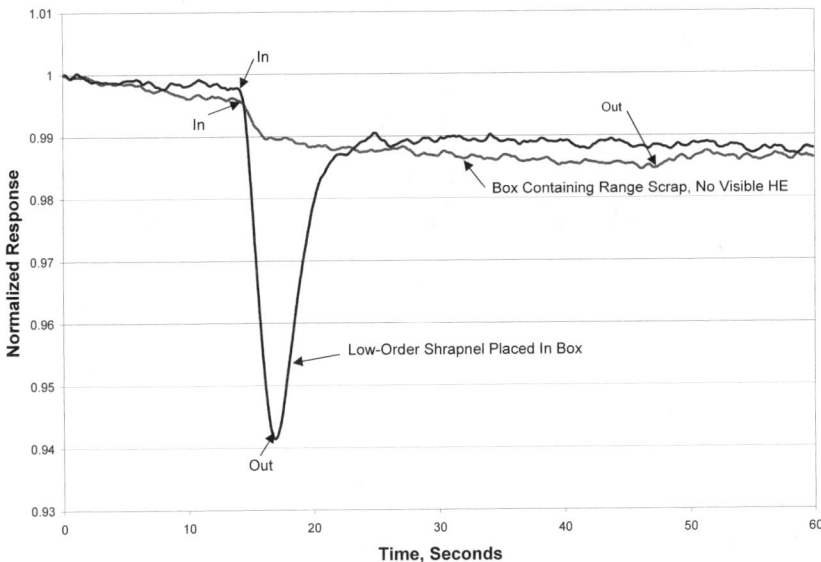

Figure 5. Response to a detonable mass of TNT placed in scrap metal.

The results of this test have implications for security and force protection applications. The test results suggest that it may be possible to use Fido technology to screen packages, shipping containers, vehicles, and other large objects for explosive devices. Even if detonable quantities of TNT are not present in the item, trace contamination in the form of vapour or explosive particulates may also be detected because of the extreme sensitivity of the device. Evidence of this type could help tie suspects who may not have the explosive device in their possession to previous activities in which they manufactured or handled a device.

6.2   Detection of underwater unexploded ordnance

The U.S. Navy performed field tests which successfully demonstrated that an underwater version of the Nomadics sensor can detect a plume of TNT in real-time in the ocean over 100 metres from its source. In these tests, the SeaPup sensor was integrated with the REMUS (Remote Environmental Monitoring UnitS) unmanned underwater vehicle (Figure 6) and deployed in a series of successful missions in the Atlantic Ocean.

Figure 6. The Nomadics sensor deployed on an autonomous underwater vehicle.

The Office of Naval Research (ONR) conducted the Chemical Sensing in the Marine Environment (CSME) Program to provide a new tool for detecting underwater UXO. Over the last few years, a number of TNT sensors have been developed and evaluated.

In June 2003, the Nomadics sensor and two other systems were given the opportunity to detect TNT in a test bed set up off the Atlantic coast. Nomadics first tested the SeaDog system which has previously detected TNT in trials in Panama City, San Diego Harbor, and San Clemente Island. This system successfully passed the tests and the newer, smaller SeaPup version of the sensor was then given mission time on the REMUS. The SeaPup sensor showed almost an order of magnitude improvement over the SeaDog due to design enhancements incorporated into the system.

One interesting aspect of the data is that data filtering allows the REMUS to monitor a threshold value for finding TNT. This system can feed information to an adaptive mission planner so that the REMUS can follow a plume upstream, to its source. The filtered data channel also allows rapid determination of mission success.

The time-domain data are integrated with the longitude-latitude positioning data from the REMUS vehicle to develop a concentration map of the area. This concentration map clearly shows a plume of TNT emanating from a source. The plume was detected over 100 metres from the source as shown in Figure 7.

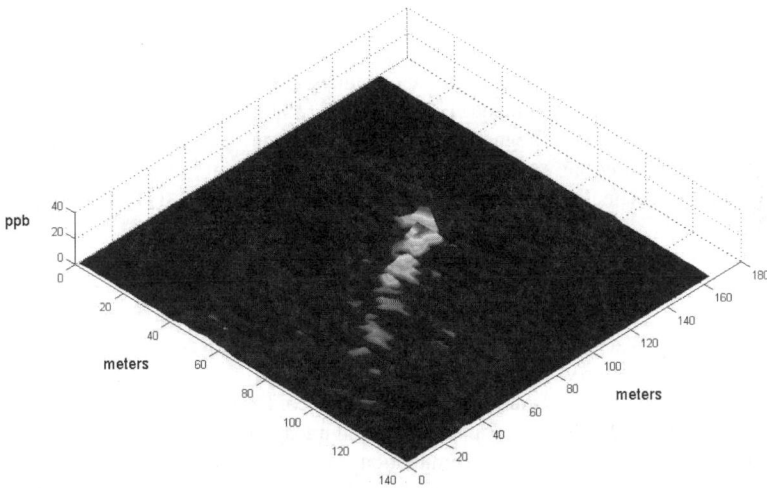

Figure 7. Detection of a TNT plume in the marine environment with the sensor mounted on an underwater autonomous vehicle.

7. FUTURE WORK

The Fido sensor is an extremely sensitive and selective detector for nitroaromatic explosives such as TNT. The sensor has also been shown to detect most smokeless powders and black powders. Work is now underway to develop polymers that enable detection of nitramine explosives, such as RDX and HMX. Recently, a new polymer was tested that shows promise for detection of the taggant dimethyldinitro butane (DMNB). Development of polymers for detection of peroxide-based explosives is also planned.

Other applications that are being pursued are detection of chemical and biological weapons, and toxic industrial chemicals (TICs). Environmental screening applications, including a groundwater monitoring probe for TNT, and a rapid screening tool to assess the extent of environmental contamination by explosives on live firing ranges and formerly used defense sites (FUDs) are also being developed. The use of Fido for forensics and security applications is ongoing.

Sensor improvements including improved sampling hardware and 'hit detection' algorithms are also being developed. These should expand the utility of the sensor in the near future.

REFERENCES

1 R.V. Nolan and D.L. Gravitte, Mine detecting canines, Report 2217, US Army Mobility Research and Development Command, 1977.
2 V. George, T.F. Jenkins, D.C. Leggett, J.H. Cragin, J. Phelan, J. Oxley and J. Pennington, Progress on determining the vapor signature of a buried landmine, Proc. SPIE, Detection and Remediation Technologies for Mines and Minelike Targets IV, 3710, Part 2 (1999) 258-269.
3 M.E. Walsh and T.F. Jenkins, Identification of TNT transformation products in soil, SR92-16, US Army Corps of Engineers, Cold Regions Research and Engineering Laboratory (1992).
4 T.F. Jenkins, M.E. Walsh, P.H. Miyares, J.A. Kopczynski, T.A. Ranney, V. George, J.C. Pennington and T.E. Berry, Analysis of explosives-related signature chemicals in soil samples collected near buried landmines, ERDC Technical Report 00-5 (2000).
5 J.M. Phelan and S.W. Webb, Environmental fate and transport of chemical signatures from buried landmines - screening model formulation and initial simulations, Sandia Report SAND97-1426, Sandia National Laboratories (1997).
6 V. George, T.F. Jenkins, J.M. Phelan, D.C. Leggett, J. Oxley, S.W. Webb, P.H. Miyares, J.H. Cragin, J. Smith and T.E. Berry, Progress on determining the vapor signature of a buried landmine, Proc. SPIE, Detection and Remediation Technologies for Mines and Minelike Targets V, 4038, Part 1 (2000) 590.
7 C. Cumming, C. Aker, M. Fisher, M. Fox, M. laGrone, D. Reust, M. Rockley, T. Swager, E. Towers and V. Williams, Using novel fluorescent polymers as sensory materials for above-ground sensing of chemical signature compounds emanating from buried landmines, IEEE Transactions on Geoscience and Remote Sensing 39 (2001) 1119-1128.
8 J.S. Yang and T.M. Swager, Porous shape persistent fluorescent polymer films: an approach to TNT sensory materials, J. Am. Chem. Soc. 120 (1998) 5321-5322.

Chapter 9

ELECTROCHEMICAL SENSING OF NITROAROMATIC EXPLOSIVES

Joseph Wang
Department of Chemistry, New Mexico State University, Las Cruces, New Mexico, USA.

Abstract: Electrochemical devices are uniquely qualified for meeting the portability, speed, cost and power demands of field detection of nitroaromatic explosives. The inherent redox activity of these compounds makes them ideal candidates for electrochemical detection. The coupling of modern electrochemical detection principles with recent advances in microelectronics and microfabrication has led to powerful, compact and 'user-friendly' explosive detection devices. Recent activity in our laboratory has led to disposable sensor strips, submersible sensors (on a cable or underwater-vehicle platforms), a voltammetric flow detector, and amperometric detectors for microchip ('Lab-on-Chip') devices for on-site electrochemical measurements of organic and inorganic explosives. The unique features of such electrochemical monitoring systems make them particularly attractive for addressing related security and environmental problems.

Key words: Nitroaromatic explosives, electrochemical detection, voltammetry, sensors, microchips.

1.   INTRODUCTION

The detection of explosive compounds has received considerable attention for national security and environmental applications [1]. Such security and environmental have generated major demands for effective field-deployable tools for detecting organic explosives in a fast, simple, sensitive, reliable and cost-effective manner.

J.W. Gardner and J.Yinon (eds.),
Electronic Noses & Sensors for the Detection of Explosives, 131-142.
© 2004 *Kluwer Academic Publishers. Printed in the Netherlands.*

This article highlights recent advances, primarily from the author's laboratory, aimed at developing electrochemical devices for detecting explosive compounds. Electrochemical sensors offer unique opportunities for addressing the needs for field screening and identification of various explosives. The inherent redox activity of nitroaromatic explosives [2], namely the presence of easily-reducible nitro groups, makes them ideal candidates for electrochemical (voltammetric) monitoring. Voltammetric techniques deal with the study of charge transfer processes at the electrode/solution interface. Here, the electrode potential is being used to derive an electron-transfer reaction and the resultant current signal is measured. The advantages of electrochemical systems include high sensitivity and selectivity, a wide linear range, minimal space and power requirements, and low-cost instrumentation. Both the sensor and the controlled instrumentation can be readily miniaturized to yield hand-held meters. Common examples are hand-held glucose meters that are widely used for self testing of diabetes.

The past two decades have seen enormous advances in electroanalytical chemistry, including the development of ultra-microelectrodes, the design of tailored interfaces, the coupling of biological components with electrical transducers, the microfabrication of molecular devices and the introduction of *smart* sensors and sensor arrays. Advances in voltammetric techniques have greatly increased the ratio between the analytical and background currents to allow quantitation down to the ppt range [3]. Particularly suited for rapid testing of explosives is square-wave voltammetry (SWV) owing to its high sensitivity, fast scan rates, and compact low-power instrumentation [4]. Readers are referred to a recent book and a review for comprehensive information on electrochemical systems [3, 5]. In the following sections we describe several recently developed powerful electrical devices for on-site monitoring of explosive compounds.

2.   TOWARDS EASY-TO-USE DISPOSABLE ELECTRODE STRIPS

In order to address the needs of field sensing of explosives, it is necessary to move away from traditional bulky electrodes and cells (commonly used in research laboratories). The exploitation of advanced microfabrication techniques allows the replacement of conventional ('beaker-type') electrochemical cells and electrodes with easy-to-use sensor strips. Both thick-film (screen-printing) and thin-film (lithographic) fabrication processes have thus been used for high-volume production of highly reproducible, effective and inexpensive electrochemical sensor strips. Such strips rely on planar working and reference electrodes on a plastic or silicon substrate. These strips can thus be considered as self-contained electrochemical cells onto which the sample droplet is placed. The thin-film fabrication route facilitates also the development of cross-reactive electrode arrays [6] that are promising for multi-explosive e-nose detection (in connection to advanced signal-processing algorithms).

We demonstrated a thick-film TNT sensor, based on the coupling of fast SWV and the screen-printing fabrication process [7]. The screen printing technology is a well established technology for the mass production of low-cost (disposable) electrochemical sensors [3, 8]. Such process involves printing patterns of conductors and insulators on surfaces of planner substrates. The screen-printed TNT sensor, developed in our laboratory, offers an attractive analytical performance, including high sensitivity (down to 200 ppb TNT), a wide linear range, fast response, and high selectivity. Such high sensitivity and selectivity were documented in connection to direct TNT measurements in ground and river water samples. Applicability to vapour detection is anticipated in connection to closely-spaced electrodes covered with a conducting permeable film.

3. REAL-TIME ELECTROCHEMICAL MONITORING

The use of sensors and detectors to continuously measure important chemical properties has significant analytical advantages. By providing a fast return of the analytical information in a timely, safe and cost effective fashion, such devices offer direct and reliable monitoring of explosive compounds. Such real-time explosive monitoring capability has been accomplished in our laboratory through on-line [9] or submersible [10] operations.

3.1 On-line continuous monitoring of TNT

We also developed an electrochemical flow system for on-line monitoring of trace TNT in marine environments, based on a square-wave voltammetric operation at a carbon-fibre based detector [9]. The flow system offers selective measurements of sub-ppm concentrations of TNT in untreated natural water matrices (with a detection limit of 25 ppb). It responds rapidly to sudden changes in the TNT concentration with no observable carry-over. About 600 runs can be made every hour with high reproducibility (e.g., RSD=2.3%, n=40) and stability. The system lends itself

134

to full automation and to possible deployment onto various stationary mobile platforms (e.g., buoy, underwater vehicles, etc.).

A flow detector, with four electrodes coated with different permselective film, is currently developed for flow injection measurements of multiple nitroaromatic explosives. The resulting array response (Figure 1) offers unique *fingerprint* of such explosive compounds.

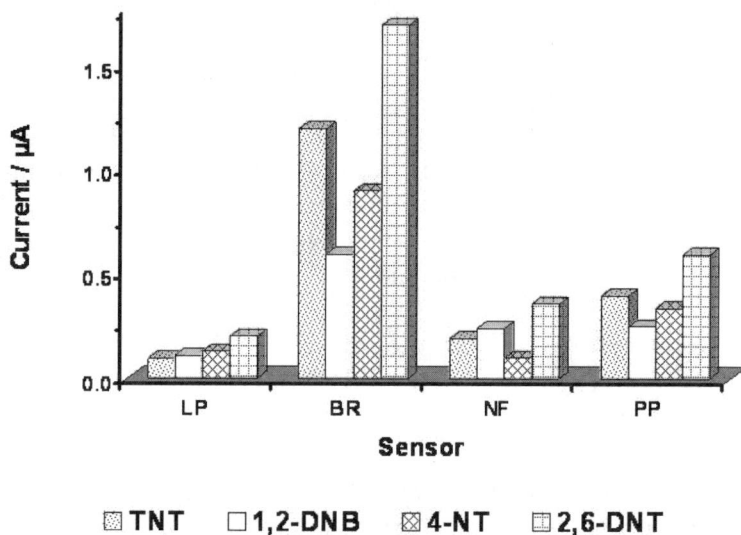

Figure 1. Amperometric array response to various nitroaromatic explosives at a flow detector with four electrodes coated with different films. LP: Lipid, BR: Bare, NF: Nafion, PP: Polyphenol.

3.2   Remote sensing of TNT

We developed a submersible electrode assembly, connected to a 50 ft (24 m) long shielded cable (via environmentally-sealed rubber connectors), for the real-time in-situ monitoring of the TNT explosive in natural water [10]. Such sensor assembly consists of the carbon-fibre, silver, and platinum working, reference, and counter electrodes, respectively, and operates in the rapid square-wave voltammetric mode. Such a remote/submersible probe circumvents the need for solution pumping and offer greater simplification and miniaturization. The facile reduction of the nitro moiety allowed convenient and rapid square-wave voltammetric measurements of ppm levels of TNT. Lower (ppb) concentrations have been detected using a background subtraction operation. Such high sensitivity is coupled to good selectivity and stability, and absence of carry-over effects. The latter reflects the absence of recognition/binding events. These capabilities were illustrated using various natural water environments. For example, Figure 2 displays voltammograms for seawater samples containing increasing levels of TNT in 250 ppb steps.

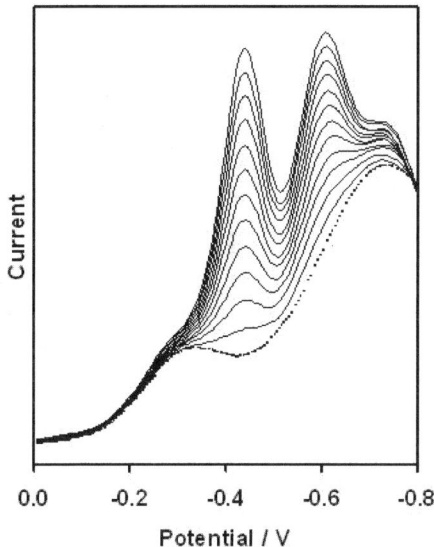

Figure 2. Square-wave voltammograms recorded with remote/submersible electrochemical probe for sea water samples containing increasing levels of TNT in 250 ppb steps.

3.3   In-situ electrochemical monitoring of TNT using underwater vehicle platforms

The voltammetric sensor has been integrated onto an Unmanned Underwater Vehicle (UUV). This integration was a part of the ONR Chemical Sensing in Marine Environment Program which had a goal to demonstrate the effectiveness, in the field environment, of advanced technology sensors for detecting explosives in coastal regions of the ocean. With such a sensor an UUV can be used to detect and prosecute plumes containing trace amounts of explosives that leach into the sea water from man-made objects, such as unexploded ordnance. The TNT sensor was added to the vehicle as a separate module and it serially sent data values to

136

the REMUS UUV. The electrode assembly was mounted on the cone-nose of the vehicle, and was connected to the internal microanalyzer (Figure 3).

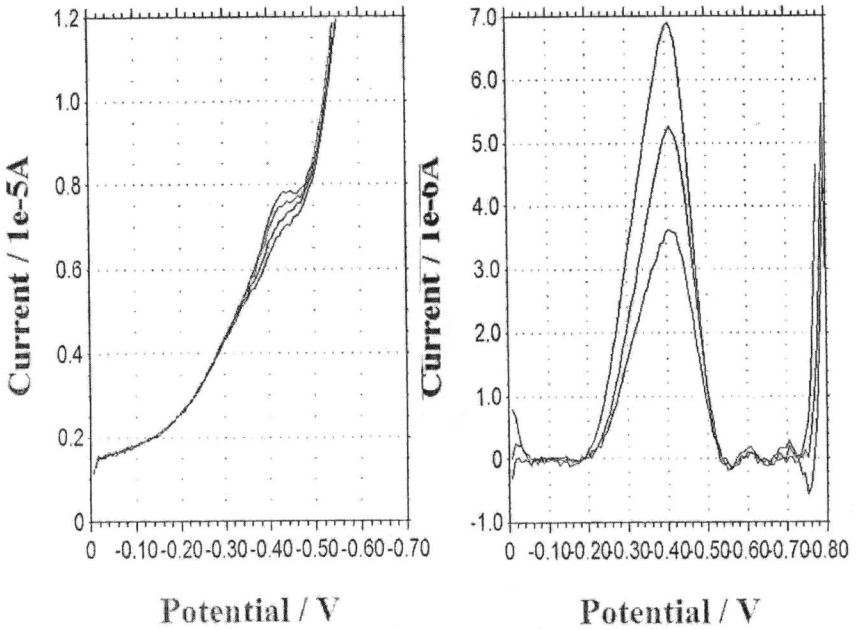

Figure 3. Square-wave voltammograms for sea water samples containing increasing levels of TNT in 20 ppb steps (Right and Left, with and without background correction, respectively).

Major attention was given to the optimization of variables of the square-wave waveform (including the frequency, potential step, and amplitude) essential for attaining high speed (1 to 4 second runs) without sacrificing the sensitivity. A computerized baseline subtraction was developed to compensate the oxygen background contribution and hence to facilitate the detection of low ppb levels of TNT (see Figure 4). The optimal device offered high sensitivity and selectivity, fast response, excellent precision/stability, and absence of matrix effect, and hence meets the demands for underwater sensing of TNT. A typical stability data involving 100 repetitive runs are displayed in Figure 5. The UUV-deployed sensor was tested for tracking TNT plumes in several field missions lasting over periods of 2–3 hours.

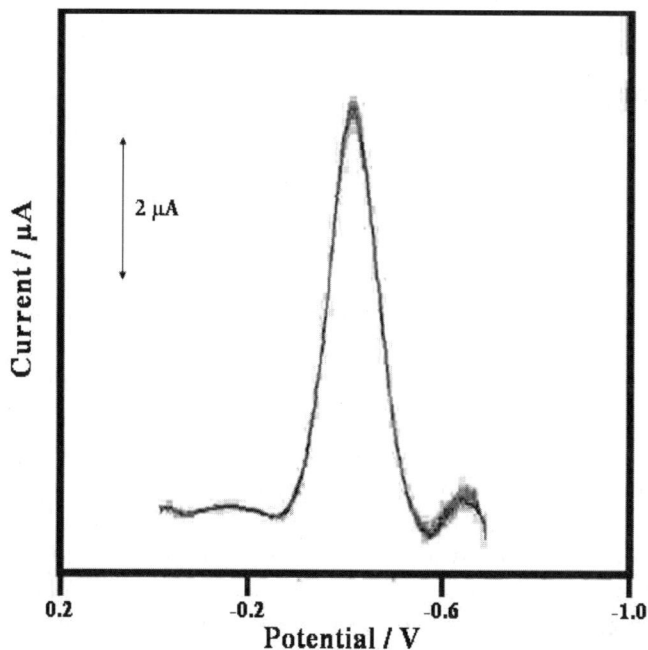

Figure 4. Stability of 100 repetitive measurements of 3 ppm TNT at submersible carbon fibre electrode assembly.

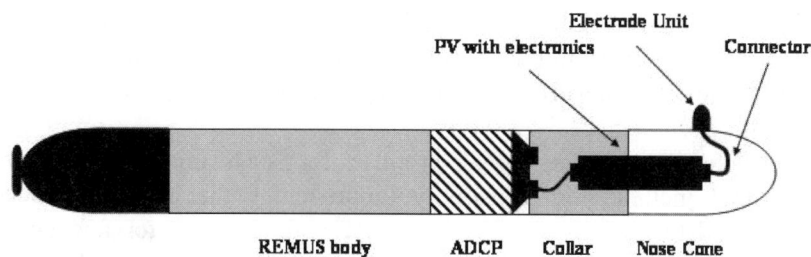

Figure 5. Electrochemical explosive sensor mounted on the Unmanned Underwater Vehicle (UUV).

3.4 Diver-held TNT sensing unit

The carbon-fibre based voltammetric sensor has also been integrated with a diver-held unit. Such integration relied on the incorporation of the PalmSense hand-held voltammetric analyzer within a pressure vessel, with the electrode assembly sticking out of the vessel surface. This configuration allows the diver to observe (in real time) the corresponding voltammogram. Repetitive square-wave voltammograms are thus performed at 2-4 sec intervals. Particular attention has been given to the adaptation of the PalmSense software to meet the requirements of continuous TNT detection. The new software was successfully tested, and the unit was deployed during a field test. Preliminary underwater testing has been very encouraging.

4.   LAB-ON-A-CHIP FOR MEASURING EXPLOSIVES

Microfabricated microfluidic analytical devices, integrating multiple sample handling processes with the actual measurement step on a microchip platform are of considerable recent interest [11,12]. For obvious reasons, such devices are referred to as *Lab-on-a-Chip* devices. Complete assays, involving sample pretreatment (e.g., preconcentration/extraction) chemical/biochemical derivatization reactions, electrophoretic separations, and detection, have thus been realized on single microchip platforms. Such analytical microsystems rely on electrokinetic fluid 'pumping' and obviate the needs for pumps or valves. Highly effective separations combined with short assay times have been achieved by combining long separation channels and high electric fields. The dramatic downscaling and integration of chemical assays make these analytical microsystems extremely promising for faster and simpler on-site monitoring of explosives. Particularly attractive for on-site security or environmental applications is the small dimensions/portability, minimal solvent/reagent consumption and waste production (to the nL level), high degree of integration, efficiency, speed and disposability.

Electrochemistry offers great promise for such microsystems, with features that include high sensitivity (approaching that of fluorescence), inherent miniaturization and integration of both the detector and control (potentiostatic) instrumentation, independence of optical path length or sample turbidity, low-power and cost requirements, and compatibility with advanced microfabrication and micromachining technologies [13]. Such properties make electrochemical detection extremely attractive for creating truly portable (and possibly disposable) stand-alone microsystems.

Our group and that of Luong have developed several effective CE/amperometric microchip protocols for detecting nitroaromatic explosives down to the ppb level [14,15]. Such amperometric detection relies on the application of a fixed (negative) potential at the working electrode, and monitoring the reduction current as a function of time. The current response thus generated reflects the concentration profiles of these explosives as they pass over the detector. For example, Figure 6 displays microchip-based electropherograms for mixtures containing increasing levels of TNT and DNB in 200 ppb steps. A surfactant, such as sodium dodecyl sulfate (SDS) is commonly added to the run buffer to facilitate the separation of the neutral nitroaromatic explosives. Hilmi and Luong [15] employed a gold working electrode, formed by electroless deposition onto the chip capillary outlet, for highly sensitive amperometric detection of nitroaromatic explosives (with a detection limit of 24 ppb TNT). Analysis of a mixture of four explosives (TNT, 2,4-DNT, 2,6-DNT and 2,3-DNT) was accomplished within 2 min, using a borate/SDS run buffer a detection potential of –0.8 V.

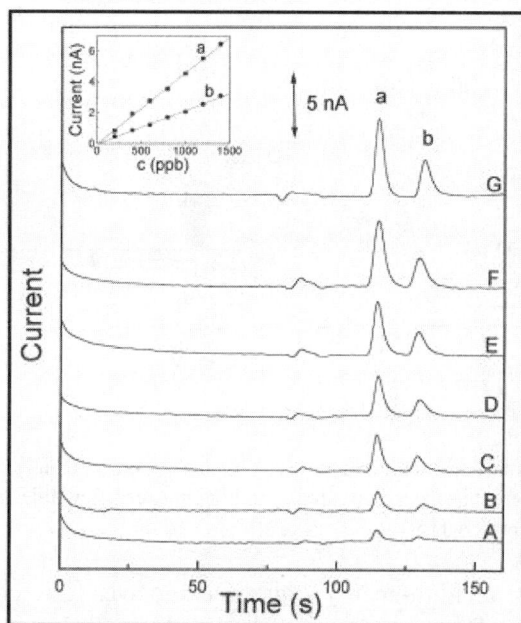

Figure 6. Microchip electropherograms for mixtures containing increasing levels of DNB and TNT in 200 ppb steps, along with the resulting calibration plots. Thick-film amperometric carbon detector held at – 0.7 V; Borate buffer (15 mM, pH 8.7) containing 25 mM SDS.

Convenient distinction between *total* and *individual* explosive compounds has been accomplished in connection to chip-based *flow-injection* (fast-screening) and the *separation* (fingerprint-identification) operation modes [16]. The realization of such dual (screening/identification) mode protocol using a single microchannel chip manifold involved a rapid switching from a run buffer that did not contain sodium dodecyl sulfate (SDS) to a SDS-containing buffer (Figure 7). As desired for various security scenarios, this allowed repetitive fast screening assays of the *total* content of explosive compounds, and switching to the detailed fingerprint identification once such substances were detected. Figure 7 illustrates such total and individual measurements for a mixture of nitroaromatic organic explosives.

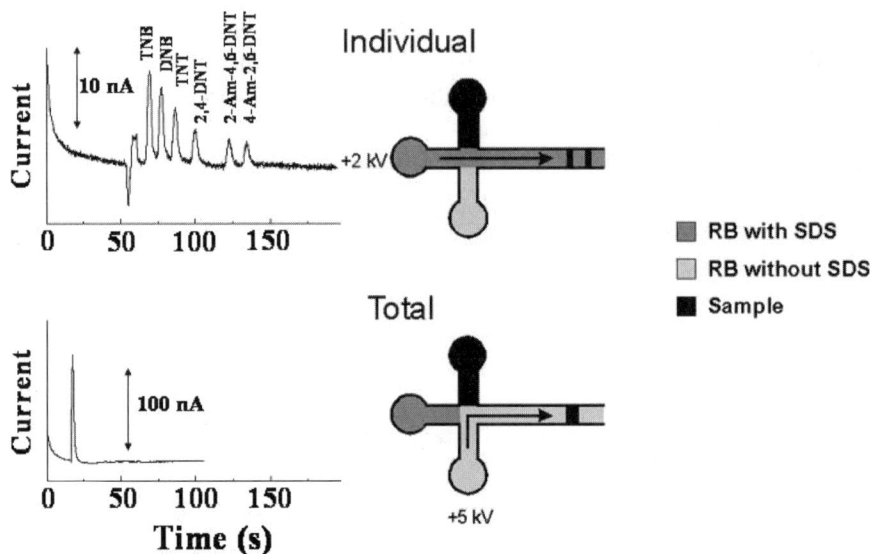

Figure 7. 'Total' and 'individual measurements' of nitroaromatic organic explosives using CE microchip (based on rapid switching between flow-injection and separation modes, respectively [16])

In addition to high-energy organic explosives, microchip devices offer great promise for separating and detecting ionic explosives [17]. A contactless-conductivity detection system has been particularly useful for this task. Such detector can sense all ionic species having conductivity different from the background electrolyte. The low electroosmotic flow (EOF) of the poly(methylmethacrylate) (PMMA) chip material facilitated the rapid switching between analyses of explosive-related cations and anions using the same microchannel and run buffer; this led to a rapid (<1 min) measurements of seven explosive-related cations and anions down to the low micromolar level. Addition of an 18-crown-6 ether modifier has been used for separating the peaks of co-migrating potassium and ammonium ions. In addition to sequential injection of anionic and cationic explosives, it is possible to use a special chip-based dual-end opposite injection protocol for simultaneous measurements of explosive-related cations and anions [18]. For this purpose, mixtures of cations and anions were injected simultaneously from both sides of the chip to the separation channel, so that the cations and anions migrated in opposite directions and detected in the centre of the separation channel by a movable contactless-conductivity detector. Simultaneous measurements of explosive-related ions and nerve-agent degradation products were also demonstrated.

5.   CONCLUDING REMARKS

The examples described in this chapter illustrate the power and versatility of modern electrochemical devices for detecting explosives. These developments would allow field testing for major explosives to be performed more rapidly, sensitively, inexpensively, and reliably, should greatly facilitate the realization of in-situ detection of explosive compounds. The resulting real-time monitoring capability should thus have a major impact on the way explosive materials are monitored and upon the prevention of terrorist activity.

ACKNOWLEDGEMENTS

The author thanks the coworkers and collaborators who have contributed to the research on electrochemical sensors, detectors and microsystems. This work was supported by grants from the EPA, DOE, ONR, and DHS.

REFERENCES

1   J. Yinon, Field detection and monitoring of explosives, Trends Anal. Chem. 21 (2002) 292-301.
2   K. Bratin, P.T. Kissinger, R. Briner and C. Bruntlett, Determination of nitroaromatic, nitramine and nitrate ester explosives by liquid chromatography and reductive electrochemical detection, Anal. Chim. Acta. 130 (1981) 295-311.
3   J. Wang, Analytical Electrochemistry, J. Wiley, New York, (2000).
4   J. Wang, E. Ouziel, C. Yarnitzky and M. Ariel, A flow detector based on square-wave polarography at the dropping mercury electrode, Anal. Chim. Acta 102 (1978) 99-999.
5   J. Wang, Real-time electrochemical monitoring: toward green analytical chemistry, Acc. Chem. Res. 35 (2002) 811-816.
6   K. Albert, N.S. Lewis, C. Schaurer, G. Soltzing, S. Stitzel, T. Vaid and D.R. Walt, Cross reactive chemical sensor arrays, Chem. Rev. 100 (2000) 2595-2626.
7   J. Wang, F. Lu, D. MacDonald, J. Lu, M.E. Ozsoz and K.M. Rogers, Screen-printed voltammetric sensor for TNT, Talanta 46 (1998) 1405-1412.

8   S. Wring and J. Hart, Screen printing carbon electrodes, Analyst 117 (1992) 1281-1285.

9   J. Wang and S. Thongngamdee, (2003) On-line electrochemical monitoring of (TNT) 2,4,6-trinitrotoluene in natural waters, Anal. Chim. Acta 485 (2003) 139-144.

10   J. Wang, R.K. Bhada, J. Lu and D. MacDonald, Remote electrochemical sensor for monitoring TNT in natural waters, Anal. Chim. Acta 361 (1998) 85-91.

11   D.R. Reyes, D. Iossidis, P.-A. Aurox and A. Manz, (2002) Micro total analysis systems. 1. Introduction, theory, and technology, Anal. Chem. 74 (2002) 2623-2636.

12   P.-A. Aurox, D.R. Reyes, D. Iossidis and A. Manz, (2002) Micro total analysis systems. 2. Analytical standard operations and applications, Anal. Chem. 74 (2002) 2637-2652.

13   J. Wang, Electrochemical detection for microscale analytical systems: a review, Talanta 56 (2002) 223-231.

14   J. Wang, B. Tian and E. Sahlin, (1999) Micromachined electrophoresis with thick-film electrochemical detectors, Anal. Chem. 71 (1999) 5436-5440.

15   A. Hilmi and J.H.T. Luong, Electrochemical detectors prepared by electroless deposition for microfabricated electrophoresis chips, Anal. Chem.72 (2000) 4677-4682.

16   J. Wang, M. Pumera, M.P. Chatrathi, A. Escarpa and M. Musameh, Single-channel microchip for fast screening and detailed identification of nitroaromatic explosives or organophosphate nerve agents, Anal. Chem. 74 (2002) 1187-1191.

17   J. Wang, M. Pumera and G.E. Collins, A chip-based capillary electrophoresis-contactless conductivity microsystem for fast measurements of low-explosive ionic components, Analyst 127 (2002) 719-723.

18   J. Wang, G. Chen, A.Jr. Muck and G.E. Collins, Electrophoretic microchip with dual-opposite injection for simultaneous measurements of anions and cations, electrophoresis, (2004) in press.

Chapter 10

IMPROVED SENSITIVITY FOR EXPLOSIVE VAPOURS AND ICAO TAGGENTS DETECTION USING ELECTROCHEMICAL (EC) SENSORS

Stewart Berry, Shirley Locquiao, Phuong Huynh, Qiaoling Zhong, Wen He, David Christensen, Lin Zhang, and Charnjit Singh Bilkhu
Scintrex Trace Corporation, 300 Parkdale Ave., Ottawa, ON K1Y 1G2, CANADA

Abstract: Improvements in electrochemical (EC) cell design, sample preconcentration, sample desorption, pyrolysis and sample transport have led to significant increases in sensitivity towards explosive detection using electrochemical NO_2 sensors. Our EC sensors have been modified to allow faster sample permeation through the sensor membrane giving sharper response peaks. Sample trapping and preconcentration has been optimized for low concentrations of explosive vapours and ICAO taggents. The trap has a low mass and achieves desorption temperatures in approximately $1/5^{th}$ of a second. Pyrolyzation uses an efficient high temperature filament, removed from but in close proximity to the desorber element. Transport of vapours to the detector is through a moisture exchanger and removes any effect of fast sampling from an overly wet or dry environment. Near zero dead volumes in the sensor housing force almost complete transmission of gases through the membrane for detection. Detection limits for most vapours is less than a ppb.

Key words: Nitrogen dioxide, electrochemical, ICAO.

1. INTRODUCTION

Electrochemical detectors have traditionally been used for sampling simple gases of interest, whether it is carbon monoxide, ammonia, or oxygen. These sensors are used in the automotive industry, medical field, and for emission monitoring. In the vast majority of cases, the sample gas concentration varies slowly. For these applications a fast sensor response (1-

J.W. Gardner and J.Yinon (eds.),
Electronic Noses & Sensors for the Detection of Explosives, 143-147.
© 2004 *Kluwer Academic Publishers. Printed in the Netherlands.*

5 seconds) is not required and the sensors generally achieve 90% of their signal in 30 seconds to several minutes. Dead volumes within the instrument have no detrimental effect as all volumes within achieve and maintain the same concentration of gas as the surrounding sampled environment. Minimum detection levels are generally no lower than a ppm. Sub-ppm levels are usually measured with a more expensive or complicated technology.

Many explosives pyrolyze to produce NO_2. Since electrochemical sensors have been used for quite some time to reliably monitor NO_2 they are a logical choice for detecting explosives. However, sensitivity in the low ppb to ppt range is essential for this application. Preconcentration is usually employed and the explosive vapours must be released from the trap in a pulse which passes through and over the sensor in 2-3 seconds. The sensor has little time to respond before the NO_2 spike is flushed from the system with clean air. The following work describes the variety of techniques employed to achieve an acceptable response at such low explosive vapour concentrations.

2.   SUMMARY OF WORK

First, a compact trap was developed. This trap needed a small cross section and efficient trapping; it needed to act like as a heat sink but cool quickly, and to be long lasting without any changes in performance. During development it was shown that smaller traps produced less background, allowed for sharper peaks and in addition the sample desorbed faster. Larger traps would collect and deliver more sample to the sensor but took more time to desorb with consequent peak broadening. Since diffusion through the membrane of the sensor is concentration driven, and smaller trap cross sections were able to deliver the highest concentration in the shortest time, the smaller cross section proved to work better. The metal mesh used to hold the trap medium was treated to prevent aging due to close proximity to the flash heater. Since the mesh tended to oxidize, a treatment was applied to prevent further oxidation. The oxidized coating enhanced heat absorption from the radiant heater. The trap medium is a proprietary powder that Scintrex Trace Corporation has been using to trap high vapour concentration explosives. It was initially developed for use in GC/IMS systems and very low levels of vapours were successfully collected and detected. With focused heating, the trap cleans out quickly even under high load conditions. For high use instruments, the trap needed to shed heat fast. Heat buildup in the trap from continuous sampling would lead to a loss in trapping efficiency and made it impossible to detect sub ppb levels. The low mass radiant heat desorber was designed to pulse for less than 250ms and to achieve complete desorption under these conditions. Testing of this trap revealed consistent trapping efficiency over thousands of samples.

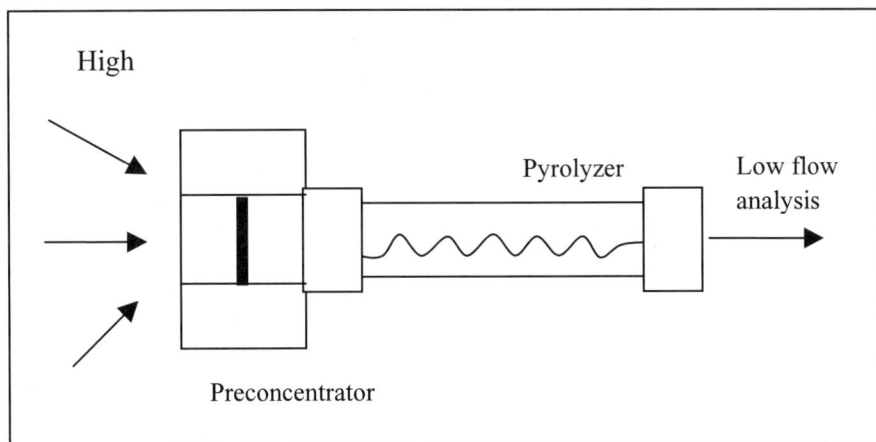

Figure 1. Desorption and pyrolysis in a single unit.

Previous systems for desorption and pyrolysis combined the two processes into one unit. The heater used to desorb the trapped vapours would continue heating to temperatures needed for pyrolysis. A portion of the sample was lost during this process. For this reason a separate pyrolyzing filament was added downstream of the trap (see Figure 1).

The trap has a tendency to preconcentrate water vapour. An example of this would be samples collected from a suitcase containing a wet towel. The sampled air would be more humid than the air outside the suitcase. This would result in a positive baseline shift when the sample is desorbed into the sensor. The opposite is also possible. It is possible to 'preconcentrate' dryness. To prevent this possibility from occurring, a length of proprietary tubing was placed between the trap and sensor (see Figure 2). The proprietary tubing acts as a moisture exchanger. It conditions the sample to have the same moisture content throughout analysis. If a relatively dry pulse passes through the tube it will humidify. The reverse is true for a relatively humid pulse of air.

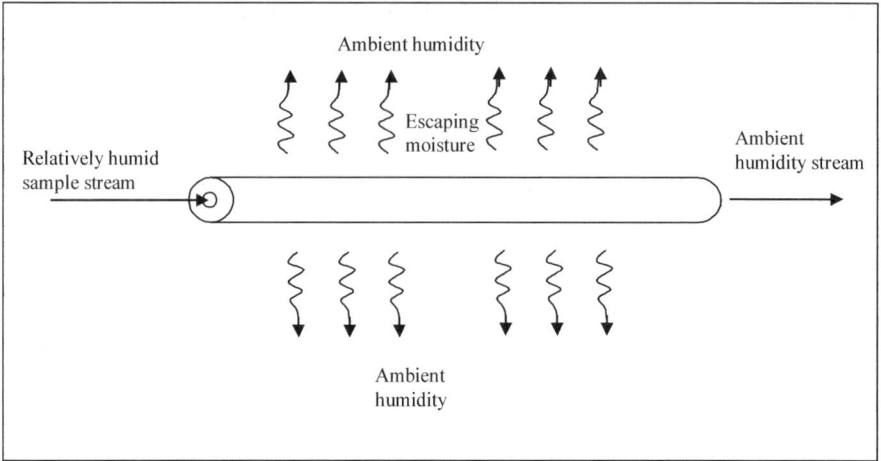

Figure 2. Moisture exchanger.

A simple change in sensor housing brought about a significant change in sensitivity. Previous designs incorporated sensor housing with a cavity above the sensor membrane. This cavity would have no impact on sensitivity for a sample stream with a constant NO_2 concentration. However in situations where the NO_2 concentration rapidly peaks (Scintrex Trace Corp. applications) the cavity causes significant dilution of sample. Without the cavity, the sample is forced across the entire surface of the membrane. By placing two sensors in line, it was shown that the first sensor now consumes all of the NO_2. In the original configuration the entire sample was not consumed. A proposed idea was to use two or more sensors in line so that all of the NO_2 generated would be detected. Simply by modifying the traditional style of housing we were able to eliminate the requirement for more sensors.

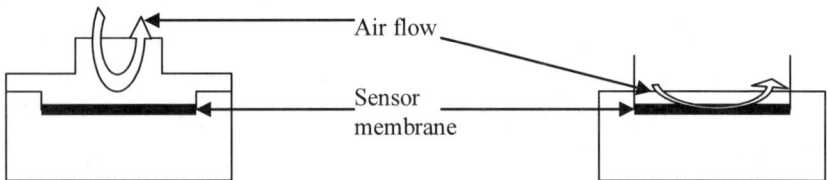

Figure 3. Two designs of sensors.

Finally the sensors themselves have been modified to a small degree (Figure 3). A thinner membrane allows for faster passage of the NO_2 through to the working electrode of the sensor. The thicker membranes normally were quite fast but did lend to some peak broadening. Further modifications also resulted in less noise. This allowed for a higher gain adjustment of the signal amplifier, and higher sensitivity.

3. CONCLUSIONS

These changes, while not necessarily increasing the sensitivity for traditional applications, have resulted in large improvements in sensitivity for Scintrex Trace explosive detectors which employ NO_2 sensors. We have taken a technology that normally detects NO_2 at low ppm levels and have developed it into a method for detecting explosive vapours in some situations to as low as 50ppt, a level suitable for taggants detection. These changes are being implemented in our product line and have allowed us to develop low cost, sensitive, and robust handheld instruments for explosives.

Chapter 11

ELECTROCHEMICAL TRACE DETECTION AND PATTERN RECOGNITION

Michael Krausa, Kartsen Pinkwart, Peter Rabenecker, and Marc Kehm
Fraunhofer-Institut für Chemische Technologie Joseph-von-Fraunhofer-Str. 7, D-76327 Pfinztal, GERMANY

Abstract: Today electrochemical sensors are widely used for personal security, environmental and process control and for medical applications (e.g. diabetes). The major advantages of electrochemical sensors are their high sensitivity, easy handling and low price compared to other sensors. Most of the electrochemical sensors are based on the amperometric method. For some years cyclic voltammetry has been used for the detection of explosives. In comparison to the amperometric method, systems based on cyclic voltammetry deliver qualitative information beside quantitative information [1]. Experimental results will be presented which show that TNT vapour concentrations down to approx. 30 ppt could be detected with our experimental setup. Moreover cyclic voltammetry could easily be used for pattern recognition like an electronic tongue. First preliminary results will be shown which underline the possibilities of cyclic voltammetry for pattern recognition.

Keywords: Electrochemical sensors, pattern recognition.

1.   ELECTROCHEMICAL SETUP

Figure 1 shows the common setup of electrochemical sensors. A membrane serves as an interface between the vapour phase and the liquid phase inside the sensor vessel. The working electrode normally is located directly on the membrane. In opposite to this common setup we use a different setup which is presented in Figure 1 for the detection of TNT. In

149

J.W. Gardner and J.Yinon (eds.),
Electronic Noses & Sensors for the Detection of Explosives, 149-158.

150

this case the electrolyte is in direct contact to the atmosphere under investigation. The reason for that is that we have not found any membrane material which is permeable for TNT, especially for the expected small vapour concentrations. Most materials we have tested adsorb TNT strongly and TNT could not be detected

A   **B**

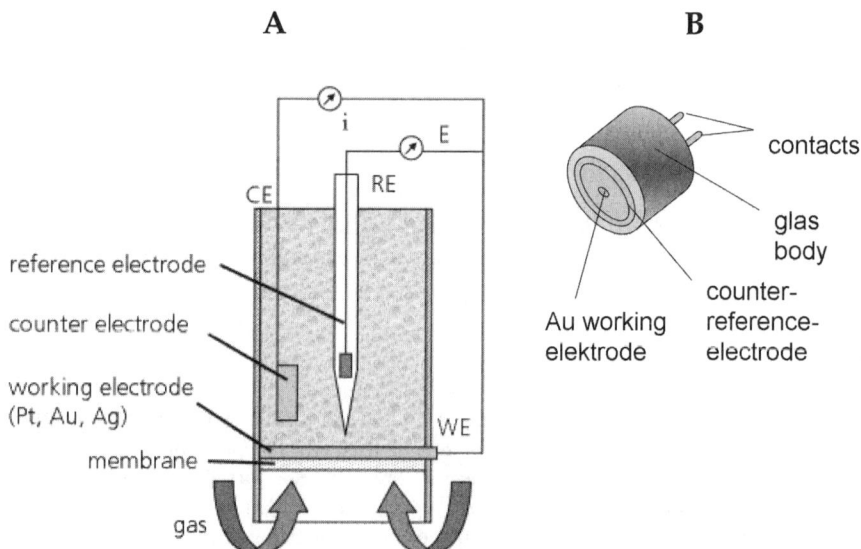

Figure 1. A: Common electrochemical sensor setup. B: Setup for TNT vapour detection

Normally large electrodes are used for the common sensors. In our case we use 25 µm diameter microelectrode (see Figure 1B) for two reasons: The first reason is that we can directly combine the complete electrochemical setup, namely working, reference and counter electrode on top of a small tip. Because of the low currents (some nA) at microelectrode, counter and reference electrode can be combined and the potential of the counter/reference electrode is nearly stable and is not distorted by the small currents. The second positive effect of the use of microelectrodes is that the ratio between Faradayic and double layer current is increasing with decreasing active surface. The reason for that is the different diffusion mechanism compared with 'large' electrodes. At microelectrodes the diffusion takes place in a spherical way like to the surface of a drop [2].

For the calibration of our sensor respectively for the further development of the sensor it was necessary to adjust definite TNT-concentrations in the vapour phase. For that we constructed a setup which allows us to adjust TNT concentrations in the vapour phase down to *ca.* 30 ppt. The setup consists of a vessel which is filled by small beads on which TNT was adsorbed. This vessel is tempered in a water bath and a constant amount of TNT desorbs from the beads to the vapour phase. A constant gas streams flows through the vessel and is saturated by TNT. This gas stream is diluted afterwards by pure air or carrier gas. The setup was calibrated by a gas chromatograph. Actually we try to automate the setup by the use of an integrated special gas chromatograph detector.

For all measurements H_2SO_4 serves as supporting electrolyte. All chemicals were p.a. grade. For all vapour phase experiments the gold ring serves as reference electrode.

2. TNT DETECTION FROM THE VAPOUR PHASE

Some years ago we conducted electrochemical measurements in view of detection of TNT in soils and liquids [3]. Figure 2 shows cyclic voltammogramms of the pure electrolyte (H_2SO_4) and a TNT containing solution. As can be seen TNT shows a large reduction peak at approx. 0.15 V vs. saturated calomel electrode (SCE). These measurements normally were conducted with a conventional electrochemical setup (three electrodes). Because information about the electrochemical behaviour of TNT is only rare we assayed different electrode materials in view of their sensitivity. Because boron doped diamond microelectrodes were not available at that time, we used electrodes of approx. 1 cm^2 geometric surface for comparison. In Figure 3 cyclic voltammogramms (CVs) of Pt and Ag electrodes in TNT containing solutions are presented. It is well known that Pt is one of the best electrocatalyst for organic substances. But beside of that we have to take the expected small TNT concentrations into account. Pt-electrodes, in comparison to other materials (BDD, Ag, Au) feature a large capacitive current (i_C) which may cover the small expected Faradayic currents (i_F) of the TNT reduction.

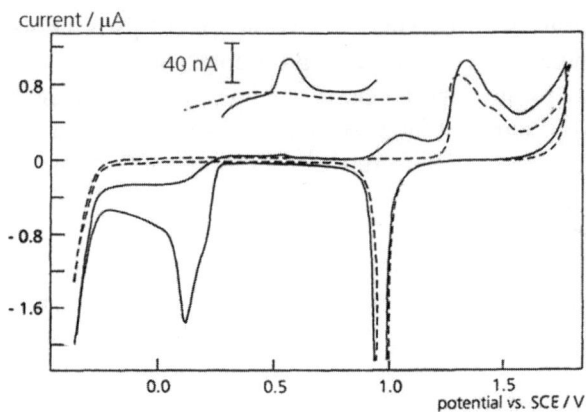

Figure 2. Cyclic voltammogramms of a Au-electrode in pure supporting electrolyte (dotted line) and in TNT containing solution (solid line), v = 100 mV/s, 0.5 M H_2SO_4.

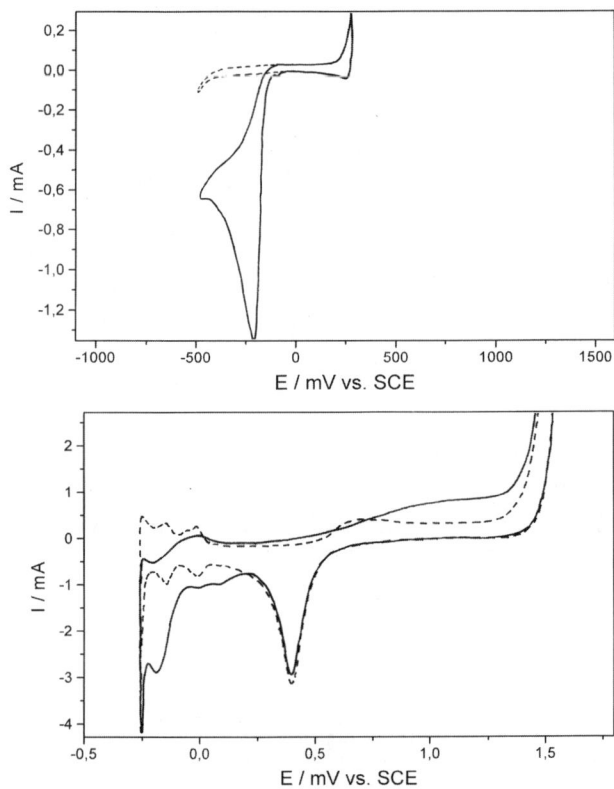

Figure 3. Cyclic voltammogramms of an Ag (top) and Pt-electrode (bottom) in pure supporting electrolyte (dotted line) and in TNT containing solution (solid line), v = 100 mV/s, 0.5 M H_2SO_4.

Therefore the ratio between the real Faradayic current (i_F-i_C) and the pure double layer current (i_C) serves as a measure for the sensitivity of the special materials. Figure 4 shows the result of these measurements. It can be clearly seen that BDD electrodes show the best ratio and therefore the best sensitivity. Ag shows a very high sensitivity also, but the use of Ag-electrodes for onsite-measurements is restricted because of their sensitivity against sulfuric substances. Actually we use Au-electrodes but we hope to repeat these measurements with BDD-microelectrodes in the near future.

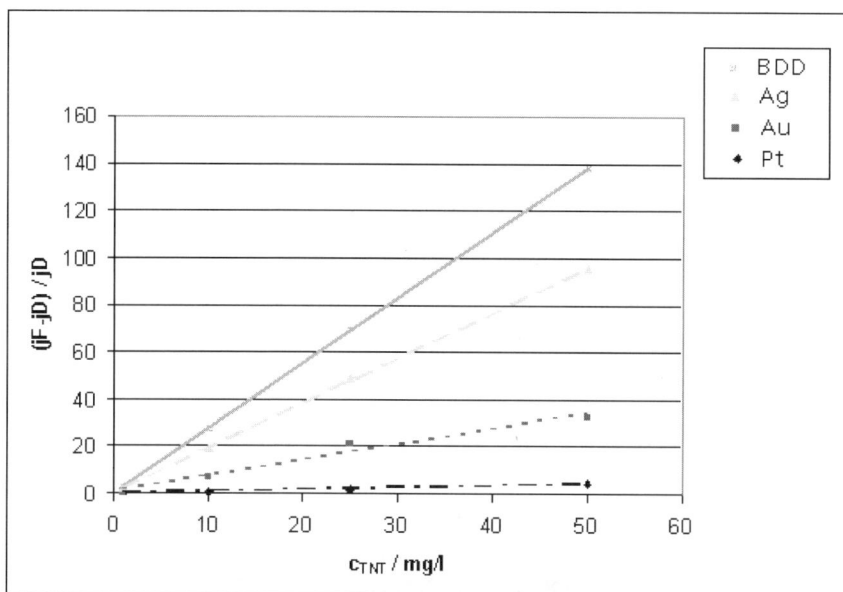

Figure 4. Ratio of (j_F-j_D)/(j_D) as a function of the TNT-concentration in the solutions at different electrode materials. Symbols j and D can be substituted by I and c.

As was stated earlier, the use of a conventional electrochemical sensor setup (Figure 1) for vapour measurements was not successful because of the adsorptive behaviour of TNT. Therefore we changed the setup and used the setup shown in Figure 1B. The electrodes tip was placed in a distance of some cm above the TNT sample and cyclic voltamogramms were recorded. As can be seen from Figure 5 the CVs show the same run of the curve like in the solution (Figure 2). This result is not fully unexpected because the probe tip is covered with a thin film of H_2SO_4 as supporting electrolyte. Obviously TNT is dissolving fast into this film. Figure 5 shows different CV for the same electrochemical setup at different adjusted TNT-vapour concentrations. Actually the lowest TNT-concentration which was measured at the Au-electrodes was 34 ppt. We are sure that it will be possible to measure smaller concentrations than the actual measured ones by varying the electrochemical procedure, the setup and the electrode materials.

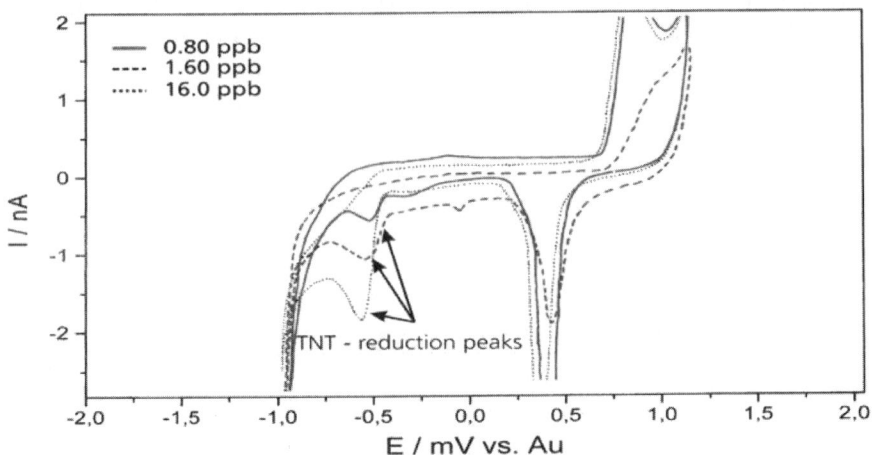

Figure 5. Cyclic voltammogramms of an Au-electrode in a TNT-containing gas stream, v = 100 mV/s, electrolyte film: H_2SO_4.

3. PATTERN RECOGNITION IN LIQUID PHASE

Up to now it is not clear what dogs really smell or which substance of an odour bouquet dogs use to identify the smell of e.g. explosives. There are several hints that dogs do not use the pure explosive but some odour bouquet of the explosive [4]. In contrast most chemical sensors systems available or under research and development focus normally on the pure explosive alone. Apparently the dog's strategy is more successful than the technical one. On the other hand today exist several systems called 'electronic noses' which simulate the smelling strategy by combining sensors of the same type, differ in composition or operated at different temperatures [5,6]. Each of these sensors will react in an unpredicted and unspecific way with an odour composed of several substances. Mathematical algorithm will combine the single signals of the sensors and deliver a specific value for different odours. It was shown for several applications that this strategy of a sensor system delivers fast and reproducible results. Moreover this proceeding is easier than the way to analyze all substances which produce a specific odour or taste and to find or develop specific sensors for single substances. Therefore pattern recognition seems to be a promising tool for explosive detection especially in consideration of the fact that we do not know what the successful sniffer dogs really smell.

So far we recorded cyclic voltammogramms for each substance to detect and we analyzed the CVs in view of oxidation or reduction reactions which are specific for the single substance under study. This means for the interpretation of the CV single values for the verification of substances and/or tastes or odour bouquets were only used. The questions arose if it is possible to extract more information from cyclic voltammogramms. For a preliminary test we tried to distinguish three explosives which contain TNT: pure TNT, Composition B, and a substance called X. If only the TNT-reduction peak is used as a measure of the explosive it is not possible to distinguish these three substances on the basis of the CV. For the conducted experiments the question is then: how do explosives taste?

Cyclic voltammogramms were taken in pure electrolyte and solutions of the three explosives. But in this case the analysis of the voltammogramms does not take place at the TNT reduction potential alone. In this preliminary case we measured the current at eight different potentials of the cyclic voltammogramm, see Figure 6, and analyzed these values with software of a MOSES-System. This means we use values which might simulate eight unspecific amperometric sensors during one potential scan at one electrode. Figure 7 shows the result of the experiments. Clearly the three different explosives can be distinguished on the basis of this analysis. No specific peak current or potential, with the exception of the TNT reduction peak is used for this analysis. Apparently cyclic voltammetry seems to be a powerful and in comparison with other methods easy tool for pattern recognition. It is planned to continue and optimize these measurements and to develop a tool for the discrimination of much more explosives by this method.

Stimulated by this result we considered other applications of this method and we tested the possibility to discriminate different apple juices. Therefore eight different apple juices were bought and without any more analysis directly assayed with cyclic voltammetry at Au-electrodes. We were joyfully surprised about the result of this preliminary and quick test. The result is presented in Figure 8. All different apple juices could be clearly distinguished. This result strengthen our opinion that cyclic voltammetry seems to be a valuable tool for pattern recognition. More investigations especially for the explosives detection in the vapour phase are planned for the future.

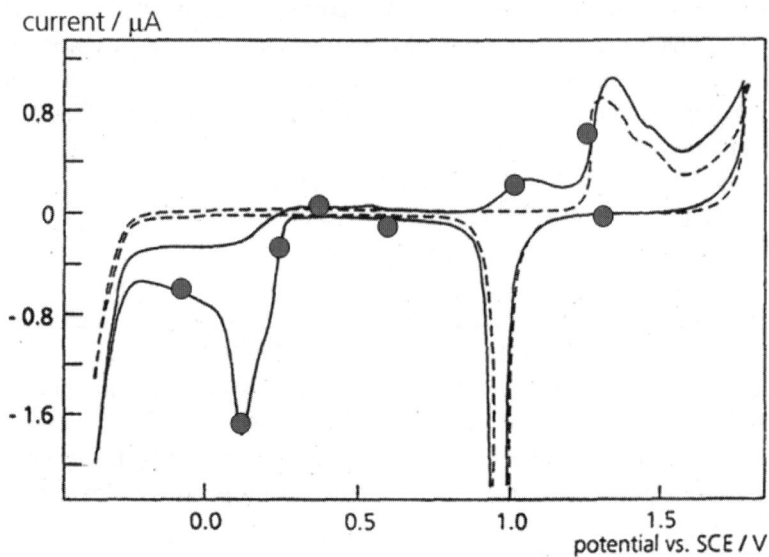

Figure 6. Cyclic voltammogramms of a Au-electrode in pure supporting electrolyte (dotted line) and in TNT containing solution (solid line), v = 100 mV/s, 0,5 M H_2SO_4, The red dots mark the potentials were the current was taken for the analysis.

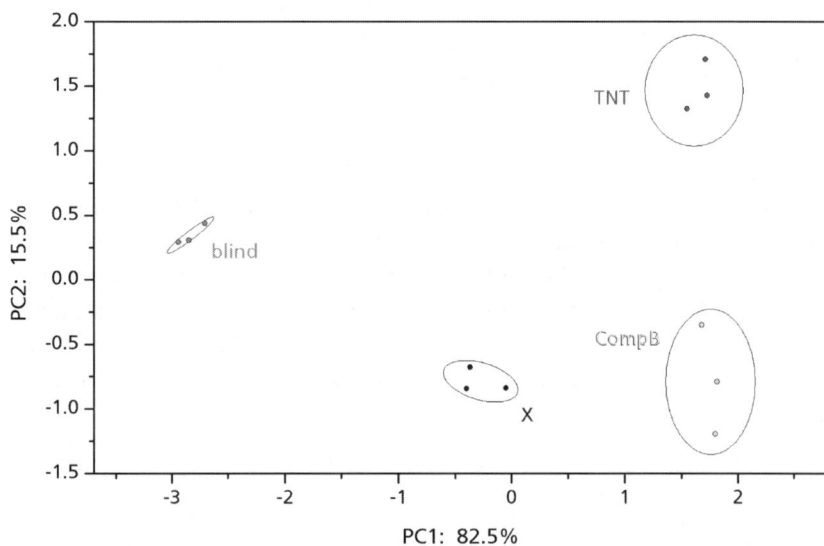

Figure 7. Results of principal components analysis (PCA) on three TNT containing different explosives. Plot shows that the compounds are linearly separable in measurement space.

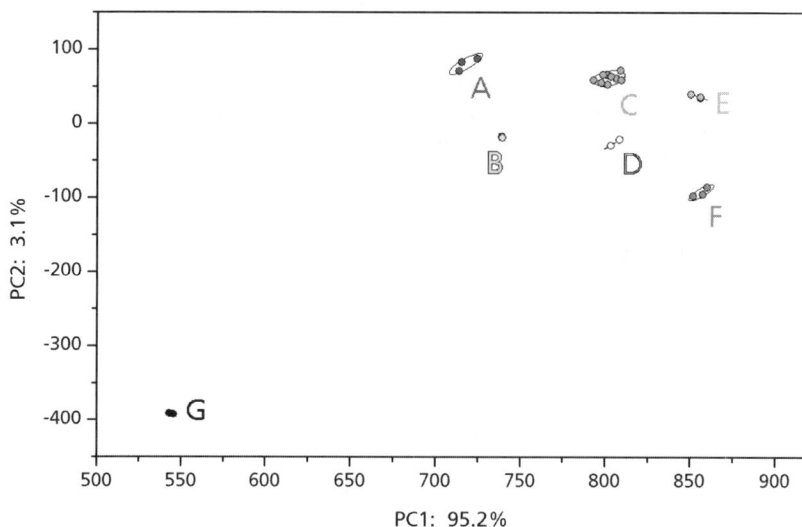

Figure 8. Results of PCA on eight different apple juices labeled A to G.

4.   CONCLUSION

Cyclic voltammetry was applied for the detection of TNT from the vapour phase. At present TNT concentrations of approx. 30 ppt can be reproducibly detected by cyclic voltammetry with a special electrochemical setup. On the basis of these results it can be stated that by varying the electrochemical procedure, the setup and especially by the use of BDD electrodes smaller TNT concentrations might be electrochemically detectable.

Preliminary results are presented for the use of cyclic voltammetry as a method for pattern recognition. The current at eight potentials of a cyclic voltammgramm was analyzed and these values served as a simulation of eight amperometric sensors. It is shown that on the basis of this procedure, different explosives and apple juices are easily distinguished by the use of one electrode alone! These results clearly show that cyclic voltammetry could easily be used for a so called electronic tongue. In the future it is planned to use cyclic voltammetry as a method in combination with different sensor methods and also for the vapour detection.

158

ACKNOWLEDGEMENTS

Financial support by the Bundesamt für Wehrtechnik und Beschaffung (BWB) is gratefully acknowledged. We like to thank DRUKKER International to provide us with BDD-electrodes.

REFERENCES

1 C.H. Hamann and W. Vielstich, Electrochemistry, Wiley-VCH, Weinheim, 1999.
2 R. Brina, S. Pons and M. Fleischmann, J. Electroanal. Chem. 244 (1988) 81.
3 M. Krausa, J. Doll, K. Schorb, W. Böke, and G. Hambitzer, Propellants Explos. Pyrotech. 22 (1997) 156.
4 M. Krausa in: H. Schubert and A. Kuznetsov, 'Bulk Detection of Explosives: Advanced techniques against Terrorism', NATO Sciences Series, in press.
5 http://www.lennartz-electronic.de/Pages/MOSES/MOSES_home_e.html
6 P. Althainz, A. Dahlke, M. Frietsch-Klarhof, J. Goschnick, and H. J. Ache, Sens. Actuators B, 24 (1995) 366.

Chapter 12

DETECTION OF DYNAMIC SMELL INTENSITY
Characterisation Of Explosives By The Response Of Gas Sensor Array To Modulated Intensity Of The Smell

Arūnas Šetkus
Semiconductor Physics Institute, A.Gostauto 11, Vilnius, LITHUANIA

Abstract: A method increasing the rate of the detection of complex volatile compounds by solid-state gas sensors is proposed. An original approach includes special modulation of the smell intensity and analysis of the kinetics of the sensor response. Essential aspects of the method are theoretically described considering dynamics of the processes in the model of the response of solid-state sensors to gas. The experimental transients of the response of metal oxide thin film sensors to a step change in gas composition of the atmosphere are analysed considering controlled injection of volatile compounds of pure TNT and explosive mixtures containing TNT.

Key words: Tin oxide, gas sensors, kinetics, volatile compounds, TNT.

1.   INTRODUCTION

Detection of explosives is an important problem in various applications, such as finding hidden explosives in airports, tracing and elimination of landmines, etc. [1-3]. The effectiveness of well-known detection methods based on electromagnetic induction and geophysical techniques, such as ground penetrating radar, is significantly limited in finding novel types of mines and explosives. Analysis of chemical composition is currently proposed as an advanced approach in development of novel methods for detection of mines and explosives [4,5]. Activities in the chemical approach is mostly initiated by the knowledge that traces of characteristic volatile compounds evapoured from the explosives or from the remaining of the explosives on the explosive's container surface can be measured and

J.W. Gardner and J.Yinon (eds.),
Electronic Noses & Sensors for the Detection of Explosives, 159-179.
© 2004 *Kluwer Academic Publishers. Printed in the Netherlands.*

recognized by suitable method (e.g. [6,7]). An artificial odour recognition system is one of the tools acceptable for the implementation of a chemical detection method to locate explosives.

A great variety of sensors and sensor systems have been developed for detection of chemical compounds in air and in liquids (e.g. [8-10]). Resistive sensors are amongst the simplest types of the sensors considering handling, capabilities to modify and production terms. In spite of the basic limitations of the selectivity to gases in single sensors, systems based upon these sensors (usually called an Electronic Nose (EN)) were effective in identification of target gas in a given mixture (see e.g. [11-13]). Standard ENs are based on the measurement of the response signals at equilibrium and processing of these signals by sophisticated methods. In these systems, each individual sensor is associated with one signal in the database. Based on numerous studies, technological methods were developed for the production of different sensors aiming to build up the systems capable to generate large databases of the response signals that are used for the featuring of a smell. Some of these systems were probed for recognition of explosives in various laboratories.

In different studies, it was recognised that mine detection by EN in the field conditions is limited by the parameters of the sensors. An increase of the sensitivity of the devices seems to be the primary task of most current developments, because very small amounts of the characteristic volatile compounds (mostly in ppb or even sub-ppb intervals) have to be traced and recognised in gas mixture. Another limitation for the EN applications is related to comparatively long response time of the sensors when small amounts of the chemicals are to be detected. Low rate of adsorption and desorption process for complex volatile compounds seems the main problem in reduction of the detection time in EN especially when it includes solid-state gas sensors.

This report deals with the dynamic response of the solid-state gas sensors when the gaseous composition of the atmosphere is intentionally varied. The study is focused on the featuring of the chemical composition of the atmosphere using the information extracted from the transients of the response. Theoretical analysis of the surface processes in the sensors is aimed at a description of the response kinetics by a set of the parameters acceptable for the featuring of an odour. The method was tested by experimental investigations of the response kinetics under exposure to various volatile compounds including that from TNT based explosives.

2.   THEORETICAL CONSIDERATIONS

A response of various solid-state sensors to gas mainly depends on the coverage of the surfaces by gas species. An amount of the gas bound to the surfaces is determined by the characteristics of adsorption and desorption and the gas concentration in the atmosphere. Presence of the gas species on the surfaces leads to some changes in the properties of the sensitive material compared to the clean surface state. Therefore, a description of a variation of the gas coverage on the surfaces is important for understanding of the sensor functioning and might be useful for development of novel gas detection methods or optimisation of the sensor parameters (see e.g. [14-16]). In current section, the gas coverage is theoretically described for metal oxide gas sensors. Dynamic aspects of this coverage are analysed.

In an oxygen rich atmosphere (e.g. normal room air), the surfaces of the sensors are initially covered by the oxygen species. Therefore, an injection of a target gas results in a bimolecular interaction between the gas and the oxygen on the surfaces in addition to the adsorption and desorption processes. The rates of all three processes define the kinetics of the coverage. An explicit form of this definition is obtained by solving the rate equations (similar to the Langmuir-Hinshelwood) for the gases involved in the surface chemical processes. Quantity of the gas types determines the number of the related rate equation similar to that

$$\frac{d\Theta_G}{dt} = \alpha_G S_{0G} P_G (1 - \Theta_G) - \beta_{dG} \Theta_G - v_{Gg} \Theta_G \Theta_g , \qquad (1)$$

In Equation (1), the parameters P_G, α_G, S_{0G}, β_{dG} and v_{Gg} define the partial gas pressure, the adsorption rate, the sticking coefficient, the desorption rate and the rate of the bimolecular interaction, respectively.

In our analysis, two coupled rate equations for oxygen (substitute Ox for G in Eq. (1)) and a reducing target gas (R for G) are solved. Assuming a change in the atmosphere composition at $t = 0$ (the air with a constant amount of a target gas is substituted for the clean air) we obtain the rise of the gas coverage on the surfaces of the metal oxide. Details are described in [14].

It is interesting to note, that the time constant included in the solution characterises not only the kinetics of the gas coverage but also the fastness of the signal rise in response to gas injection. The response time is defined as

$$\tau_{resp}^{-1} = ((\alpha_R S_{0R} P_R + \beta_{dR} + v_{ROx} \Theta_{Ox0})^2 - 4 \cdot \alpha_R S_{0R} P_R v_{ROx})^{1/2} . \qquad (2)$$

Based on the Equation (2), the speed of the sensor response (and hence time of the signal takes to fall back to the clean air value) might be explained in terms of fundamental characteristics of the surface properties.

Considering the possibilities of practical application, unique method of the gas detection is possible to develop using dynamic features of the surface processes. It is assumed that the partial pressure of the target gas is intentionally varied like a periodic function of time:

$$P_R(t) = P_{R\,max} \cdot (1 - \cos(\omega \cdot t))/2 \quad .$$

(3)

For such injection, the gas coverage of the surfaces will also periodically vary with time. It followed from the theoretical model that the variation of the coverage is similar to a monotonous increase superimposed by an oscillation like

$$\Theta_R(t) \cong 1 - \frac{Y}{Y+Z} \cdot \exp\left(\frac{Y}{\omega} \cdot \sin(\omega \cdot t)\right) \exp(-(Y+Z) \cdot t)$$

$$- \frac{Z}{Y+Z} \cdot \exp\left(\frac{Y}{\omega} \cdot \sin(\omega \cdot t)\right) \quad ,$$

(4)

In Equation (4), the following substitutions are used

$$Y = 1/2 \cdot \alpha_R S_{0R} P_R ,$$
$$Z = \beta_{dR} + V_{ROx} \Theta_{Ox0} .$$

(5)

In general, the oscillations cease with time as it follows from Equation (4). A decrease factor of the amplitude is proportional to $\exp(-(Y+Z)\,t)$ with $Y+Z$ as a sum of the rates for adsorption, desorption and the chemical reaction.

According to this model the amplitude of the oscillations and the decrease factor of the amplitude are the primary characteristics of the response that might be evaluated during a few periods of such response. Since these characteristics are directly related to the fundamental parameters characterising the surfaces and gas, it seems reasonable to suppose that the novel approach will increase the selectivity of the gas detection and decrease the detection time because the saturation of the signal is not required.

3.   EXPERIMENTS AND METHODS

The experiments were carried out for tin and indium oxide thin film sensors. The sensors were manufactured using an original prototype structure consisting of planar heater, thermometer and electrical contacts shaped on a substrate of Si with SiO_2 insulating layer. The temperature of the structures could be fixed at any value from 300 K up to 700 K. On the contacts, tin and indium oxide thin films were grown by dc-magnetron sputtering. The sensors were individually modified by an original method of post-growth doping. Metallic catalysts Pt, Ag and Cu were used for it (see [15,16] for the details).

During the tests, the sensors are mounted in a chamber with strictly controlled atmosphere. Synthetic air is flowing through the chamber at a constant rate (about 150 ml per minute). The composition of the air is controlled by the system of the flow-controllers. Two flows were produced in separated channels by the system. Clean synthetic air was flowing in one of the channels while synthetic air containing a constant amount of the target volatile compounds was in another channel. Special switching valve changed the connection of the inlet of the sensor chamber from one channel to the other one. This switching produced a very step change in gas composition in the chamber (the transition time about 2 ms). The transients of the resistance response to this change were measured by computerized system. Details of the flow control are described in [15,16].

Samples of a smell source were set in the container that was at the room temperature. Small amounts (about 2-10 mg) of explosive materials reduced to powder namely TNT, the TNT based mixture called 'JURA' and unknown mixture was used in the tests. Volatile compounds were blown from the glass container by a flow of dry synthetic air.

Transients of the response signals were analysed by the method of multi-exponential decays described in [14]. In the limit of small response signals (the resistance change is $R_{gas}/R_0 \ll 1$ in the layers), the resistance response might be expanded into series of exponents like

$$R(t) = a_0 + \sum_{i=1}^{N} a_i \exp(-t/\tau_i) \ . \tag{6}$$

Weighting coefficients a_0 and a_i might be related to the rate parameters of the surface chemical reaction. This relationship is not the subject of present study and will be analysed elsewhere. It is important to note that the time constants τ_i in Equation (6) are proportional to the time constants determined by Equation (2). The fitting of the theory to the experiment was accomplished and a set of the time constants was obtained for each of the transients in our work (details e.g. in [14,15]).

The response to periodic variation in the gas composition was analysed theoretically because the parameters of the set-up were not acceptable for the implementation of the theoretical conditions in the experiment at present stage of the investigation.

4.   RESULTS AND DISCUSSIONS

Typically, the electrical parameters of a metal oxide based sensor gradually change with time after an injection of a target gas into the surrounding air. If there is no an intentional control of the injection the change in the composition of the atmosphere depends on the accidental processes and unknown set of the parameters. This situation is typical for a standard electronic nose. Therefore, it is extremely important to measure the saturated magnitudes of the response signals. The target smell is characterised by the database of the sensor outputs in which quantity of the signals is equal to the number of the sensors in an array of the electronic nose. In current section it is demonstrated how the featuring of a smell is improved by a strict control of the injection. Additional information is extracted from the transient part of the response corresponding to: (1) step change in composition of the atmosphere and (2) to periodic variation of the composition.

4.1   Step change in the gas composition

Typical resistance responses to a step change in the gas composition are illustrated in Figure 1. The experimental results (open circles, labels 1, 2, 3) and the theoretical calculations (dash-dot lines, labels 1*, 2*, 3*) are plotted in this figure. The resistance response to gas was measured in several sensors in which the rate parameters of the surface chemical reaction were made different by modification with the metallic catalyst. The temperature of the sensors was fixed at about 600 K long before the tests.

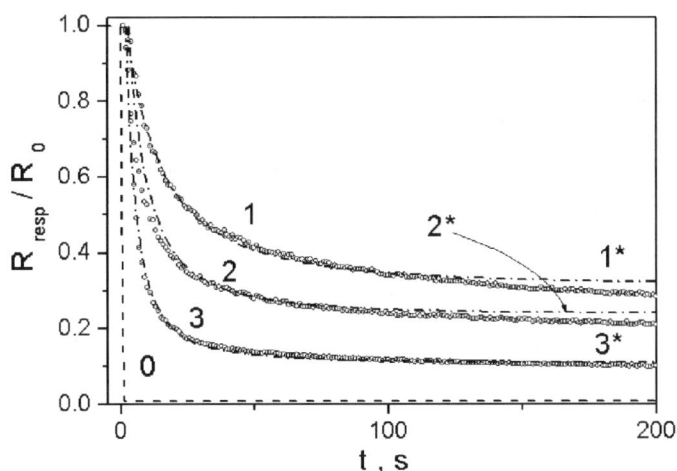

Figure 1. The resistance response to a pulse of H_2-concentration in the air measured in SnO_2-based sensors modified by Pt (1), Au (2) and Pt+Ag (3) and calculated from the phenomenological model (1*, 2* and 3*).

The transients in Figure 1 were approximated by multi-exponential decays (6) as described in Section 3. Details of the approximation accuracy were described in our recent report [15]. Up to three exponential components were typically obtained from the approximation in present study but this is not general rule. The weighting coefficients a_i and the time constants τ_i were quantitatively defined for each component. A set consisting of a few pairs of the coefficients $\{\tau_i , a_i\}$ completely describes the kinetics of the rise (or the fall) of the signal after a step change in gaseous surrounding. Though these parameters are much more useful for a theoretical study of the dynamic picture of the sensor response, the coefficients of the multiexponential decay can also be used as the elements of a database of the smell features. Two possible methods were proposed for characterisation of a smell by these coefficients until now.

First, the time constant described by Equation (2) is analysed and the phenomenological rates of the surface chemical interaction between a target gas and the surface are evaluated. A set of the reaction rates specific to a target gas and the sensor surfaces is supposed as the smell features.

Second, the coefficients of the multiexponential decay are included into a database of the sensor outputs. Original graphical representation of the database is composed for visual inspection and identification of the smell.

Some essential details of the methods are described in Sub-sections 4.1.1 and 4.1.2 in the text below.

4.1.1 The reaction rates

The reaction rates are evaluated from analysis of the experimental dependencies of the time constants on gas concentration. The phenomenological relationship (2) is used for the calculation of the theoretical dependencies. The phenomenological rates are varied until the best fit between the theory and the experiment is obtained.

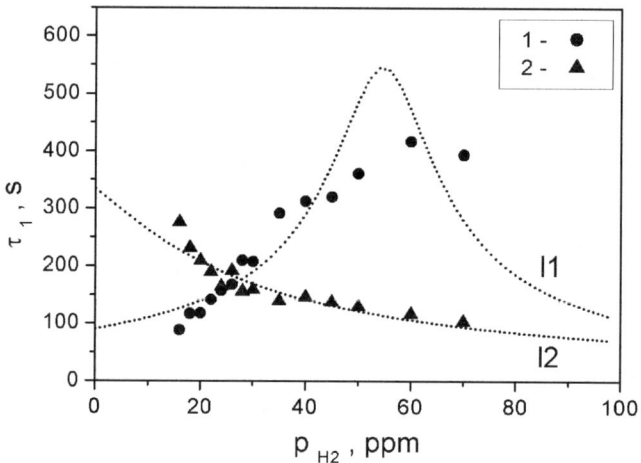

Figure 2. Dependencies of the response time constant τ on H_2-concentration in the atmosphere for the SnO_2- sensors modified by Mo using different technologies (1, 2).

Typical dependencies of the time constants on gas concentration are illustrated for two types of SnO_2 sensors with modifying impurity Mo in Figure 2. The two types of the sensors are different in the methods of the modification by the impurity metal. Each of the points in Figure 2 represents an individual transient of the sensor response to the appearance of fixed partial pressure of H_2-gas (p_{H2}) in the air. The lines illustrate the calculated dependencies that correspond to the best fit between the theory and the experiment. Individual sets of the rate parameters $\alpha_R S_{0R}$, β_{dR} and ν_{ROx} were obtained for the sensors. These parameters are listed in Table 1.

It followed from the investigations that the rate parameters are dependent not only on the properties of the sensor surfaces but also on the gas to which the surfaces are exposed. The rate parameters of these investigations are summarized in Table 1. The values of the rate parameters presented in the table are the averages of several equivalent tests for each of the sensors.

Table 1. The rate parameters of the surface reaction evaluated by fitting of the phenomenology (2) to the experimental dependencies of τ on gas concentration for various sensors modified by individual metallic impurity and exposed to H_2 and CO gases in the air of the relative humidity RH= 33 %.

Sensor	H_2			CO		
	$\alpha_B S_{0B} \cdot 10^4$	$\beta_B \cdot 10^3$	$v_{AB} \cdot 10^3$	$\alpha_B S_{0B} \cdot 10^4$	$\beta_B \cdot 10^3$	$v_{AB} \cdot 10^3$
SnO_2-pure	1.23	1.2	1.7	-	-	-
SnO_2-Ag	1.4	2.9	6.7	0.58	0.7	8
SnO_2-Mo	1.2	3.5	4.8	0.2	2.6	3
SnO_2-Pt	0.94	3.1	3.3	0.65	0.55	28
SnO_2-W	0.8	2.3	3.4	0.97	0.9	18

Detailed analysis of the rate parameters in Table 1 reveals the role of individual metallic impurity in each of the steps of the surface interaction. For example, it can be concluded that the chemisorption of H_2 gas is nearly independent of the metallic catalyst because the rates $\alpha_R S_{0R}$ of the chemisorption are practically the same on the surfaces of all the sensors in our experiments. On the contrary, the desorption rate β_{dR} and the rate of the bimolecular interaction v_{ROx} are evidently higher for the surfaces modified by the metals than for pure SnO_2 sensor. Such detailed comparison of the rates is important for the fundamental description of the sensor.

If omitting the fundamental aspects from the comparison of the rates in Table 1, it can be concluded that unique collection of the rate parameters can be associated with individual sensor in a certain atmosphere. In general, such a collection can be accepted as an additional database of the smell features extracted from the transient part of the sensor response.

An influence of co-adsorption effect on the reaction rates is illustrated in Figure 3 by typical dependencies of the rate of bimolecular interaction v_{ROx} on presence of H_2O vapour in the atmosphere. Presence of water vapour in the atmosphere significantly affects the rate of the bimolecular interaction only on the surfaces modified by Pt (3) or composed of mixed oxides of tin and molybdenum (1). On these surfaces, the rate v_{ROx} is comparatively high in dry atmosphere but drastically decreases if relative humidity of the atmosphere increases. It must be noted that the other rate parameters are also dependent on the co-adsorption effect. Consequently, a gaseous mixture will be distinguished from a single gas by a change in the parameter set.

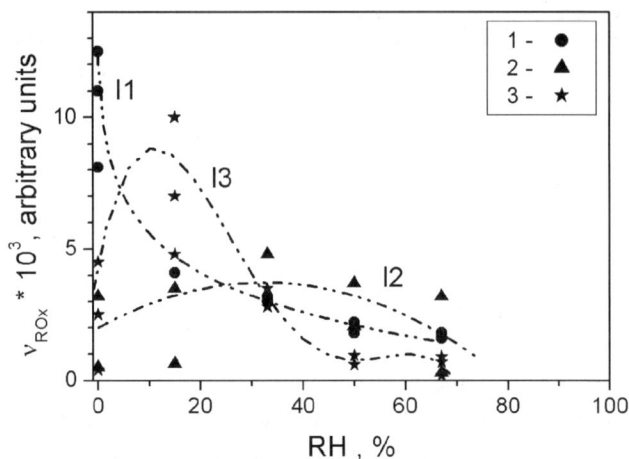

Figure 3. Dependencies of the rate of the bimolecular interaction on the relative humidity of the air on the surfaces of SnO_2 sensors modified by the bulk Mo (1), the surface Mo (2) and the surface Pt (3). The lines l1, l2 and l3 are related eye-guides.

4.1.2 Time constant based featuring of a smell

The time constant based characterisation of a smell includes an approximation of the response kinetics by multiexponential decay similar to the description (6). A collection of several pairs of the parameters $\{\tau_i; a_i\}$ is usually obtained for each of the sensors in the array from this approximation. The characteristic parameters, namely the time constants τ_i and the weigh coefficients a_i, are used for the composition of a graphical representation of the sensor outputs that might be visually inspected. The graphical representation is build up by an original method.

The pairs of the related parameters, such as time constant τ_i and weight coefficient a_i are supposed as the co-ordinates of a point in a two-dimensional plot. Reciprocal time constant was assumed to be the length of the vector while the weigh coefficient defined the sine of the rotation angle ϕ for the vector. The points of separate sensors from an array are displayed in individual segments of the circular plot.

In general, the time constants τ_i are the phenomenological parameters related with the rate of the chemical reactions on the surfaces of the sensors. Consequently, a change of a dominant compound in the target smell must be displayed by a change in the τ-constants. Such a change will be visualised by a change in the distance from the centre at which the corresponding point is plotted. The weighing coefficients a_i are dependent on the intensity of the target smell, i.e. on the concentration of the volatile compounds in the smell. A change in the concentration will be related to a change in the rotation angle for the vector in the plot. Taking into account that several pairs of the coefficients $\{\tau_i;\ a_i\}$ describe the transient of the response, graphical representation of the outputs of even one sensor will include a few points.

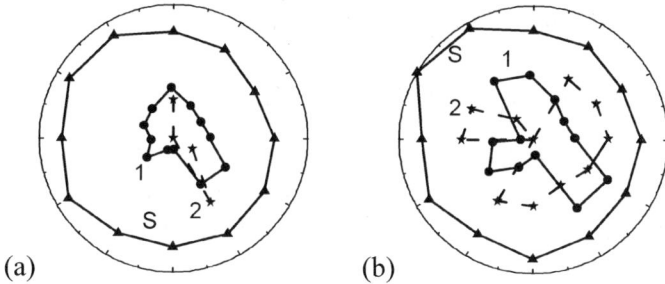

(a)

(b)

Figure 4. Visualised response transients for the array of 12 sensors exposed to the growth substances with different bacteria: (a) *Streptococcus progenes*; (b) *Staphylococcus aureus*.

In Figure 4, typical 2Dτ-images of the smell are illustrated for the samples with bacteria from the human wounds. In the images, the output data were collected from an array of 12 sensors. One or two exponential terms were usually obtained for the transients of the responses of the sensors to the smell. The exponential terms of the same order (the magnitude of the time constants is considered) are displayed for all the sensors of the array by the same type of the points in the images. These points are connected by a closed line, which outlines some shape. Consequently, the exponential terms of the same order are summarised for all the sensors of the array as a profile of some specific shape in the two-dimensional images. Several profiles based on the individual collection of the terms of the same order were laid one on top of the other in the same image. As a result, three layer images were composed of the output data collected from the array of 12 sensors in these experiments. In Figure 4, the profiles (or layers) marked with labels 1 and 2 correspond to the exponential terms of the first and the second order while the layer with label S represents the stationary signals. The multilayer arrangement of the profiles is supposed as the 2Dτ-images of the smell.

The 2Dτ-images are compared for two types of the bacteria in Figure 4. The profiles that visualise the exponents of the same order (the same label in Figure 4) might easily be distinguished one from another. The shapes with labels 1 and S include the points attributed to all the sensors in the array while only some of the sensors are represented by the points in the layer 2 in Figure 4. The 2Dτ-images seem to be individual for each type of bacteria.

The differences between the smells of the related bacteria cultures are clearly seen by visual inspection of all the layers in the 2Dτ-images in Figure

4. If only the stationary response signals are considered, the layers labelled by *2* and *3* should be omitted in the analysis of the plots. Some difference between the layers S can be seen. If the layer *1* is analysed, the differences between the target odours are much more evident. However, the recognition of the odour is much more reliable if the parameters of all the layers are considered in the characterisation of the odour.

4.1.3   The dynamic response to explosives

The transients of the sensor responses to a smell of explosives are analysed in this sub-section. The array of the sensors collected for the experiments with bacteria (see previous sub-section) was used in this experimental study. A step change in the composition of the atmosphere in the sensor chamber was produced by the controlled injection of the headspace air from a glass container with explosive materials. Thought the flow rates of the mixed streams were permanently controlled and the proportions between the carrier gas and the headspace air were well defined, the amount of the volatile compounds emitted from the explosive material was not defined.

Typical responses of several types of the sensors to the headspace of explosives are compared in Figure 5. The complete transients (including the rise and the fall of the signal) were measured for the TNT-based explosives.

The experimental results revealed that typically there is no saturated part in the transient of the response even for comparatively long (about 10-15 minutes) exposure to the headspace air. In general, a sophisticated curving shape of the transients displays both the rise and the fall of the response signals. It is clearly seen that the response to the headspace does not correspond to the pure multi-exponential decay. Moreover, most of the transients can be split into a linear combination of two curves, each of which is described by a pure multi-exponential decay.

One of these constituent curves is attributed to a fast response to one of the volatile compounds emitted from the TNT based explosives. Some traces of saturated signal can be related to these fast responses. In addition to the fast term, the curve representing much slower response to the smell is also visible in the measured signal. This constituent curve superimposes the fast response. There is no tendency towards the saturation in this constituent curve representing the slow response term. Monotonous increase of this term indicates a poisoning of the surfaces by some of the volatile compounds.

Based on the results of other authors (e.g. [4,5]) the response signals might be attributed to the two compounds that are typically found in the headspace air of TNT. These compounds are 2,4,6-trinitrotoluene and 2,4-dinitrotoluene. At this stage of the study it is not possible to attribute any of these compounds to a certain part of the response kinetics of the metal oxide thin film based sensors. Moreover, the experimental results prove that the sensors also detect some other volatile components in the headspace of the TNT based explosives.

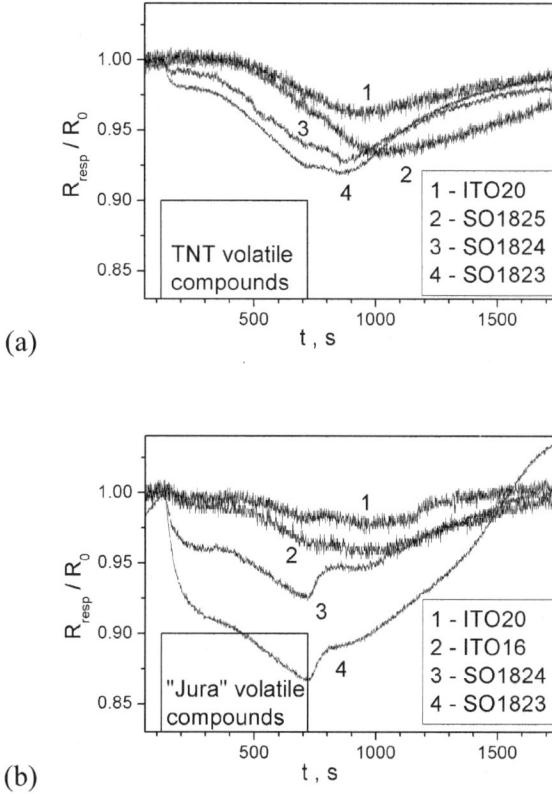

(a)

(b)

Figure 5. The resistance response of several types of SnO$_2$ and InSnO based sensors to a pulse of the volatile compounds blown from a glass container with pure TNT (a) and the TNT-based mixture called 'Jura'.

In Figure 6, the responses to three types of the TNT based explosives are compared for one sensor from the array.

Only the presence of compounds that are not related to the volatiles of TNT can explain the difference between the responses labelled by 2 and 3 in Figure 6. Supposing that the sensors detect only the volatile compounds specific to TNT the largest response must be expected for the pure TNT. It also is considered that the method of the exposure was the same and that the amount of the powder in the glass container was analogous in all the tests. In any mixtures, partial amount of pure TNT is reduced and the intensities of the expected components have to be comparatively lower. In general, the

172

transient of the response to the mixture 1 in Figure 6 supports such the premise. In contrast to this, the mixture called 'Jura' (label 3) produces significantly larger response in the metal oxide based gas sensors than pure TNT. It seems reasonable to suppose that it is necessary to consider some other volatile compounds not specific to pure TNT. At present stage of the investigation, these volatiles are not identified.

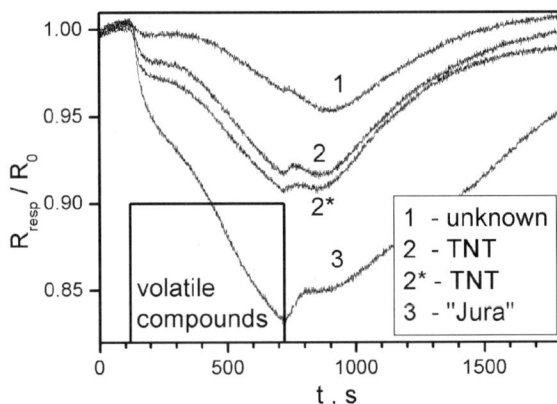

Figure 6. The resistance responses of a SnO$_2$ thin film based gas sensor to a step-like change in the composition of the atmosphere produced by controlled injection of the volatile compounds from a glass container with unknown explosive mixture (1), pure TNT (2) and the TNT-based mixture called 'Jura' (3)).

A complex shape of the curve displays the complete transient of the response to the volatile compounds of the explosive materials. The possibility to apply the method of the multi-exponential decay is highly complicated. Therefore, an analysis of the graphical images of the sensor responses to the headspace of the explosives has been omitted from this study.

4.2 Periodic modulation of the smell intensity

A novel method for the evaluation of the kinetic parameters of the sensor response to a gas is described in this section. The theoretical study of the phenomenological model described in Section 2 made it possible to propose this method. The core of the method is the controlled periodic variation of the gas composition in the atmosphere. It was theoretically demonstrated by [4] that the oscillations of the gas coverage on the surfaces are determined by the rate parameters of the chemical reaction. Since main characteristics of the oscillations can be evaluated within a few initial periods, the gas detection time can significantly be reduced without a loss of the information.

The transition from the clean air signal to a saturated response to gas is not required for the definition of the sensor output in the novel method. An experimental implementation of the method is related to significant modification of the existing equipment. Therefore, some essential aspects of the methods are theoretically studied in present work.

The gas coverage on the surfaces is supposed as the key parameter in this study. Such approach allows extending the findings of the study to various types of the sensors. This is because the gas coverage on the surfaces determines not only the functioning of the resistive gas sensors but also is essential in any other types of the sensors in which the response depends on the chemisorption of gas.

The key description of the oscillations of the gas coverage (4) was derived within the phenomenological model supposing that the clean air is replaced by an atmosphere in which an amount of a target gas is periodically varied. According to (4), simple periodic variation of the gas concentration in air (3) produces complicated oscillations of the gas coverage on the surfaces of metal oxides. Two components are distinguished in these oscillations. One of the components defines the oscillations of the coverage with decreasing amplitude while another component represents oscillations characterised by constant amplitude. The weight of the components depends on the rate parameters of the surface chemical reaction.

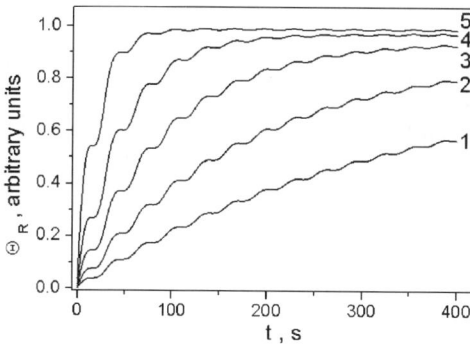

Figure 7. Dynamic response of the gas coverage on the metal oxides to the modulated concentration of a target gas chemisorbed on the surfaces characterised by different chemisorption rates α_R (10^{-4}):1 – 0.5; 2 – 1; 3 – 2; 4 – 4; 5 – 10.

In Figure 7, the component of the oscillations that mainly depends on the chemisorption rate is illustrated. The dependencies in this figure represent the response of the gas coverage to the modulated exposure of metal oxide surfaces to target gas in air. These dependencies are calculated in accordance with (4) in which negligible weight was assumed for the second time dependent term. The assumption means that chemisorption of a target gas is the dominant process in the formation of the gas coverage. An influence of the chemisorption rate on the response was investigated by calculation of a series of the curves defined by individual set of the parameters. The parameters are comparable with that defined in Table 1.

Since the chemisorption is the dominant process, only the disappearing oscillations are clearly seen in the coverage responses in Figure 7. The decrease of the amplitude is much more significant if the chemisorption rate is higher. Only very few waves are visible at the beginning of the response labelled by 5 while a decrease in the amplitude is hardly seen in the curve 1 in Figure 7. Moreover, the transient part of the response is clearly shorter if the chemisorption rate is higher.

The second time dependent term in the coverage definition (4) is mostly determined by the processes that 'consume' the chemisorbed species of the particles. The sum of the desorption rate and the rate of the bimolecular interaction determines the character of this type of the oscillations. Assuming that the bimolecular interaction is the dominant process in the surface chemical reaction, these oscillations are revealed in Figure 8. It is interesting to note that the amplitude of this type of the oscillations is constant with respect of time.

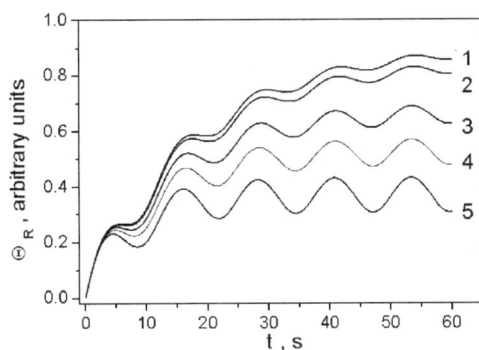

Figure 8. Dynamic response of the surface coverage by gas to the modulated concentration of a target gas chemisorbed on the surfaces characterised by the rates of the bimolecular interaction v_{ROx}: $1 - 0.4$; $2 - 4$; $3 - 20$; $4 - 40$; $5 - 80$.

The coverage response to modulated partial pressure of a target gas was calculated at different rates of the bimolecular interaction. As it follows from Figure 8, an increase in the 'consumption' of the target gas species on the surfaces results in a decrease of the amplitude of the oscillations. In addition, the transient part of the average of the response is longer for the surfaces on which the rate of the bimolecular interaction is higher (e.g. compare the curves 1 and 5 in Figure 8). Analogous changes in the coverage response are obtained if an influence of the desorption on the coverage is analysed by a variation of the desorption rate in the calculations using Equation (4).

If, in the theoretical study, the surfaces characterised by the rate parameters listed in Table 1 are considered, both types of the oscillations are revealed by the calculations in the coverage response to the modulated pressure. Depending on the proportions between the rate parameters for the real surfaces, the oscillations of the response will hardly be revealed in the signal under the conditions defined for the study that is described above in the text. From the practical point of view it is important to note that the oscillations of the gas coverage can be enhanced for the same surfaces by a change of the enforcing modulation of the partial pressure of a target gas.

Capabilities to optimise the conditions for the investigation of the response oscillations are demonstrated by the results in Figure 9. In this figure, the coverage responses were calculated using constant rate parameters of the surface reaction and different frequencies of the enforcing modulation of gas pressure. It follows from Figure 9 that the amplitude of the oscillations is dependent on the frequency of the pressure modulation. The most significant changes in the amplitude are obtained if the frequency of the pressure modulation is comparable with the reciprocal response time of the sensor defined by Equation (2). At such frequencies the amplitude increases significantly (see curves 1 and 2 in Figure 9). The reciprocal response time $(2\pi/\tau)$ is equal to 0.125 for the surfaces characterised by Figure 9. If the difference between the frequency of the pressure modulation and the reciprocal response time is huge the oscillations (e.g. curve 4) can practically be ignored due to comparatively low amplitude of the response oscillations.

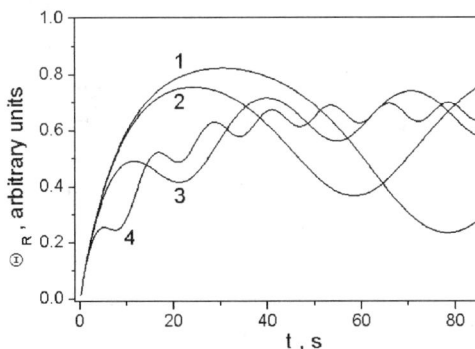

Figure 9. Dynamic response of surface coverage by gas to modulated concentration of a target gas for different frequencies ω: 1 – 0.06; 2 – 0.08; 3 – 0.2; 4 – 0.5.

In general the results in Figure 9 mean that the reaction rates can be extracted from the coverage oscillations by conscious tuning of the external influence, namely the frequency of the pressure modulation. The measurements at different frequencies will define the solution for the rates.

The partial pressure of a target gas can also be evaluated from the amplitude of the oscillations. A dependence of the oscillating response to the modulated pressure on the gas concentration in air is illustrated in Figure 10. As it follows from these results the amplitude of the oscillations proportionally increases with an increase in the partial pressure. It is interesting to note that the response time related to the average signal also depends on gas partial pressure in the atmosphere. If the pressure is comparatively low (label 1 in Figure 10) an increase in the average signal is slow while this increase is comparatively fast (label 7) if the partial pressure is comparatively high. In general, the relationship between the response time and pressure is analogous to dependence defined by our phenomenological model for the response to a step change in the gas composition.

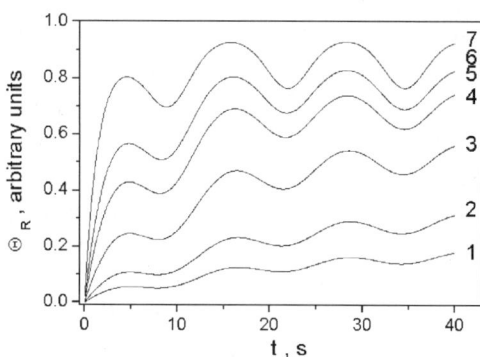

Figure 10. Dynamic response of the surface coverage by gas to the modulated concentration of a target gas for different concentrations of the gas in the atmosphere P_{gas}: 1 – 100; 2 – 200; 3 – 500; 4 – 1000; 5 – 1500; 6 – 2000; 7 –3000.

The capability to optimise the modulation frequency with respect to the parameters of metal oxide surfaces is extremely valuable for a selective detection of certain target gas. The selectivity of the method with respect to gas is demonstrated by the calculated results in Figure 11. The coverage response was calculated for different sets of the rate parameters. These parameters are listed in Table 2.

The frequency of the pressure modulation was chosen to be optimal for the detection of gas 1. It means, that the oscillations of the coverage response to the modulated pressure are easily detected and analysed (curve 1 in Figure 11). An injection of another gas instead of gas 1 in the atmosphere is simulated by a change is simulated by a change of the rate parameters in the calculations.

It follows from the results in Figure 11 that the oscillations are clearly visible for gas 1 selected as the main target gas. If other gases are considered (sets 2, 3 and 4 in Table 2), the coverage response to modulated pressure is significantly different from the reference signal of the key gas (compare curves 2, 3 and 4 with 1 in Figure 11). It must be noted that the differences between analogous parameters in different sets are comparatively low. In spite of this, the change of the gas is easily recognised by comparison of the detected response with the reference. As it can be seen in Figure 11 the oscillations are practically not detected if the rate parameters do not correspond to the characteristics of the key gas.

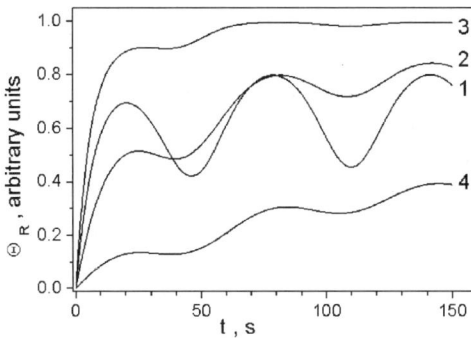

Figure 11. Dynamic response of the surface coverage by gas to the modulated concentration of different gases defined by individual set of the rate parameters (see Table 2).

It must be added that the explanation of the results in Figure 11 in terms of the selectivity to gas is not the only one possible. Strictly speaking, not only the type of target gas determines the rate parameters listed in Table 1 and used in the calculations. These reaction rates highly depend on the properties of the surfaces on which the reaction occurs. Therefore, the same results can be attributed to the modification of the surfaces. This approach means that different gas sensors are exposed to the same gas under the same conditions.

Table 2. Phenomenological rate parameters (chemisorpion rate $\alpha_R S_{0R}$, desorption rate β_{dR} and the rate of bimolecular interaction ν_{ROx}) of the surface reaction typical for SnO_2 based thin films gas sensors exposed to H_2 gas (see Figure 11.).

No.	$\alpha_R S_{0R}$	β_{dR}	ν_{ROx}
1	0.20	5.0	20.0
2	0.10	3.0	3.0
3	0.30	0.2	0.5
4	0.02	3.0	3.0

178

In spite of two possible explanations of the results in Figure 11 both approaches are applicable. The choice of the approach depends on the basic assumptions and the conditions selected for the study. Assuming the same gas, different sensors must be considered in the study. If the same sensor is used in the tests, the selectivity of this sensor to gas must be discussed.

5. CONCLUSIONS

Taking into account the absence of any saturated signals, several parameters evaluated from strictly controlled oscillating response seem more acceptable for the featuring of the smell than a signal of the terminated response. These parameters are related to commonly known characteristics of the sensors namely the response (signal rise) and the recovery (signal fall) time. A unique set of the surface reaction rates determines the rise and the fall of the response of metal oxide sensors during the well-controlled temporary exposure to the gas.

The parameters of the oscillating response are basically related to the features of the key volatile compound while the presence of other compounds instead of the key one will significantly change the appearance of the oscillations. The detection time might be reduced to the duration of the two initial periods of the oscillating response without a loss of the information about the gas mixture.

Periodic variation of the gas concentration induces an oscillating sensor response; the amplitude of the oscillations depends on gas. 'Resonance' in the response oscillations is an indicator that the gas 'fits' the 'preferences' of the system that are determined by an intentional choice of the modulation conditions for the gas injection. In general, the method seems acceptable to any type of the sensors in which the response signal is proportional to surface coverage by the gas particles.

ACKNOWLEDGEMENTS

The work was supported by Lithuanian State Science and Studies Foundation. The author acknowledges the support related to the attendance at the NATO ARW on Electronic Noses/Sensors for Detection of Explosives provided by NATO Science Program.

REFERENCES

1 F. Cremer, K. Schutte, J.G.M. Schavemaker and E. den Breejen, A comparison of decision-level sensor-fusion methods for anti-personnel landmine detection, Information Fusion 2 (2001) 187-208.

2 J. Yinon, Field detection and monitoring of explosives, Trends in Analytical Chemistry 21 (2002) 292-301.

3 S. Singh and M. Singh, Explosives detection systems (EDS) for aviation security (Review), Signal Processing 83 (2003) 31-55.

4 K.G. Furton and L.J. Myers, The scientific foundation and efficacy of the use of canines as chemical detectors for explosives, Talanta 54 (2001) 487-500.

5 T.F. Jenkins, D.C. Leggett, P.H. Miyares, M.E. Walsh, T.A. Ranney, J.H. Cragin and V. George, Chemical signatures of TNT-filled land mines, Talanta 54 (2001) 501-513.

6 X. Yang, X.X. Du, J. Shi and B. Swanson, Molecular recognition and self-assembled polymer films for vapour phase detection of explosives, Talanta 54 (2001) 439-445.

7 D. Heflinger, T. Arusi-Parpar, Y. Ron and R. Lavi, Application of a unique scheme for remote detection of explosives, Optics Communications 204 (2002) 327-331.

8 U. Weimar and W. Goepel, Chemical imaging: II. Trends in practical multiparameter sensor system, Sens. Actuators B 52 (1998) 143-161.

9 K.J. Albert, N.S. Lewis, C.L. Schauer, G.A. Sotzing, S.E. Stitzel, T.P. Vaid and D.R. Walt, Cross-reactive chemical sensor arrays, Chem. Rev. 100 (2000) 2592-2626.

10 J.R. Stetter, S. Strathmann, C. McEntegart, M. Decastro and W.R. Penrose, New sensor array and sampling systems for a modular electronic nose, Sens. Actuators B: Chem. 69 (2000) 410-419.

11 W. Goepel, Chemical imaging: I. Concepts and visions for electronic and bioelectronic noses, Sens. Actuators B 52 (1998) 125-142.

12 P.C. Jurs, G.A. Bakken and H.E. McClelland, Computational methods for the analysis of chemical sensor array data from volatile analytes, Chem. Rev. 100 (2000) 2649-2678.

13 C. Distante, M. Leo, P. Siciliano and K.C. Persaud, On the study of feature extraction methods for an electronic nose, Sens. Actuators B 87 (2002) 274-288.

14 A. Šetkus, Heterogeneous reaction rate based description of the response kinetics in metal oxide gas sensors, Sens. Actuators B 87 (2002) 348-359.

15 A. Galdikas, Ž. Kancleris, D. Senulienė and A. Šetkus, Influence of heterogeneous reaction rate on response kinetics of metal oxide sensors to gas: application to the recognition of an odour, Sens. Actuators B 95 (2003) 244-251.

16 A. Galdikas, A. Mironas, D. Senulienė and A. Šetkus, Specific set of the time constants for characterisation of organic volatile compounds in the output of metal oxide sensors, Sens. Actuators B 68 (2000) 335-343.

Chapter 13

CHEMOTACTIC SEARCH IN COMPLEX ENVIRONMENTS
From Insects To Real-world Applications

Tim Pearce[a], Kwok Chong[a], Paul Verschure[b], Sergi Bermudez i Badia[b], Mikael Carlsson[d], Eric Chanie[c] and Bill Hansson[d]

a) NeuroLab, Centre for Bioengineering, Department of Engineering, University of Leicester, University Road, Leicester, UK
b) Institute of Neuroinformatics, ETH-Universitat Zürich, Zürich, SWITZERLAND
c) Alpha MOS, 10 Av Didier Daurat, Z1 Montaudran, Toulouse,FRANCE
d) Department of Crop Science, Swedish University of Agricultural Sciences, Sundsvagen 14, Alnarp, SWEDEN

Abstract: Searching for trace quantities of predefined chemical compounds in natural environments, as in the case of explosives detection, is a difficult and challenging technological problem. Much chemical interference is likely to be present and the environment itself is often complex, nonstationary and unpredictable. Here we discuss nature's ultimate solution to this problem, in the form of pheromone mediated chemotactic search of the moth. We argue here that this organism provides an ideal model for solving the technological problem and we discuss in detail the exploitation of specific sensory processing and behavioural mechanisms relevant to the task. Finally, we discuss the project AMOTH in which the authors work towards a fully implemented, real-world instantiation of these sensorimotor mechanisms in the form of an unmanned aerial vehicle.

Key words: Pheromone detection, chemotaxis, computational neuroscience, moth, olfaction, odour plumes.

1.   INTRODUCTION

Chemical sensing technology based upon arrays of chemosensors, often called Electronic noses (ENs), have been actively developed over the past 22 years since the seminal paper of Persaud & Dodd [1]. In their most common

J.W. Gardner and J.Yinon (eds.),
Electronic Noses & Sensors for the Detection of Explosives, 181-207.

instantiation these systems rely upon (usually small numbers of) broadly tuned chemosensors comprising an array. The array may potentially generate a wide variety of responses that are then interpreted by some pattern classification scheme. The arrangement has been shown to function adequately for distinguishing a large variety of complex (multi-component) odours, provided that there is sufficient diversity of tunings to encode the relevant differences in the components.

The architecture of EN systems has close analogies to sensory pathways for chemical sensing in biology, in that the broadly tuned chemosensor responses on which these systems employ similar to odour-evoked responses seen in the receptor neurons of the olfactory system [2]. The broad tuning of these responses imparts some beneficial properties to these systems (both biological and technological). In particular, the precise chemical make-up of the stimulus does not necessarily need to be well defined beforehand in order to detect it as being distinct from learnt odours. By possessing a large repertoire of receptors/sensors with sufficient diversity in their tunings, different complex chemical stimuli are likely to lead to distinct receptor/sensor population responses. It appears that this approach is followed in biology to solve what we shall call the 'general odour coding problem', *i.e.* how to discriminate between large numbers of chemical components in combination by assessing some change in the chemoreceptor/sensor population response.

To illustrate the coding capacity of such broadly tuned, population coded, chemosensor arrays, consider a very small array of just 6 chemosensors, each with different tunings to the chemical world. If we imagine that the signal-to-noise ratio provided by each of these sensors to some pattern classification scheme is adequate to support just 5 distinct levels of activation (a very conservative estimate) then such an array would, in principle be capable of encoding 5^6, or just over 15,000 different patterns of input, each potentially corresponding to a discrete and separate complex odour stimulus. From this argument it is clear that even modest sensor arrays are capable of enormous combinatorial coding of stimuli. In practice many of these combinations would not be used by the system but are a latent resource for capturing new, previously unexposed, combinations of chemicals (for details on how chemical stimuli are represented by these arrays see Pearce & Sanchez-Montanes [3]).

The problem of explosives detection has not been a typical domain of electronic nose research, mainly since it requires extreme sensitivity to specific chemical compounds (usually a small number) which often occur in trace amounts below the detection limit of available sensors. However, as new and more sensitive technologies for chemical sensor instrumentation are discovered, we should consider how these can be exploited individually and in combination for specific detection tasks, such as finding explosives. ENs are typically exposed to highly complex stimuli, the chemical components of which are not necessarily known and well defined beforehand. Yet in the explosive detection case the problem is a very well defined one, in that, the task is to identify the presence of specific compounds (albeit at extremely low levels) in a very wide variety of interfering 'background' compounds. Under such operating conditions the broadly-tuned, combinatorial coding capacity of an EN system does not work in its favour, since background odours of no interest to the system have to be factored out of the population/array response in order to assess the compound(s) of interest. This may become a very hard problem when we consider that the interfering compounds are likely to exist in the environment at far higher levels than the compounds of interest and may also vary over time. This we shall call the 'specific odour coding problem', that is how to find the chemicals you are interested in whilst ignoring everything else.

When we look to the biology, we see that there are counterparts to both of these technological problems – specific and general odour coding. Both appear to have been solved by the biology, in that the main olfactory system of animals is known to be responsible for coding a very wide variety of complex odours through its broadly tuned receptor population, whilst the accessory olfactory system produces responses which have been closely correlated with highly specific chemicals of special behavioural relevance to the animal [4]. We believe that the accessory olfactory system provides an ideal model for the solution of the specific odour coding problem since it is tuned to identify highly specific (predefined) compounds in trace amounts, in the context of a variety of different chemical and operating conditions.

As such we (the authors) are engaged in an ambitious project to understand, model, and instantiate the neuronal mechanisms which underlie the specific odour coding problem and chemotactic search to provide a new technology for chemical source localisation which we discuss in Section 5. We do this since this process is likely to inform us of general principles deployed in the biology which are likely to be relevant to performing robust chemotactic search in a variety of unpredictable and time varying environments. For this project the moth is our model system of choice, due to its exquisite pheromone sensitivity and the fact that the stimulus and its associated pathways in the nervous system are well understood [5]. As we shall see in Section 4 the moth also has a highly effective search strategy for chemical source localisation which makes it ideal for the purposes of our project.

The main purpose of this chapter, then, is to consider these parallels between the specific odour detection problem and its biological solution, in particular the pheromone coding pathway of moths. We will firstly overview the main olfactory system in Section 2, highlighting its broadly-tuned properties and their function relevance, demonstrating its ability for solving the general but not necessarily the specific odour coding problem of importance in the case of explosives detection. In Section 3 we will then consider how pheromones are detected by one of their prime exponents, the moth, and consider its associated chemotactic search behaviour in Section 4. We will conclude with a discussion on the potential for chemical sensing and chemical search technology to exploit what is known about pheromone detection in these animals.

2.   GENERAL ODOUR DETECTION IN BIOLOGY - MAIN OLFACTORY SYSTEM

The main olfactory system in most animals responds to an extremely wide variety of molecular stimuli. Humans are said to be able to distinguish more than 10,000 different odours. Where this precise number comes from is unclear, but if you consider that any arrangement of molecular structure under mass weight 300 Daltons can potentially give rise to a unique odour sensation then it is clear that there are likely to be many undiscovered, yet to be synthesized, chemical compounds which may produce olfactory responses.

This diversity in the response of the system to single compounds mirrors the complexity of chemical stimuli we see in natural environments. Most naturally occurring odours we recognize consist of very high numbers of chemical compounds. For instance, the smell of coffee consists of 400-600 different chemical elements, including 10 ketones; 60 aromatic benzenoid compounds, including 16 phenols; 300 heterocyclic compounds, including 74 furans, 10 hydrofurans, 37 pyrroles, 9 pyridines, 2 quinolines, 70 pyrazines, 10 quinoxalines, 3 indoles, 23 thiophens, 3 thiophenones, 28 thiazoles, and 28 oxazoles [6]. The recognition of a complex odour such as coffee is a feat of information processing performed by the main olfactory pathway. Chemical compounds that diffuse through the air enter the mucus layer in our nasal cavity and bind with the receptor proteins hosted by Olfactory Receptor Neurons (ORNs). The properties of these receptor neurons are fundamental to this exceptional diversity of odour coding in the main olfactory system. For discrimination of general odours (as opposed to specialised pheromones) the main olfactory system relies upon the input from a large population of olfactory neurons, the signals from which provide a representation of a multicomponent odour stimulus. This means that for general odours, each of the receptor types responds to a large number of odour compounds (since they are broadly tuned), but each has preferences for certain groups of compounds. The extent of the broad tuning underlying olfactory perception of general odours was vividly demonstrated by Sicard & Holley (1984) after taking electrophysiological recordings from olfactory receptor neurons from the frog during exposure to different chemical compounds [7]. Figure 1 shows the high degree of broad tunings observed across a sample of the neuron population, the spot size relating to the activity produced by the cell in response to a single chemical compound. The results showed conclusively how odour perception is supported by neurons that have overlapping sensitivities of varying degrees to groups of compounds, each here with distinct tunings.

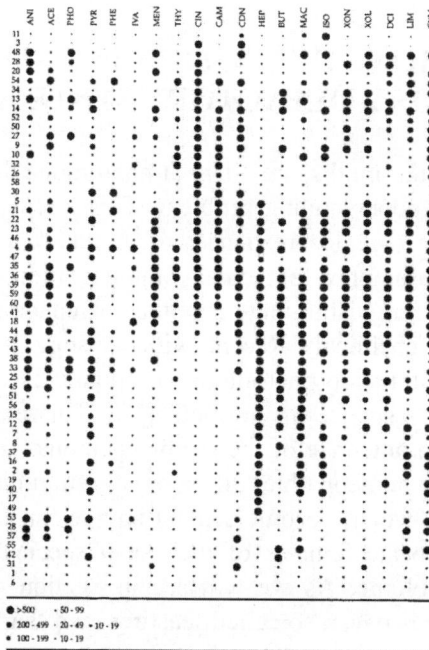

Figure 1. Diagrammatic representation of olfactory receptor cell activity during odour stimulation. The spot size is roughly proportional to spike frequency (spike/min). Receptor cells taken at random from the epithelium of a frog are identified by a serial number in the left column (60 in all). ACE - acetophenone, ANI - anisole, BUT - n-butanol, CAM - DL-camphor, CDN - cyclodecanone, CIN - cineole, CYM, p-cymene, DCI- D-citronellol, HEP - n-heptanol, ISO - isoamylacetate, IVA - isovaleric acid, LIM - D-linonene, MAC - methyl-amylketone, MEN - L-menthol, PHE - phenol, PHO - thiophenol, PYR - pyridine, THY - thymol, XOL - cyclohexanol, XON - cyclohexanone. (From Sicard & Holley [7]).

Importantly, for the case of general odour detection none of the receptors in Figure 1 show highly specific responses to the odours being investigated. So in order to solve the general odour detection task, it seems the main olfactory pathway must be able to decode and recognize the specific patterns of activity across the receptor population. Not only must the system be capable of recognising an individual compound occurring in isolation but it must do so for mixtures occurring from many hundreds of compounds acting in combination. This is truly an impressive feat of information processing but is far away from the specific odour coding problem we consider next, where the task is to pick out individual compounds of relevance in varying backgrounds.

3. PHEROMONE DETECTION IN ANIMALS – ACCESSORY OLFACTORY SYSTEM

Insects, in particular moths, are adept at pheromone detection – regularly solving the specific odour coding problem. This is clear from behavioural studies regarding the complex responses of insects to olfactory stimuli [8-12]. Moths using pheromones for sexual communication show that males are able to detect and distinguish minute amounts of female pheromones from a background of other chemicals. Recent evidence suggests that as little as an amount of 10^{-18} g of major pheromone compound is enough to elicit an increased heart-beat response in the moth [13]. This equates to just a few molecules hitting the antennae of the moth – demonstrating a physiological response to almost homeopathic dilutions of stimulus. Calling females release pheromones, which become a point source for a pheromone plume. Males are able to locate females of their own species by identifying the correct pheromone plume. As we will see in Section 4 males do this by switching behaviour between so called 'casting' and 'surging' depending on the presence, quality, and time dynamics of pheromone plumes [8]. The fact that the moth has a well defined pheromone stimulus and stereotypical behavioural response makes it ideal as a model system to study chemical source localisation.

As shown in Figure 2, the olfactory system plays a central role in the sensorimotor pathways of the moth during interaction with its environment. Chemical, and to some extent anemometric, information is routed via the antenna to the Antennal Lobe (AL) and hereon to the higher centres of the mushroom body calyces. ORNs at the antenna are responsible for chemical detection. The role of the AL might be to amplify and encode this olfactory information, preserving the aspects that are relevant, and encapsulating this for higher brain processes and motor functions. The relevant coding

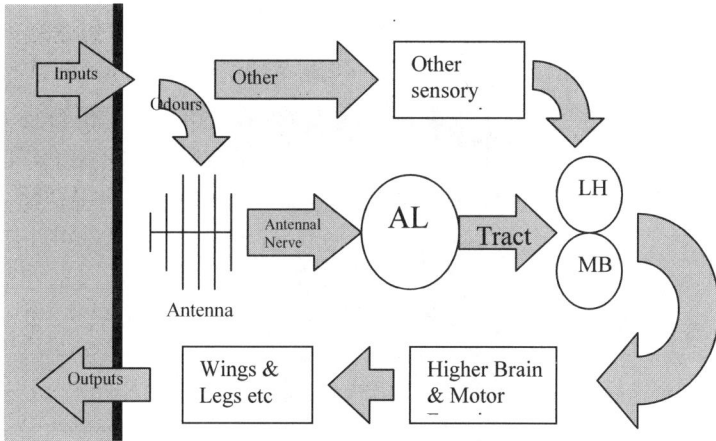

Figure 2. Overall system context of the pheromone detection system in moths (shown in centre). AL is the antennal lobe, MB the mushroom bodies and, LH the lateral horn. This schematic shows the animal in an 'input/output' diagram with respect to its environment.

parameters include odour identity, quality, concentration, component ratios and temporal aspects. The AL must be able to encode some or all of these parameters using a spatiotemporal output, which is then passed onto the lateral horn (LH) and mushroom body calyces (MB) for sensory integration and memory. How this is achieved in the nervous system of the moth is potentially very revealing of the key processing principles required to solve the specific odour coding problem, particularly as it relates to extracting relevant information (both spatial and temporal) for the detection task. For this reason we next look at the different steps of the pheromone detection pathway in the moth in order to appreciate it in some more detail and uncover some general principles of pheromone detection.

3.1 The olfactory detectors

How are these extremely low amounts of pheromone detected and processed? As mentioned above, the detection takes place on the antenna. Here ORNs are housed within hairlike cuticular structures, called sensilla, shown clearly in Figure 3. Each sensillum houses between one and three ORNs which send a dendrite up into the hair. On this dendrite receptor proteins act as a lock for a certain odour 'key.' It is thus the receptor protein that is the identification site for the odour molecule.

188

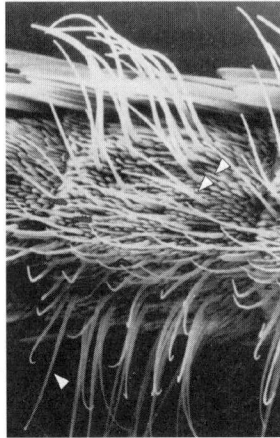

Figure 3. Scanning electron micrograph of a single segment of a male *S. littoralis* antenna. The longer sensilla trichodea type I (single arrow head) are primarily found along the ventral and lateral margins whereas the shorter type II (double arrow heads) are found on the medial surface. Type I sensilla contain pheromone detecting neurons and are far more abundant in males than in females. Type II sensilla contain mainly cells tuned to plant-associated compounds. The distal segment of the antenna, the flagellum, contains the olfactory sensilla. A sensillum houses the dendritic ends of the olfactory receptor neurons (ORN). In *S. littoralis* each sensillum generally houses two ORNs (Ochieng' *et al.*, 1995 [31]). Three major types of olfactory sensilla can be discriminated, the basiconic, the coeloconic and the trichoid (Ljungberg *et al.*, 1993 [14]). The *sensilla trichodea* can further be divided into two distinct size classes, the larger type I and the smaller type II (Figure 3). There is a clear sexual dimorphism when it concerns the number of type I sensilla. Males have about 50-60 type I sensilla on each annulus whereas females have less than 10 (Ljungberg *et al.*, 1993 [14]).

Pheromone communication between moths usually involves more than one molecular cue, since these typically occur in blends. Moreover the chemical composition, number of components and ratio of concentrations of these blends appears to be important to the pheromone detection task since different species often use similar chemicals in a different context, and interfering with the blends can block appropriate behavioural responses.

As such, ORN responses within the sensilla are often highly specialised to individual components of the blend, having highly specific response patterns. For example, the two major pheromone components of *S. littoralis* do not excite each other's respective ORN even at very high concentrations meaning that the response can be highly selective and specific (Ljungberg *et al.* 1993 [14]).

Therefore an important and obvious feature of pheromone reception in the moth is that receptors are deployed which are highly specific to *each* individual component of relevance to the detection task. Contrast this to the main olfactory system and EN systems which deploy broadly tuned chemosensors across a large population. That is not to say that the biological solution we see is here the only strategy to solving the specific odour coding problem, but that it is likely to be an effective one.

Olfactory receptor neurons not only have to be selective to a spectrum of molecules, they must also report the quantity or concentration of molecules in the environment and also temporal properties. Receptor neurons respond to fluctuations in stimulus concentration with changes in action potential firing frequency. We also see very clear diversity in the temporal responses of ORNs responsible for pheromone detection in *S. litoralis* (J. Mackenzie *et al.*, unpublished) which may suggest that the front end of the pheromone detection system in the moth is tuned to different frequency information which is presumably rich in turbulent odour plumes (see Section 4). In solving the problem of chemical source localisation the capture of temporal information has already been demonstrated to be important [15].

When a pheromone molecule interacts with the receptor protein a transduction cascade triggers the formation of a nervous signal. This signal travels to the ORN cell body, located just below the base of the sensillum. There, action potentials are triggered that travel down the ORN axon into the AL the primary olfactory centre of the moth brain.

3.2   Antennal Lobe

In the antennal lobe ORN axons terminate in structures known as glomeruli (Figure 4). The ORNs innervate the glomeruli – spheroids of tightly packed dendritic and axonal branches. These first-order synaptic neuropil are where ORNs, and the two major classes of AL neuron, principal neurons (PNs) and local interneurons (LNs) interconnect. The moth *M. sexta* (male) has around 66 glomeruli, 360 LNs and 900 PNs [16]. There are many different receptor types, since most of the ORNs are tuned to respond to a variety of chemicals.

It was for the first time shown in invertebrates that ORNs expressing a certain receptor protein project to a common glomerulus. Functionally characterised ORNs were found to project in a chemotypic manner [17-19]. Further support for this is seen in *Drosophila melanogastor*, where the number of receptor gene types equals the number of glomeruli [20]. Thus, it is thought that, although different ORN types are intermingled on the antennae, ORNs of the same type send axons to the same glomerulus. Thus each glomerulus is a site of convergence for ORNs of the same type. This convergence has important consequences for signal detection in the system, since such spatial averaging is likely to boost the signal-to-noise ratio by $\sqrt{n}$

(n being the number of convergent ORNs) [21]. By deploying this convergent architecture it appears that the moth is able to enhance its pheromone response to a high level. However, this appears to be one of many tricks that it uses to do so since sensitivity at the periphery of the system is around 10^{-9} g whereas physiological responses to pheromones have been demonstrated at far lower levels [13].

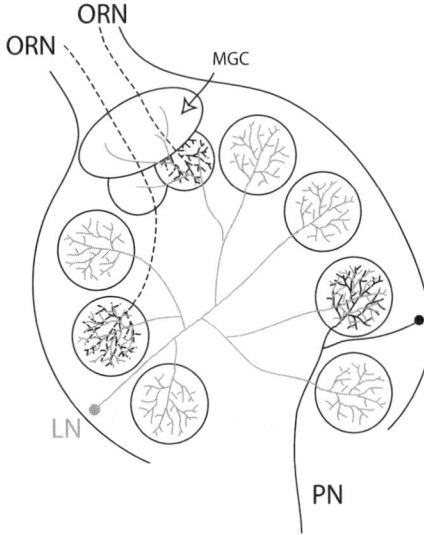

Figure 4. The ramifications of ORNs, a PN and a LN. ORN input (dashed black lines) and PNs (solid black lines) typically innervate single glomeruli, whereas LNs (solid grey lines) tend to innervate many glomeruli. Also in the AL this sexual dimorphism is manifested, here a large cluster of male-specific glomeruli, the macroglomerular complex (MGC) are a key feature. The primary olfactory brain thus has an area found only in the male where the female pheromone signal is received. Information regarding plant odours and other so called 'general' odours is received in sexually isomorphic 'ordinary' glomeruli (black circles).

In moths the AL glomerular setup displays a pronounced sexual dimorphism [16,22,23]. Males have large glomeruli at the top of the AL called the macroglomerular complex (MGC) which are responsive to sex-pheromones (actually, males and females possess homologous glomeruli, but males have become morphologically different) [24]. Pheromone specific ORN axons project exclusivelyto the MGC (Figure 5). In the moth *Spodoptera littoralis*, the MGC is composed of three glomeruli. The largest glomerulus is the Cumulus, which sits atop two smaller satellite ones. The MGC is the pheromone processing subsystem and is largely separate to the rest of the olfactory system. Also the LNs and PNs within the MGC are sexually dimorphic [25,26].

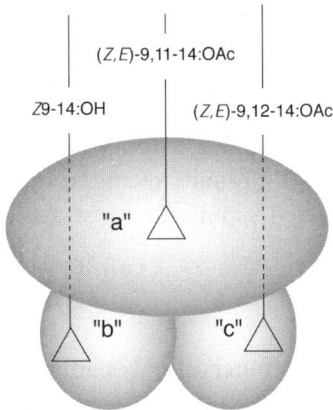

Figure 5. A schematic illustration of ORN projections in the MGC of S. littoralis. Three physiologically identified types of ORNs send axons to specific glomeruli within the MGC. Modified from Hansson 1997 [35].

Consequently we see that the evolutionary pressure to detect pheromones with as high sensitivity as possible has produced a detected pathway in the olfactory system for this purpose. This specialisation is a potential clue to the solution of the specific odour coding problem.

MGC innervation

Stainings of single pheromone sensitive ORNs have shown a specific innervation pattern with respect to the neural tuning in several moth species [18,19,27,28]. In *S. littoralis*, neurons responding to the major pheromone component Z9, E11-14:OAc arborised in the largest of the MGC glomeruli, the cumulus or 'a' glomerulus (Figure 6) [29]. The second neuron in this type of sensillum, with unknown physiology, sent axons to a glomerulus outside the MGC. Two satellite glomeruli termed 'b' and 'c' received axons from ORNs tuned to the behavioural antagonist, Z9-14:OH, and to the secondary component Z9, E12-14:OAc, respectively.

In the female, most ORNs tuned to Z9, E11-14:OAc arborised in a glomerulus located at the antennal nerve entrance [31]. This glomerulus is far smaller than the corresponding glomerulus in the male. Furthermore, no cluster of glomeruli equivalent to the male MGC has been observed in the female. It is, however, likely that the single glomerulus in females and the MGC in males have a common origin but selective pressure on the males has initiated an evolution of an enlarged and subdivided organ.

Optical imaging

In order understand the coding properties of neurons within the AL investigators have adopted optical imaging techniques. In a series of experiments in *S. littoralis* studied spatial representation of odour information among glomeruli both with respect to chemical structure and to intensity of the stimulus [30,31]. A topical application of a calcium sensitive dye to the AL reports mainly ORN activity [32]. As all sensory cells expressing the same receptor protein seem to terminate in the same glomerulus [33,34] the activated glomeruli constitute a map of responding ORN types.

Responses to pheromones

Due to the established innervation patterns of physiologically characterised neurons in the MGC, optical measurements of pheromone-evoked activity provide an excellent control for the validity of the method. Increased calcium evoked by pheromone exposure was restricted to the MGC in *S. littoralis* (Carlsson *et al.* 2002 [17]). Plant-associated compounds, on the other hand, evoked no or only weak responses in the MGC. Hence, there is a rough division of labour in the AL, which is in accordance with other studies, and virtually no overlap between the two subsystems. Optical responses to sexual pheromones were found to corroborate earlier stainings of individual physiologically characterised ORNs (see Figure 6) [16,31,35-37].

Beyond this first step of encoding individual pheromone components, specific neurons in the male moth AL have been shown to exhibit so-called blend specificity. These neurons will only respond to a stimulus containing all pheromone components produced by the conspecific female, and displaying the correct ratio between these components. These kinds of neurons have been found in a number of noctuid moths. Consequently we see evidence for a type of ratiometric processing of pheromone blend information. Networks of neurons appear to be performing a computation to identify specific blends of individual molecular ligands occurring in appropriate ratios. In the project we describe in Section 5, we are beginning to build models of the ratiometric detection which are likely to be relevant to solving the specific odour coding and closely related blend identification problem.

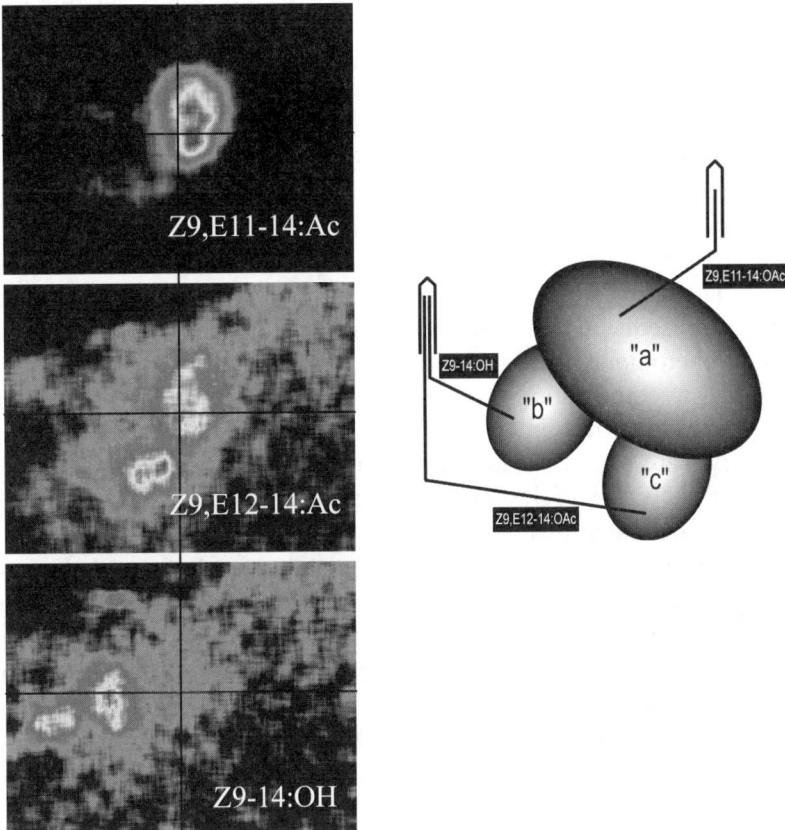

Figure 6. The principal pheromone component, Z9, E11-14:OAc, activated a large glomerulus close to the entrance of the antennal nerve, which most likely corresponds to the identified 'cumulus', or 'a' glomerulus [17]. Likewise, the secondary pheromone component, Z9, E12-14:OAc, showed highest activity in a medio-ventrally located glomerulus (with respect to the former), which likely corresponds to the 'c' glomerulus, whereas a behavioural antagonist, Z9-14:OH, activated the 'b' glomerulus A third putative pheromone component, Z7-12:OAc, activated a fourth glomerulus. This glomerulus was located proximal to the others but does not belong to the MGC. Accordingly, Anton and Hansson (1995) found that single projection neurons responding to Z7-12:OAc arborised in an 'ordinary' glomerulus [37]. (Reproduced from Carlsson et al., 2002 [36])

Within the AL it is clear that a considerable degree of processing takes place. Local interneurons connect more or less all glomeruli, allowing lateral interactions. In fact, synapses in the AL are almost entirely restricted to glomeruli (Figure 4). ORNs synapse directly onto both LNs and PNs, and are largely excitatory. The most common is dyadic synapsing where one pre-synaptic bouton contacts two post synaptic elements. Moreover, inputs from higher brain areas can modulate the activity of the entire glomerular ensemble, depending perhaps on the appetitive state of the animal. The final 'product' of AL processing is delivered to higher brain areas via projection neurons. These neurons receive input from a single or few glomeruli and lead the signal to higher brain areas as the mushroom bodies and the lateral brain. Some of these antennal lobe neurons in the male moth have been shown to exhibit so-called blend specificity. These neurons will only respond to a stimulus containing all pheromone components produced by the conspecific female, and displaying the correct ratio between these components. These kinds of neurons have been found in a number of noctuid moths [38]. Consequently we see evidence for a type of ratiometric processing of pheromone blend information. Networks of neurons appear to be performing a computation to identify specific blends of individual molecular ligands occurring in appropriate ratios. In the project we describe in Section 5, we are beginning to build models of the ratiometric detection which may are likely to be relevant to solving the specific odour coding and closely related blend identification problem.

In the higher processing sensory integration and memory formation takes place. From the brain the signal is delivered via the ventral nerve cord to different ganglia. For pheromone behaviour the most important ganglion is the thoracic, where wing muscle innervating neurons interact with odour information-carrying neurons from the brain. When the right odour is there, the wings beat and an upwind flight commences

4.   PHEROMONE MEDIATED CHEMOTACTIC SEARCH IN MOTHS

Insects must interact with their environment in order to achieve key objectives, be it locating a mate, food or other activities. Critically, its environment determines its behaviour, but its behaviour is also designed to change its environment to maximise its ability to attain its goals.

Therefore, what is relevant to the animal behaviourally and how it interacts with the environment determines its biological setup (Figure 2). The odour stimuli in which moths are primarily interested are plumes with fast temporal features and complex blends. This has implications for the form and function of the AL and its inputs. For instance, it has been suggested that there are ORNs specifically designed for detection of flux in pheromone components.

Male moths are capable of finding mates over large spatial scales (hundreds of metres) using extremely low concentrations of pheromones. This behaviour depends essentially on the quality of the pheromones released by the female but is also gated by the visual system, *i.e.* flight is only initiated in the presence of visual cues while also the actual localization of the target and landing behaviour appears to require visual cues. However, it has been demonstrated that the overall search behaviour does not require vision [39].

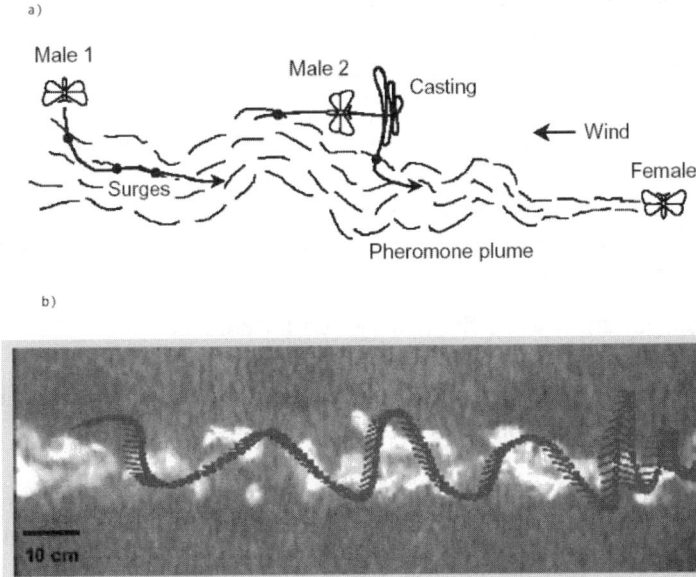

Figure 7. a) Typical flight behaviour of a male moth tracking a pheromone plume released by a female moth, b) Trace of a typical pheromone search compared to the structure of the pheromone plume.(Image on top by Ishida & Morizumi 2002 [40] bottom image by J.Hildebrand).

The female moth releases a sex attractant pheromone blend, which is transmitted downwind generating a specific plume shape – as shown in Figure 7 [40]. If there are other species nearby that release similar pheromones, they use different times at night to avoid crosstalk and inappropriate mating. Once a male moth detects its own species specific pheromone blend, it flies slowly upwind, tracking the filament of the plume that has been intercepted. The plume that guides the male moth to the female is made up of pockets with high concentrations of pheromone interspersed with clear air (Figure 7 bottom). Its structure is quite complicated, being unpredictable and following complex patterns. The chemical plume is seen to have a filamentous structure, these filaments travelling downwind. The gaps of clean air expand as the plume travels away from the source, producing a plume that consists of a sequence of bursts of pheromones of variable concentration and spacing. This structure varies over space and time, changing the concentration of its components because of the prolongation of the length of the burst and increasing the inter-pulse time [41,42] Interestingly, even under turbulent wind conditions, individual parcels of air can travel in straight lines for several meters [43].

It has been suggested that the key properties of chemical plumes that influence moth chemical search behaviour are [41]:

- Peak concentration of the odour filaments.
- Concentration gradients on the edges of these filaments.
- Intermittency (duration / frequency) of filaments.

Hence, given the complex structure and dynamics of chemical plumes male moths cannot rely on gradient-based chemotaxis alone to find a mate but need to resort to a more advanced search behaviour. Indeed, the search behaviour of male moths is described to consist of at least two stereotypic behaviours (Figure 7, top):

- *Surge*; an upwind flight triggered by contact with a pheromone filament
- *Cast*; a zigzag movement orthogonal to the wind direction in the absence of any pheromone filament.

Surprisingly, when the male moth looses the plume, it usually finds it again closer to the source than before [44]. In addition, it always maintains an altitude close to that of the plume and usually its body is not totally aligned with the wind direction (Figure 8) [45]. When the male moth is getting closer to the source the casting frequency increases while its speed decreases. This culminates in a landing close to the pheromone source and a putative reward.

4.1   Mechanisms

Male moths make use of the *olfactory system* and an *orientation system* to reach the female. Since 1940, when Kennedy conducted one of the first studies in this field, the mechanisms that govern the olfactory search behaviour of a number of insects have been studied. Kennedy discovered that the orientation of the flight trajectory depends on the speed of movement of the ambient visual cues and their angle relative to the moth [46].

At the end of the 1950s and at the beginning of the 1960s, Wright and Kellog realized that some insects fly upwind in response to food odours [47,48] and later that male moths respond to pheromones [49]. In later experiments [50,51], it was shown that during the surge phase of the search behaviour, the speed of the moth did not depend on the wind speed but that it always maintains a constant ground speed. This suggests that visual cues

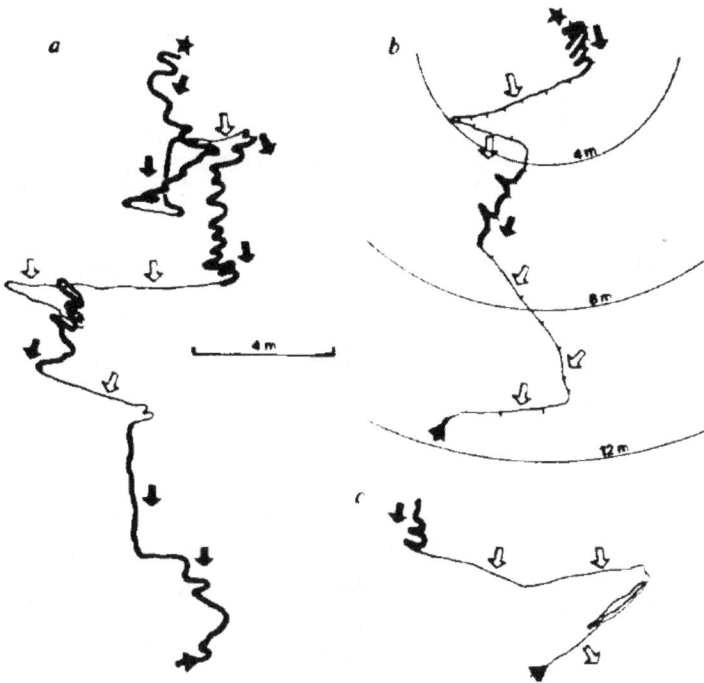

Figure 8. a) Frequency distribution of track angle vectors for male flights. a) shows an unimodal distribution when males are flying upwind in the plume (n=0), and b) shows a bimodal distribution when the same males are casting in clear air gaps (n=+-90) From Fadamiro *et al.* [45].

provide a reference for the chemical search control system. Ludlow *et al.* subsequently suggested that as opposed to using angle and velocity the moth relies on the longitudinal and transversal movement of its optic flow (Figure 9) [43].

These observations raise a number of questions on the principles and mechanisms underlying moth navigation including:

- How does the moth achieve a similar crosswind ground speed in winds of different speeds?
- Why does the moth keep the same ground speed?
- How and when does a moth turn upwind?
- How are visual and chemical cues combined to control search behaviour?

So far no conclusive evidence exists that would argue for the use of mechanoreceptors in the detection of wind-direction. Moreover, the readings

198

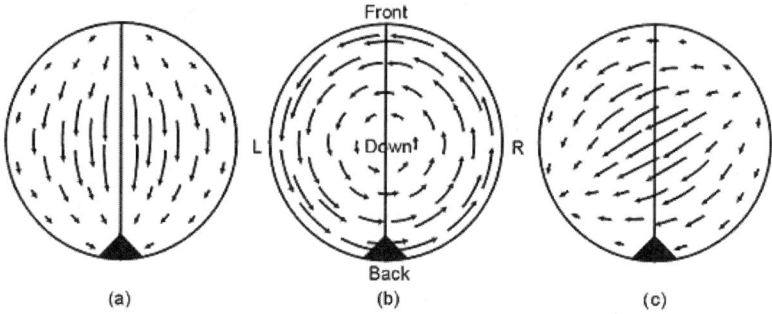

Figure 9. The direction of the visual flow field movement across the downward-facing half of an insect's visual field for forward flight (a), rotation in place (b) and side-slipping (c). The length of the arrows represents the amount of image flow at each position as viewed. Reproduced from Willis [44].

provided by such sensors would result from the displacements due to drift due to wind currents and the flight apparatus itself. Hence, in the absence of specialized sensors to determine wind direction and speed, it is believed that the moth can only calculate these key movement parameters in reference to visual flow in order to maintain a constant ground speed. Moreover, it has been shown that moths retain their estimate of the current wind direction for some time, being able to fly in the direction of the plume source in absence of any wind [52]. Because of the contributions of both vision and chemosensing to this localization behaviour it is called *optomotor anemotaxis*. The system that controls the turns and their amplitude during the search is called *self-steered counterturning* [53].

Using both sensory systems, the male moth flies slowly upwind in the direction of the source when it finds a pheromone plume, performing a number of turns in a quite regular structure when the plume filaments are lost.

4.2 Moth responses to different plumes

The moth performs different flight manoeuvres in different types of plumes. Thus, the moth does not respond identically to plumes with high frequency pulsing compared to those with a low frequency or continuous plumes. The plume pulsation frequency determines the shape of the flight. In a continuous ribbon-like plume the moth is performing some consecutive turnings, whereas in turbulent plumes it flies straight. In both cases out of the combination of the plume dynamics and the behaviour of the moth the antenna could receive a signal with similar properties [54]. In addition, it has been suggested that signal pulsation also controls for the speed of flight; the male moth shows a faster and straighter flight upwind for high frequencies than for low frequencies [55,56]. A similar effect has been observed for the concentration of the plume [57].

The pulsation of the plume has an effect on the counterturning program of the moth. This program, responsible for zigzagging, is suppressed or inhibited during straight flight. This suppression can be caused by the pheromone pulse frequency or concentration or a refractory period. It has been suggested that the intermittency of the plume could reduce the sensory adaptation, which can help to resolve pulsation rates [58]. The antenna resolves odour pulses of up to 20 Hz (*P. americana*), while the projection neurons does not appear to preserve the temporal structure of the stimulus [59]. This suggests that there could be some kind of encoding of the temporal signal characteristics of the odour at the early stages of olfactory processing [60].

In the MGC of *M. sexta* a small group of PNs has been found that appear to discriminate the duration and inter pulse of pheromone plumes at moderate frequency rates (10 Hz) [61].

4.3   Other parameters

The response of the male moth appears to also be modulated by other factors, such as *temperature* and other *chemical signals*. For instance, temperature modulates electrophysiological and behavioural responses, including an increase in the sensitivity to pheromone and the ability to resolve high frequency pulsed pheromones, a faster upwind flight and an enhanced ability to sustain the upwind flight. In some cases also the time of the emission of pheromones is modulated by temperature. In cooler temperatures the females can release the pheromone early in the day, even in sunlight while usually release occurs at night [62].

Certain chemical compounds have been observed to affect the responses from the male moth. The olfactory search to locate the source of pheromone is affected by a *behavioural antagonist*, a substance contained in pheromones of other species. In general male moths fail to track a pheromone plume in the presence of the behavioural antagonist, at concentrations as low as 1%. In these cases a decrease in the surge displacement and flight time, and also a significant increase in the latency to surge can be observed [63,64]. Antagonists are detected by receptors that are sensitive to both the secondary pheromone compound as well as the behavioural antagonist. This appears to provide for an optimal way to reduce the uncertainty about the species of the calling female since the antagonist of some species constitutes a pheromone component of other species [65].

5. TECHNOLOGY FOR CHEMICAL SOURCE LOCALISATION

In terms of the specific odour coding problem we have seen that in the case of the moth biology invests significant resources to its solution. The underlying design principles become obvious when we look closely at the biology. We have seen how the AL in the moth extracts relevant information related to the solution of this problem in the context of pheromone blend detection. A number of general principles can be established:

1. Highly specific receptors are preferred so as to remove at source interfering information related to compounds which have no relevance to the problem.
2. ORNs as well as the AL itself appear to code for stimulus time-related information, which is likely to give valuable clues in chemical source localisation.
3. A dedicated centre for solving the specific odour coding is present in males of the moth species, presumably indicating it is distinct from the general odour coding problem, which is undertaken by the main olfactory pathway.
4. The pheromone detection pathway boosts sensitivity to an extreme level, not it seems by having ultra-sensitive receptors but by virtue of the subsequent neuronal processing of the information.
5. Ratiometric processing of specific ORN responses seem to underlie the pheromone blend detection problem.

We are interested how to exploit our knowledge of the neuroscience and behaviour of the moth overviewed in Sections 3 & 4 in order to construct a new technology in the form of an chemosensing Unmanned Arial Vehicle (cUAV) capable of recognising specific odours in its chemical environment. As part of our activity the authors are involved in an ambitious project (project website http://www.amoth.org/) to understand the principles of pheromone detection and mechanisms underlying chemotactic search behaviour in the nervous system of the moth, from the point of pheromone reception at the antennae, through to initial processing at the AL, the subsequent interpretation at the MB, LAL, and the final control of motor response in the thoracic ganglion.

Detailed, biologically constrained, models of these specific nervous centres are being developed relating to olfactory, anemometric, and visual sensory inputs and integrated on real-world artefacts. Two approaches are being taken within the project. Firstly, we are testing biologically plausible models of information processing for the purposes of chemotactic search in a controlled wind tunnel environment (Figure 10). Secondly, we are attempting to transfer what we learn about the 2-dimensional problem to 3-dimensions in uncontrolled environments.

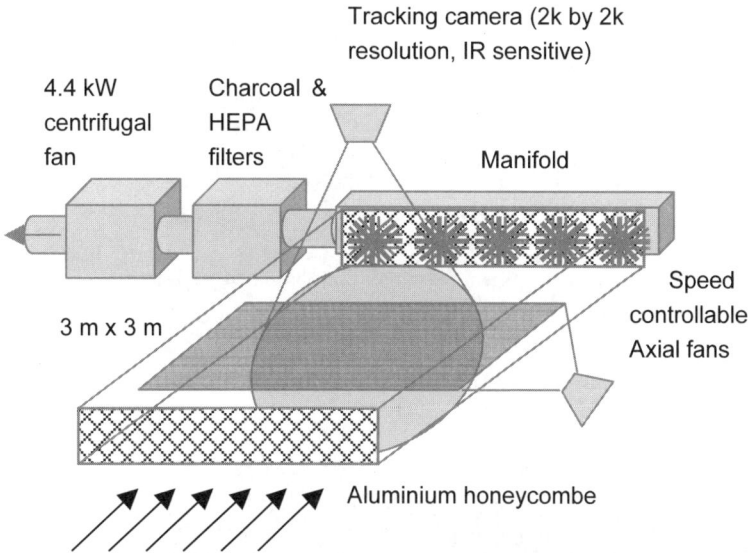

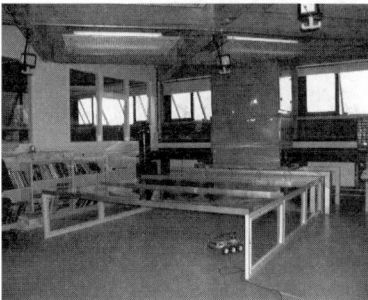

Figure 10. Projects controllable wind tunnel installation for chemical source localisation experiments and chemosensing robot with sensor array. By using arrays of speed controllable axial fans we are able to reproduce the 'S'-shaped chemical plumes that result from changing wind direction in the natural environment. Dedicated flow control enables us to test from highly laminar through to turbulent and unsteady flow conditions.

202

As was clear in Section 4 moths are able to localize chemical sources by virtue of an active sampling strategy. This approach is not incorporated in current technology that emphasizes grids of static monitoring devices. The AMOTH project aims at changing this situation by equipping unmanned aerial vehicles with arrays of chemical sensors on our cUAV (Figure 11). Each single cUAV is functionally modelled after the moth and colonies of these artefacts collectively map and monitor the chemical composition of their environment. The control system of the cUAV comprises both

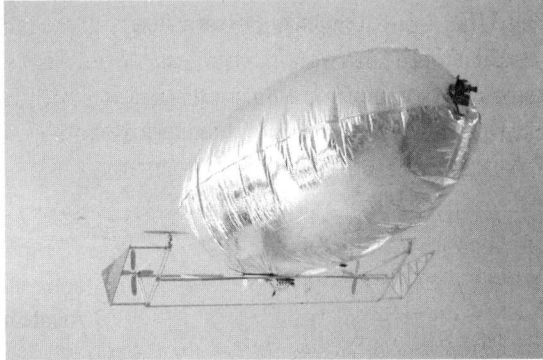

a)

b)

Figure 11. Prototypes of the cUAV as developed in the AMOTH project. a) Indoor cUAV measures about 1.2 m × 0.6 m and is equipped with 4 propellers, 2 cameras and a chemo-sensor array. b) An outdoor cUAV on a nocturnal chemical search mission. This model measures about 4 m × 1 m and has a similar sensorium as the indoor model. Both systems have energy autonomy of over 1 hour and provide full wireless communication, monitoring and control.

traditional algorithmic and biologically constrained components. The former are used for monitoring and scheduling purposes while the latter, developed also on the basis of the experiments performed on moths, support the functionality of individual cUAVs. These cUAVs are equipped with standard sensor devices such as GPS, 3D compass, altimeter and accelerometers as input devices for the monitoring and scheduling system while arrays of chemical sensors and a number of video cameras provide the input the their biological counterparts. cUAVs will individually be capable in localizing single chemical sources by active chemical search behaviour modelled on the moth pheromone system and collectively construct maps of chemical environments

In the construction of the cUAVs one has to find a compromise between the wish to define a biologically based artefact while the biological system itself is not fully understood. It has become common place to sidestep this problem by alluding to a mere biological inspiration that, as the name already suggests, does not require any fundamental sense of understanding. In contrast the AMOTH project pursues an approach where the construction of real-world artefacts directly serves our understanding of biology systems by evaluating well defined hypothesis on the structure and function of biological systems. In the AMOTH project we combine both abstract robot based models of learning and behavioural control such as the Distributed Adaptive Control architecture [66] and biologically constrained neuronal models of insect sensory processing [67,68]. The assumption motivating this approach is that the brains of insects and humans share common design principles, as illustrated by the design of their respective chemical sensing systems discussed in this chapter. It is in identifying these common design principles by combining neuroscience and engineering that we will both come to understand biology and be able to construct more advanced real-world behaving systems.

ACKNOWLEDGEMENTS

The authors are grateful to the European Commission for funding the research described in this chapter under the project 'A Fleet of Artificial Chemosensing Moths for Distributed Environmental Monitoring (AMOTH)' funded under the IST Future and Emerging Technologies Programme (awarded to TCP, PFMJV, EC, and BSH. Grant reference IST-2001-33066, project website http://www.amoth.org/).

204

REFERENCES

1 K.C. Persaud and G.H. Dodd, Analysis of discrimination mechanisms in the mammalian olfactory system using a model nose, Nature 299 (1982) 352-355.
2 T.C. Pearce, Computational parallels between the biological olfactory pathway and its analogue 'the electronic nose': Part I. Biological olfaction, BioSystems 41 (1997) 43-67.
3 T.C. Pearce and M. Sanchez-Montanes, Chemical Sensor Array Optimization: Geometric and Information Theoretic Approaches, in Handbook of Machine Olfaction, Pearce T.C., Schiffman S.S., Nagle H.T., Gardner J.W. (eds), Wiley-VCH: Weinheim, (2003).
4 C. Dulac and A.T. Torello, Molecular detection of pheromone signals in mammals: from genes to behaviour, Nature Reviews Neuroscience 4 (2003) 551-562.
5 B.S. Hansson, Insect Olfaction, Springer-Verlag: Berlin, (2002).
6 R.J. Clarke, The flavour of coffee, Dev. Food Science 3B (1986) 1-47.
7 G. Sicard and A. Holley, Receptor cell responses to odorants: similarities and differences among odorants, Brain Res. 292 (1984) 283-96.
8 C. Quero et al., Responses of male Helicoverpa zea to single pulses of sex pheromone and behavioural antagonist, Physiological Entomology 26 (2001) 106-115.
9 T.C. Baker and K.F. Haynes, Pheromone-mediated optomotor anemotaxis and altitude control exhibited by male oriental fruit moths in the field, Physiological Entomology 21 (1996) 20-32.
10 L.P.S. Kuenen and R.T. Carde, Strategies for recontacting a lost pheromone plume - casting and upwind flight in the male gypsy-moth, Physiological Entomology 19 (1994) 15-29.
11 M.O. Harris and S.P. Foster, Wind-tunnel studies of sex pheromone-mediated behavior of the hessian fly (Diptera, cecidomyiidae), Journal of Chemical Ecology 17 (1991) 2421-2435.
12 E. Hartlieb and P. Anderson, Olfactory-released behaviours, in Insect Olfaction, (Hansson, B., Ed.), Springer-Verlag: Berlin, (1999), pp. 315-349.
13 A.M. Angioy, A. Desogus, I.T. Barbarossa, P. Anderson and B.S. Hansson, Extreme sensitivity in an olfactory system, Chem. Senses 28 (2003) 279-284.
14 H. Ljungberg, P. Anderson and B.S. Hansson, Physiology and morphology of pheromone-specific sensilla on the antennae of male and female Spodoptera littoralis (Lepidoptera: Noctuidae), J. Insect Physiol. 39 (1993) 253-260.
15 T. Kikas, H. Ishida, D.R. Webster and J. Janata, Chemical plume tracking. 1. Chemical information encoding, Anal Chem. 73 (2001) 3662-8.
16 S. Anton and U. Homberg, Antennal lobe structure. in Insect Olfaction, Hansson B.S. (ed.), Springer-Verlag: Berlin, (1999), pp. 97-124.
17 M.A. Carlsson et al., Spatial representation of odors in the antennal lobe of the moth Spodoptera littoralis (Lepidoptera : Noctuidae), Chemical Senses 27 (2002) 231-244.
18 B.S. Hansson et al., Chemical Communication in Heliothine Moths. 5. Antennal Lobe Projection Patterns of Pheromone-Detecting Olfactory Receptor Neurons in the Male Heliothis-Virescens (Lepidoptera, Noctuidae), Journal of Comparative Physiology a-Sensory Neural and Behavioral Physiology 177 (1995) 535-543.
19 B.S. Hansson et al., Functional specialization of olfactory glomeruli in a moth, Science 256 (1992) 1313-1315.
20 L.B. Vosshall, The molecular logic of olfaction in Drosophila, Chemical Senses 26 (2001) 207-213.
21 T.C. Pearce, P.F.M. Verschure, J. White and J.S. Kauer, Robust stimulus encoding in olfactory processing: hyperacuity and efficient signal transmission, in Neural Computation Architectures Based on Neuroscience, Wermter S., J. Austin, and Willshaw D. (eds.), Springer-Verlag, (2001).
22 T.A. Christensen et al., Discrimination of sex-pheromone blends in the olfactory system of the moth, Chemical Senses 14 (1989) 463-477.

23 J.R. King et al., Response characteristics of an identified, sexually dimorphic olfactory glomerulus, Journal of Neuroscience 20 (2000) 2391-2399.

24 J.P. Rospars and J.G. Hildebrand, Sexually dimorphic and isomorphic glomeruli in the antennal lobes of the sphinx moth Manduca sexta, Chemical Senses 25 (2000) 119-129.

25 U. Homberg et al., Structure and function of the deutocerebrum in insects, Annual Review of Entomology 34 (1989) 477-501.

26 U. Homberg et al., Anatomy of antenno-cerebral pathways in the brain of the sphinx moth Manduca sexta, Cell and Tissue Research 254 (1988) 255-281.

27 J.L. Todd, S. Anton, B.S. Hansson B.S and T.C. Baker, Functional organization of the macroglomerular complex related to behaviorally expressed olfactory redundancy in male cabbage looper moth, Physiol. Entomol. 20 (1995) 349-361.

28 B.G. Berg, T.J. Almaas, J.G. Bjaalie and H. Mustaparta, The macroglomerular complex of the antennal lobe in the tobacco budworm moth Heliothis virescens: specified subdivision in four compartments according to information about biologically significant compounds, J. Comp. Physiol. A. 183 (1998) 669-682.

29 S.A. Ochieng, P. Anderson and B.S. Hansson, Antennal lobe projection patterns of olfactory receptor neurons involved in sex pheromone detection in Spodoptera littoralis (Lepidoptera: Noctuidae), Tissue Cell. 27 (1995) 221-232.

30 M.A. Carlsson and B.S. Hansson, Dose-response characteristics of glomerular activity in the moth antennal lobe, Chemical Senses 28 (2003) 269-278.

31 J. Meijerink, M.A. Carlsson and B.S. Hansson, Spatial representation of odorant structure in the moth antennal lobe: a study of structure response relationships at low doses, J. Comp. Neurol. 467 (2003) 11-21.

32 C.G. Galizia, K. Nägler, B. Hölldobler and R. Menzel, Odor coding is bilaterally symmetrical in the antennal lobes of honeybees (Apis mellifera), Eur. J. Neurosci. 10 (1998) 2964-2974.

33 L.B. Vosshall, A.M. Wong and R. Axel, An olfactory sensory map in the fly brain, Cell. 102 (2000) 147-159.

34 Q. Gao, B. Yuan and A. Chess, Convergent projections of Drosophila olfactory neurons to specific glomeruli in the antennal lobe, Nat. Neurosci. 3 (2000) 780-785.

35 B.S. Hansson, Antennal lobe projection patterns of pheromone-specific olfactory receptor neurons in moths, in Insect pheromone research: new directions, R.T. Cardé, A.K. Minks (eds.), Chapman & Hall: New York, (1997), pp. 164-183.

36 M.A. Carlsson, C.G. Galizia and B.S. Hansson, Spatial representation of odours in the antennal lobe of the moth Spodoptera littoralis (Lepidoptera: Noctuidae), Chemical Senses 27 (2002) 231-44.

37 S. Anton and B.S. Hansson, Sex pheromone and plant-associated odour processing in antennal lobe interneurons of male Spodoptera littoralis (Lepidoptera: Noctuidae), J. Comp. Physiol. A 176 (1995) 773-789.

38 B.S. Hansson, S. Anton and T.A. Christensen, Structure and function of antennal lobe interneurons in the male turnip moth, Agrotis segetum (Lepidoptera: Noctuidae), J. Comp. Physiol. A 175 (1994) 547-562.

39 R.E. Charlton and R.T. Carde, Orientation of male gypsy moths, lymantria dispar (l.) to pheromone sources: The role of olfactory and visual cues, Journal Insect Behavior 3 (1990) 443-469.

40 H. Ishida and T. Moriizumi, Machine Olfaction for Mobile Robots, in Handbook of Machine Olfaction, Pearce T.C., Schiffman S.S., Nagle H.T., Gardner J.W. (eds), Wiley-VCH: Weinheim, (2003).

41 J. Murlis and C. Jones, Fine scale structure of odour plumes in relation to insect orientation to distant pheromone and other attractant sources, Physiol. Ent. 6 (1995) 71-86.

42 M.A. Willis, J. Murlis and R.T. Cardé, Spatial and temporal structures of pheromone plumes in fields and forests, Physiol. Entomol. 25 (2000) 211-222.

43 A. Ludlow, J. Perry, C. David, J. Kennedy and C. Wall, A reappraisal of insect flight towards a distant source of wild-borne odour, Journal of Chemical Ecology 8 (1982) 1207-1215.

44 M. Willis, Odor-guided flight in moths, web-site http://flightpath.neurobio.arizona.edu/newindex.html.

45 H.Y. Fadamiro, C. Quero and T.C. Baker, Responses of male helicoverpazea to single pulses of sex pheromone and behavioural antagonist, Physiological Entomology (2001).

46 J.S. Kennedy, The visual responses of flying mosquitoes, Proceedings Zoological Society London 109 (1940) 221-242.

47 R. Wright, The olfactory guidance of flying insects, Canadian Entomology 90 (1958) 80-89.

48 D. Frizel, F. Kellog and R. Wright, The olfactory guidance of flying insects, Canadian Entomology 94 (1962) 884-888.

49 J.S. Kennedy and D. Marsh, Pheromone-regulated anemotaxis in flying moths, Science 184 (1974) 999-1001.

50 J.S. Kennedy, D.M. Marsh and A.R. Ludlow, Anemotactic zigzagging flight in male moths stimulated by pheromone, Physiological Entomology 3 (1978) 221-240.

51 J.S. Kennedy, D.M. Marsh and A.R. Ludlow, Analysis of zigzagging flight in moths: a correction, Physiological Entomology 6 (1981) 225.

52 T.C. Baker and L.P.S. Kuenen, Pheromone source location by flying moths: a supplementary non-anemotactic mechanism, Science 216 (1982) 424-427.

53 T. Baker, Upwind flight and casting flight: complimentary phasic and tonic systems used for location of a sex pheromone sources by male moths, Proceedings of the 10th Internation Symposium on Olfaction and Taste, (1990), pp. 18-25.

54 A. Mafra-Neto and R.T. Cardé, Influence of plume structure and pheromone concentration on upwind flight of caudra cautella males, Physiol. Entomol. 20 (1995) 117-133.

55 M.A. Willis and T.C. Baker, Effects of varying pheromone component ratios on the zigzagging flight movements of grapholita molesta, Journal of Insect Behavior 1 (1998) 357-371.

56 A. Mafra-Neto and R.T. Carde, Fine-scale structure of pheromone plumes modulates upwind orientation of flying moths, Nature 369 (1994) 142-144.

57 R.T. Cardé and T.E. Hagaman, Behavioral responses of the gypsy moth in a wind tunnel to air-borne enantiomers of disparlure, Environ. Entomol. 8 (1979) 475-484.

58 V.G. Dethier, Sniff, flick, and pulse: an appreciation of interruption, Proc. Am. Philos. Soc. 131 (1987) 159-179.

59 W. Lemon and W. Getz, Temporal resolution of general odour pulses by olfactory sensory neurons in american cockroaches, J. Exp. Biol. 200 (1997) 1809-1819.

60 W.C. Lemon and W.M. Getz, Rate code input produces temporal code output from cockroach antennal lobes, BioSystems 58 (2000) 151-8.

61 T.A. Christensen and J.G. Hildebrand, Coincident stimulation with pheromone components improves temporal pattern resolution in central olfactory neurons, J. Neurophysiol. 58 (1987) 151-158.

62 K.A. Justus Bau and R.T. Cardé, Antennal resolution of pheromone plumes in three moth species, J. Insect Physiol. 48 (2002) 422-433.

63 N.J. Vickers and T.C. Baker, Latencies of behavioral response to interception of filaments of sex pheromone and clean air influence flight track shape in heliothis virescens (f.) males, Journal of Comparative Physiology A 178 (1996) 831-847.

64 T.A. Mustaparta, N.J. Vickers and T.C. Baker, Chemical communication in heliothine moths. iv. Flight behavior of male helicoverpa zea and heliothis virescens in response to varying ratios of intra- and interspecific sex pheromone components, Journal of Comp. Physiol. A 178 (1987) 831-847.

65 J.L. Todd, A. Cossé and T.C. Baker, Neurons discovered in male helicoverpazea antennae that correlate with pheromone-mediated attraction and interspecific antagonism, J. Comp. Physiol. A 182 (1998) 585-594.

66 P.F.M. Verschure, T. Voegtlin and R.J. Douglas, Environmentally mediated synergy between perception and behaviour in mobile robots, Nature 425 (2003) 620-4.

67 J.M. Blanchard and P.F.M. Verschure, Using a mobile robot to study locust collision avoidance responses, International Journal of Neural Systems 9 (1999) 405-410.

68 C. von Planta, J. Conradt, A. Jencik and P.F.M. Verschure, A flying robot to study navigation in freely flying insects, in Proceedings of the International Conference on Artificial Neural Networks-ICANN02: Madrid, Spain, August 2002, pp. 1268-74. Lecture Notes in Computer Science. Berlin: Springer.

Chapter 14

A SPIKING NEURAL NETWORK MODEL OF THE LOCUST ANTENNAL LOBE
Towards Neuromorphic Electronic Noses Inspired From Insect Olfaction

Dominique Martinez and Etienne Hugues
LORIA, Campus Scientifique, BP 239, 54506 Vandoeuvre-Les-Nancy, FRANCE

Abstract: In analogy with the non-selectivity of gas sensors, an olfactory receptor neuron is not tuned to a specific odour and hence presents a lack of selectivity. Despite this shortcoming, insects have impressive abilities to recognize odours. Understanding how their olfactory system works could then be highly beneficial for designing efficient electronic noses. In particular, the antennal lobe, the first structure of the insect olfactory system, is known to encode odours by spatio-temporal patterns of activation of projection neurons. We propose here a simplified, but still biologically plausible, model of the locust antennal lobe. Our model is a network of single variable spiking neurons coupled via simple exponential synapses. Its reduced complexity allows a deeper understanding of the mechanisms responsible of the network oscillatory behaviour and of the spatio-temporal coding of the stimulus. In particular, we show how a stimulus is robustly encoded at each oscillation of the network by a spatial assembly of quasi-synchronized projection neurons, each one being individually phase-locked to the local field potential. Moreover, it is shown that frequency adaptation is responsible of the temporal evolution of this spatial code and that this temporal aspect of the code is crucial in enhancing the distance between the representations of similar odours.

Key words: Insect olfactory system, antennal lobe, spiking neurons, pattern recognition.

J.W. Gardner and J.Yinon (eds.),
Electronic Noses & Sensors for the Detection of Explosives, 209-234.
© 2004 *Kluwer Academic Publishers. Printed in the Netherlands.*

210

1. INTRODUCTION

The detection and localization of explosives or gas leaks in hostile environments or in public places is currently a very active area of research. There have been several attempts at developing autonomous olfactory robots capable of searching for a specific odour source [1-4]. Although these approaches have shown encouraging results (see Figure 1 for an example of a tracking experiment), they are still limited to the presence of a single odour in the environment and the use of a gas sensor sensitive to that odour. It is well known however that gas sensors are non-selective and respond to a wide variety of gases. Therefore, in an outdoor environment, the odour we would like to recognize is likely to be mixed with interfering odours and the robot may locate a source that is different from the one of interest.

Many animals, and among them insects, have the ability to recognize and track a specific odour even when it is mixed with interfering odours as it generally occurs in natural environments. In parallel with the non-selectivity of gas sensors, a biological olfactory receptor is not tuned to a specific odour and hence is not selective [5]. Because the variety of receptor proteins underlying chemoreception is not as rich as the repertoire of existing odorant molecules, different molecules may react with a particular protein and thus non-selectivity is probably an unavoidable situation an olfactory system has to face. This problem is still a challenge for electronic noses, as no really satisfactory general solution, taking into account sensor drift and poisoning and various environmental conditions, has been proposed so far. As many animals prove it by their impressive olfactory abilities, the olfactory pathway is probably fundamental in its design for facing this problem. Understanding how the biological olfactory system works could then be highly beneficial for designing efficient electronic noses.

Among all animals, insects have one of the simplest olfactory systems. In particular the first stage of this system, the antennal lobe (AL), has been shown to encode odours. Many experimental data, and in particular from the work of Laurent's group from Caltech[1], have shown that subsets of AL projection neurons (PNs) – whose activity is projected on higher structures – get synchronized in presence of an olfactory stimulus and that these subsets change in time in an odour-specific manner. Such a transient synchronization encoding scheme or spatio-temporal code has already been reproduced and studied by means of a biologically detailed model of the insect antennal lobe [6]. Still, however, because of the high level of complexity of this model, several aspects of the coding remain insufficiently understood. In particular,

[1] see at http://marvin.caltech.edu

what are the underlying mechanisms responsible of this spatio-temporal code and what are its properties and limitations? The use of a minimal model that can still reproduce in the same manner this spatio-temporal code could therefore be helpful to answer these questions.

We propose here a simple spiking neural network model of the antennal lobe that permits mathematical analysis and large scale simulations. In contrast to the model of Bazhenov *et al.* 2001 [6] that consists of conductance based neurons and biologically detailed synapses, our model is a network of single variable neurons coupled via simple exponential synapses. This reduced complexity allows a deeper understanding of the mechanisms underlying the obtained spatio-temporal coding and provides a direct input for designing data analysis methods for artificial electronic noses – for example by adapting the model so as to interface it with gas sensors.

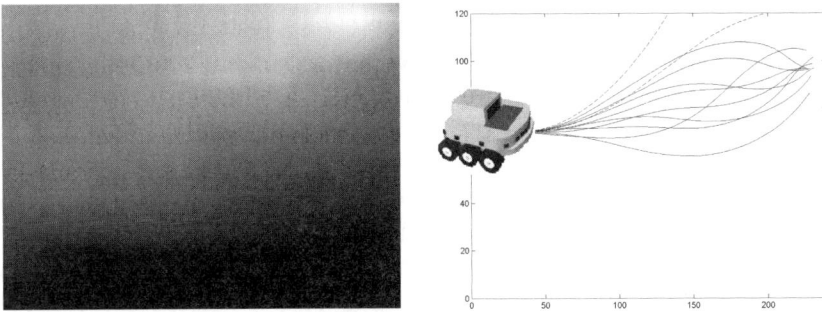

Figure 1. Tracking experiments performed with our olfactory robot: We have equipped a koala robot with two gas sensor arrays and performed tracking experiments with the robot moving in an arena. The visualisation of the odour plume can be seen at the left and robot trajectories at the right. The ethanol source was placed at $(x,y)=(240,100)$ and released at a low rate of 0.35 l/min. Over 16 runs performed with the robot starting from the same location $(x,y)=(20,70)$, 13 have successfully converged to the source location [4].

2. NEURAL CIRCUITS AND ACTIVITIES IN THE LOCUST OLFACTORY SYSTEM

In insects, odorant molecules are captured by Olfactory Receptor Neurons (ORNs) (~90,000 in the locust) distributed on their antennae. A large number of ORNs that express the same odorant receptor genes converge onto many fewer glomeruli (~900 in the locust), presumably for improved sensitivity. The odorant identity is then robustly represented in the glomerulus layer as a spatially distributed pattern. In the following we will consider this pattern of activity as input for our model. As described in Figure 2, the first two processing stages of the olfactory system of insects are: an encoding stage, called the antennal lobe (AL) and a decoding stage called the mushroom body (MB). The AL is a network of interconnected inhibitory local neurons (LNs) and excitatory projection neurons (PNs), both types receiving excitatory connections from the glomeruli. In the locust, the AL is made of about 900 PNs and 300 LNs, defining what we will refer later as the locust AL real scale. The PNs project to higher brain structures, but not the LNs which are local to the AL. The MB consists of a large number of neurons called kenyon cells (KCs) (~50,000 in the locust) which has afferent connections from the PNs : the ratio between the two population sizes shows that there is, what it is called, a divergence in the connectivity. Each PN projects to many different KCs (~600 in the locust), but a KC has only few afferent connections (~10 to 20 in the locust) from the PNs [7].

In the presence of an olfactory stimulus, subsets of PNs fire in synchrony giving rise to 20 Hz oscillations in the local field potential (LFP) that can be seen as an overall activity of the PNs. Nevertheless, no global properties of the oscillatory activity (e.g. mean frequency) have been found to convey information about the odour. However, the subset of neurons that is synchronized at any oscillation is odour-specific. It has therefore been conjectured that transient synchronization defines functional neuronal assemblies that evolve in time so as to encode odour identity. Each odour is therefore represented by a specific sequence of transiently synchronized neurons across the PN population defining a spatio-temporal code [8]. Several observations support the functional relevance of synchronization in odour coding. First, when it is pharmacologically abolished honeybees are no longer able to discriminate between similar odours [9]. Second, the KCs receiving inputs from the PNs do not seem to act as integrators but as coincidence detectors [7]. Although the spatial aspect of this code gives a synchronous population code which seems *a priori* sufficient to encode an odour, the role of the temporal aspect for encoding a constant olfactory stimulus remains, at this point, unclear: a possible role of it could be, as it has been shown for the mitral cells of the zebrafish, to decorrelate the representations of similar odorants over time [10].

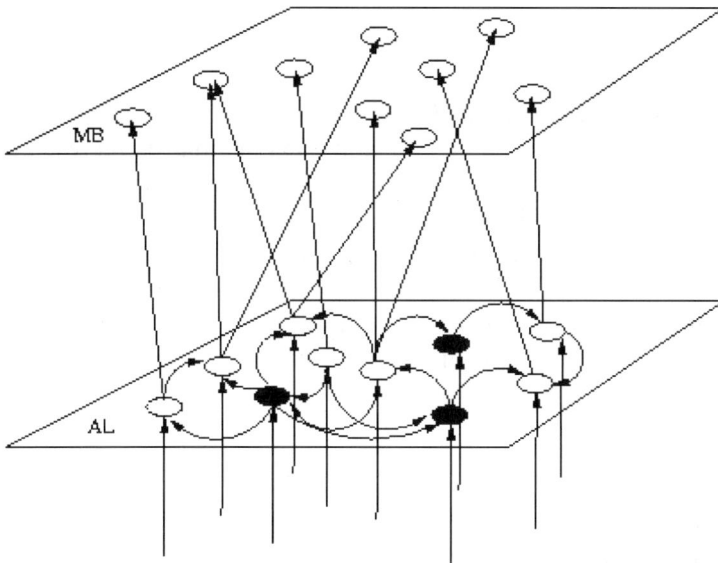

Figure 2. Schematic view of the insect olfactory system. It consists of two stages, the antennal lobe (AL) and the Mushroom Body (MB). For the Locust, the AL is a network of approximately 900 excitatory Projection Neurons (PNs) and 300 inhibitory Local Neurons (LN), represented by circles in black and white, respectively. The MB has a huge number of Kenyon Cells (KC) (~50,000 in the locust) receiving inputs from the PNs only.

3. THE ANTENNAL LOBE MODEL

The type of AL model we will consider is a sparsely connected network of N_E PNs and N_I LNs [2] (with $N_E = 3N_I$). After a careful study of the conductance based neurons of the PN and LN (see the Appendix), we found that they were type I neurons which means, to say it briefly, that their firing frequency in response to a constant input current can be arbitrarily low. Therefore, to model these neurons, we chose to use the theta neuron, also called quadratic integrate-and-fire neuron (QIF), because it has been shown to be a very good approximation of any type I neuron around the threshold [11,12].

The potential θ_j of a theta neuron j, or the equivalent potential v_j of the QIF neuron j with $v_j = \tan \theta_j/2$, obeys the following equation

---

[2] These numbers will depend on the scale of the model we consider with respect to the locust AL real scale.

$$\frac{d\theta_j}{dt} = (1 - \cos\theta_j) + \alpha_j I_j (1 + \cos\theta_j) \tag{1}$$

where I_j is the input current and α_j is a constant characterizing the neuron current-frequency response curve. Such a theta neuron j is represented by a point $(\cos\theta_j, \sin\theta_j)$ moving on a unit circle [11,12]. When $I_j < 0$ and constant, the neuron tends to be in its stable resting state $\theta_{rest}(I_j) < 0$. If it receives enough excitation, by synaptic interactions for example, it can cross the unstable threshold state $\theta_{thres}(I_j) > 0$ from which its tendency is to emit a spike. A spike occurs as soon as θ_j crosses π. When $I_j > 0$ and constant, θ_j is always increasing and the neuron emits spikes regularly. Then its firing frequency is given by

$$F_j = \frac{1}{\pi}\sqrt{\alpha_j I_j} \tag{2}$$

The current I represents actually the sum of four different currents – an external current (I^{ext}), a threshold current (I^{th}), an adaptation current (I^{adapt}) and a synaptic current (I^{syn}), such that

$$I_j = I_j^{ext} - I_j^{th} + I_j^{adapt} + I_j^{syn} \tag{3}$$

The adaptation current is given by Izhikevich 2000 [13]

$$I_j^{adapt} = g_j^{adapt} z_j^{adapt}$$

and

$$\frac{d z_j^{adapt}}{dt} = \delta(\theta_j - \pi) - \frac{z_j^{adapt}}{\tau_j^{adapt}} \tag{4}$$

in which δ is the Dirac distribution. Thus each time neuron j spikes when θ_j crosses π, the adaptation current is decreased by the quantity $g_j^{adapt} \leq 0$ and then relaxes exponentially towards zero with the time constant τ_j^{adapt}.

Similarly, the synaptic current is given by

$$I_j^{syn} = \sum_{i=1, i \neq j}^{N} g_{ij}^{syn} z_{ij}^{syn}$$

and

$$\frac{d\, z_{ij}^{syn}}{dt} = \delta(\theta_i - \pi) - \frac{z_{ij}^{syn}}{\tau_{ij}^{syn}}$$

(5)

such that each time a neuron i spikes when θ_i crosses π, the synaptic current of neuron j is increased or decreased if there exists a connection between i and j and depending on the sign of the synapse g_{ij}^{syn} between the neurons i and j. g_{ij}^{syn} is negative or positive whether neuron i is an inhibitory LN or an excitatory PN.

We have first simulated the conductance based models of the PNs and LNs and then fitted the parameters of the theta models accordingly (see the appendix). The parameters chosen for the neurons are as follows. PNs and LNs have a threshold current $I_j^{th} \approx 0.5$ and 0.8 respectively, which means that LNs are less excitable than PNs. The stimulus is applied to 33% of the neurons chosen at random as a constant external current $I_j^{ext} = 0.75$ with added Gaussian noise. With these parameters, the external current is above threshold for a PN and thus a PN receiving the stimulus is able to fire if it is not too much inhibited by the LNs. In contrast, the external current is below threshold for a LN and thus additional excitatory synaptic current coming from the PNs is needed to generate LN spikes.

The AL network is a sparsely and randomly connected network with the same probability of connection from LNs to PNs, PNs to LNs and between LNs. We did not consider here interconnections between PNs because there seem to have a negligible influence in the original model of Bazhenov *et al.* [6]. When two neurons i and j are connected, the connection strengths are $g_{ij}^{syn} = 0.05$ between a PN and a LN, $g_{ij}^{syn} = -0.5$ between a LN and a PN and $g_{ij}^{syn} = -0.1$ between two LNs. If we compare the values of the inhibitory synapses to the current $I_j^{ext} - I_j^{th}$ received by the neurons and if we do not consider the excitatory synapses, we find that a unique inhibitory spike prevents the firing of any neuron in the network for a non negligible duration. The excitatory synaptic time constant is $\tau_{ij}^{syn} = 5$ ms and the inhibitory one is $\tau_{ij}^{syn} = 10$ ms, which reinforce the dominant role of the inhibition.

4. DYNAMICAL BEHAVIOUR OF THE NETWORK

As stated in [14], it takes only 7 floating point operations to simulate 1 ms of a theta model as compared to 1,200 for a conductance based model. This reduced complexity leads more easily to large scale simulations. Thus, it is interesting to simulate a network of 900 PNs and 300 LNs that corresponds to the entire locust AL at scale 1, as in [6] only the scale 1/10 was simulated. The simulations performed below will allow us to confirm that results obtained with the reduced size were valid. In our model, we have considered a probability of connection of 0.05 for the total number of 300 LNs and 900 PNs. As mentioned above, we did not consider interconnections between PNs. The parameters for the input stimulus and for the theta neurons and the synapses are given in the appendix. The simulation of the model at scale 1 takes 20 minutes only on a Pentium 4 based PC running at 2.66 GHz. Note that the simulation is three times longer when interconnections between PNs are taken into account.

Figure 3 represents the raster plot, for an input current of 700 ms duration, of the 900 PNs (middle), the 300 LNs (bottom) as well as the time evolution of the LFP (top) estimated as the average of the state variables for the PNs, i.e.

$$LFP(t) = \sum_{j \in PN} \theta_j(t) \tag{6}$$

Synchronized activity of the PNs can be clearly seen as well as oscillations of the LFP around 16 Hz which is smaller than the 20 Hz oscillations found in experimental observations for the locust [9]. This lower frequency of the LFP oscillations can be explained by the fact that PN-PN connectivity was not considered here ($g_{ij}^{syn} = 0$ if i and j are two PNs).

In the presence of a stimulus, the network shows the following characteristic dynamical behaviour: a repeated alternance of a quasi-synchronized PN spike volley followed immediately after by a similar LN spike volley and then followed by a silent period. Without entering into details (see Section 6), we can explain this behaviour as follows: after a silent period due to the strong inhibition from the LNs, some of the stimulus-excited PNs fire and excite some LNs. After a sufficient number of PNs have fired, some LNs will start to fire and, as inhibitory connections are strong, will prevent the connected PNs and LNs to fire again.

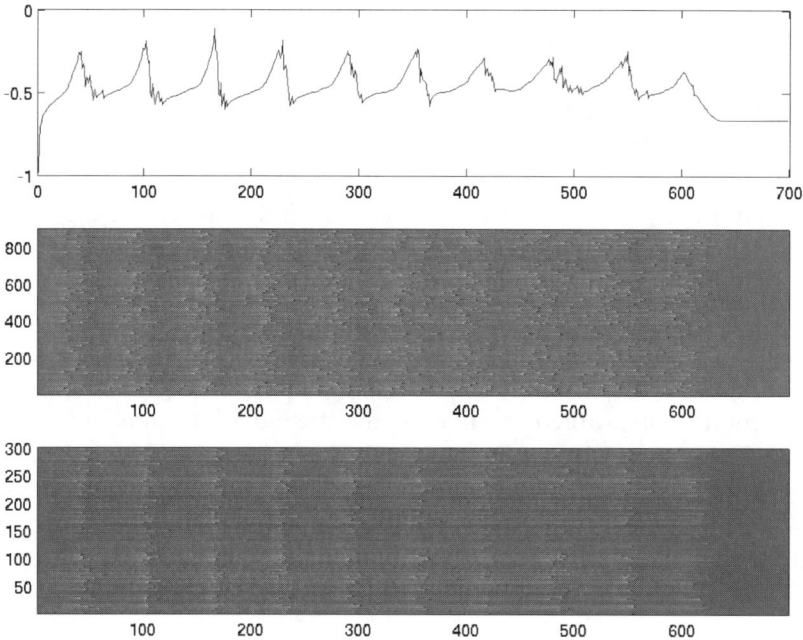

Figure 3. Simulation of the model of the locust AL at scale 1 (300 LNs and 900 PNs). Top- LFP. Middle- spike raster of the PNs. Bottom spike raster of the LNs.

In order to see if the scale of the network has an influence on its dynamical behaviour, we have simulated a size reduced network (scale 1/10). The simulation time is now of the order of 1 minute. The network architecture of 90 PNs and 30 LNs is similar to the one of Bazhenov *et al.* [6]. As before, the parameters for the theta neurons and the synapses are given in the appendix. However because the size of the network is reduced by a factor 10, the probability of connection is increased by 10 and thus is 0.5. The reason is that every neuron will then receive the same number of connections on average from excitatory or inhibitory neurons and, furthermore, the variance of this number of connections will be of the same order. This will guarantee that in presence of a similar activation of the network, i.e. same percentage of neurons excited by the stimulus, the neurons receive the same amount of excitation and inhibition independent of the size of the simulated network.

Figure 4 shows the time activity of the PNs (middle), LNs (bottom) as well as the LFP (top) for an input current of 700 ms duration. Synchronized activity of the PNs and the LNs can be clearly seen as well as oscillations of the LFP around 16 Hz which are in agreement with previous simulation results performed with the network at scale 1. Simulations using different stimuli have shown that the global synchronization of both populations of neurons and its consequence, the LFP oscillations, indicate the presence of a stimulus, but that the mean frequency of these oscillations is independent of the stimulus. We can then deduce that there is no information about the nature of the odour in any global dynamical description of the network (as the LFP). Because the number of neurons is now ten times smaller, it can be seen from Figure 4 that the PNs and LNs which do not receive the stimulus remain silent and that subsets of active LNs which receive the stimulus evolve in time. For example, LN 4 is active at the first and the fourth oscillation of the LFP but not at the second and the third. This can be explained by the fact that at the second and the third LFP oscillation, the inhibitory current received by LN 4 from the other LNs is sufficiently large to prevent it to fire. Moreover, dividing the strength of the inhibitory LN-LN and LN-PN synapses by a factor 10 resulted in a loss of PN synchronization and no more LFP oscillation (see Figure 5). Thus, the temporal evolution of subsets of active LNs, found in the intact network (Figure 4) and in [6], is likely to have an influence on the output of the AL given by the synchronization of the PNs. This will be studied in the next section.

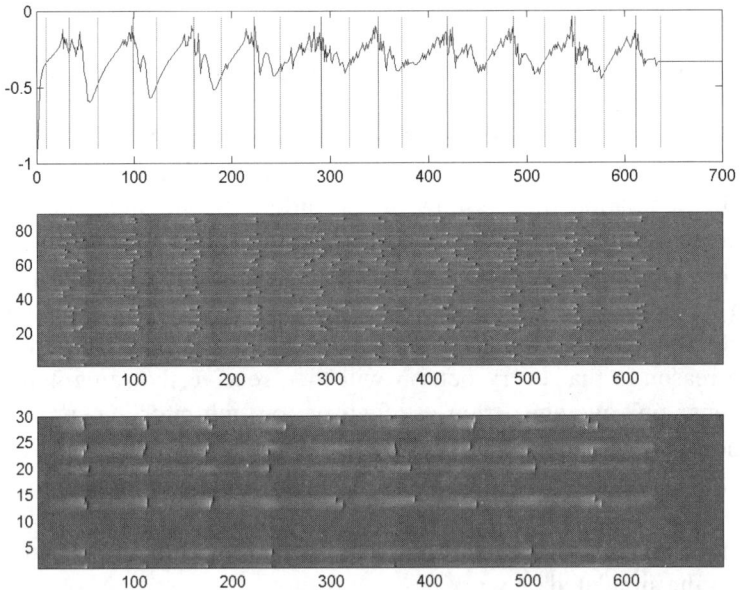

Figure 4. Simulation of the model of the locust AL at scale 1/10 (30 LNs and 90 PNs). Top: LFP. Middle: spike raster of the PNs. Bottom: spike raster of the LNs.

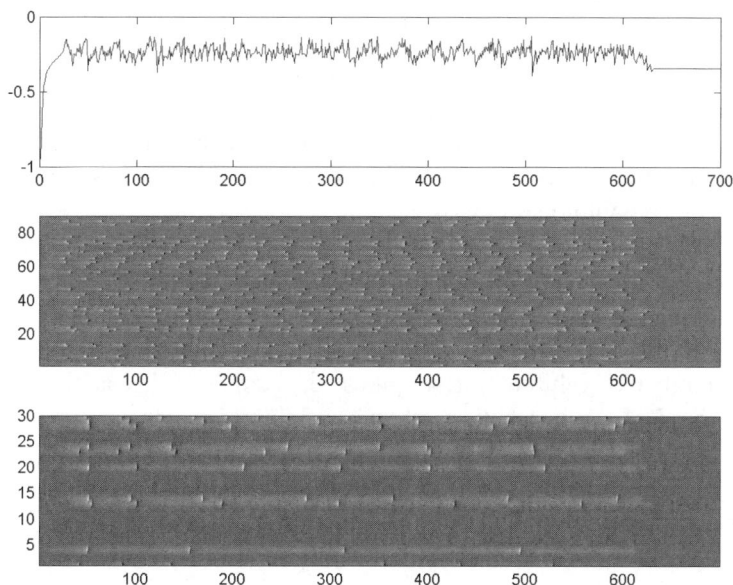

Figure 5. Simulation of the model of the locust AL at scale 1/10 with weak inhibitory synapses.

5. IS THE STIMULUS ENCODED BY THE NETWORK?

As we have seen in the previous section, the AL model at scale 1/10 exhibits a very similar behaviour to the complete one. From now on, we will therefore use for convenience the size reduced model. Furthermore, this will illustrate the capacities of this size reduced model which, for application purposes, is an attractive candidate: actually, if 3 periods of the LFP are enough to recognize the odour, the result could be obtained in about 10 s, which is a reasonable time for many applications

In order to study the transient aspect of the PN synchronization with respect to the oscillations of the LFP, we have repeated the analysis done in [6,15]. First, we have isolated the positive and negative peaks of the filtered LFP (see the vertical bars in the LFP of Figure 4). Second, we have assigned to each PN spike a phase $(-\pi, +\pi)$ according to its closest positive LFP peak, a zero phase meaning that the PN spike is perfectly synchronized with the peak of the LFP. Figure 6 at the left shows the result of this analysis for 20 different runs of the AL network and 6 particular PNs (see figure caption for details). Transient synchronization can be clearly seen in Figure 6 left. For example PN#1 (1st row) is desynchronized at the first peak of the LFP (1st box), synchronized for the next three peaks, desynchronized at the 5th peak and so on. The output of this PN could then be seen as a binary vector (0, 1,

1, 1, 0, ...) where the kth bit 1 and 0 correspond to a synchronization or a desynchronization at the kth peak of the LFP, respectively. Thus, the output as the entire AL could then be seen as a spatio-temporal binary code. If we repeat the same experiments and analysis as before but with a noise level 10 times higher, then Figure 6 at the right indicates a spatio-temporal code similar to the one found before and thus this code presents some robustness with respect to input noise.

In the previous section we mentioned that the time evolution of subsets of active LNs might have an influence on the output of the AL given by the transient synchronization of the PNs. In order to explore this, we plotted in Figure 7 the mean inhibitory drive for each of our 6 particular PNs, i.e. the average number of inhibitory LN spikes received by a PN at the previous peak of the LFP. We see that the mean inhibitory drive a PN receives changes from peak to peak. Moreover, by comparing these plots to those in Figure 6, we find a strong correlation between the amount of inhibition a PN receives and its degree of synchronization with respect to the LFP. Thus, the transient synchronization of a given PN depends on the time evolution of subsets of active LNs that are connected to it. However, the converse is also true because the LNs are not capable to fire by themselves without additional excitatory synaptic current coming from the PNs.

Is the stimulus effectively encoded by such an intertwined LN-PN dynamics? Figure 8 shows the response to a different stimulus that differs from stimulus 1 only by the identity of the LNs receiving the external current. This change is enough to produce a different PN synchronization pattern. Similarly, a different response is obtained if we change the identity of a part of the PNs that receive the stimulus.

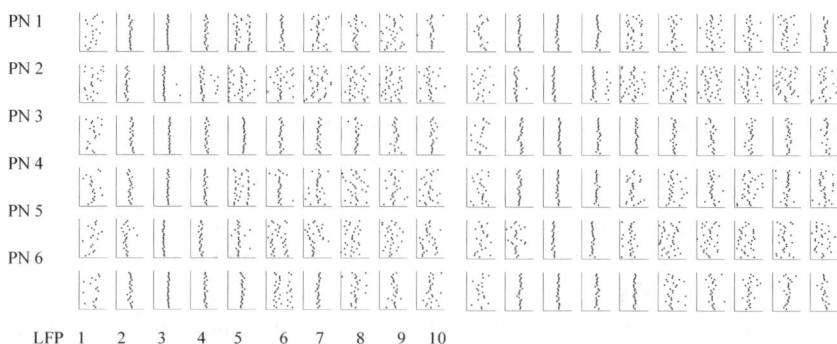

Figure 6. Phase plot for stimulus 1 and for a standard deviation of the input noise equal to 0.01 (left) and 0.1 (right). In the figures, each row corresponds to a given PN (from 1 to 6) and each column corresponds to a given peak of the LFP (from 1 to 10). The phase of a given spike fired by PN *i* with respect to the peak *j* of the LFP is plotted as a dot in the box (LFP *j*, PN *i*). Each row in each box corresponds to a different trial (20 trials in total).

PN 1
PN 2
PN 3
PN 4
PN 5
PN 6

LFP 1   2   3   4   5   6   7   8   9   10

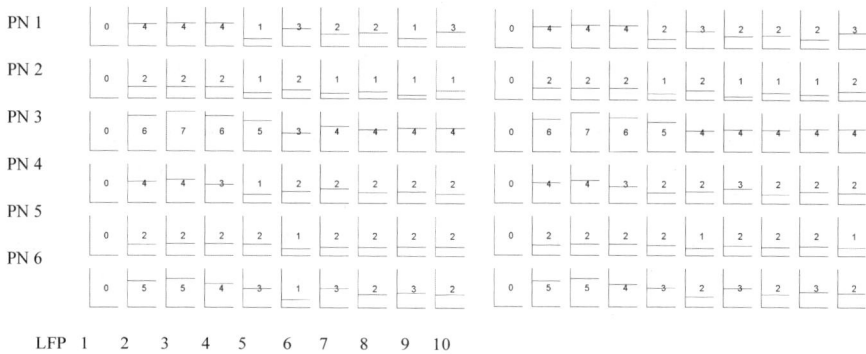

Figure 7. Mean inhibitory drive for stimulus 1 and for a standard deviation of the input noise equal to 0.01 (left) and 0.1 (right). As in figure 6, each row corresponds to a given PN (from 1 to 6) and each column corresponds to a given peak of the LFP (from 1 to 10). The mean inhibitory drive of the PN i at the peak j of the LFP is both indicated by the number and the level in the box (LFP j, PN i). The mean inhibitory drive is measured as the average over the 20 trials of the number of LN spikes received by the PNs right after the previous peak of the LFP.

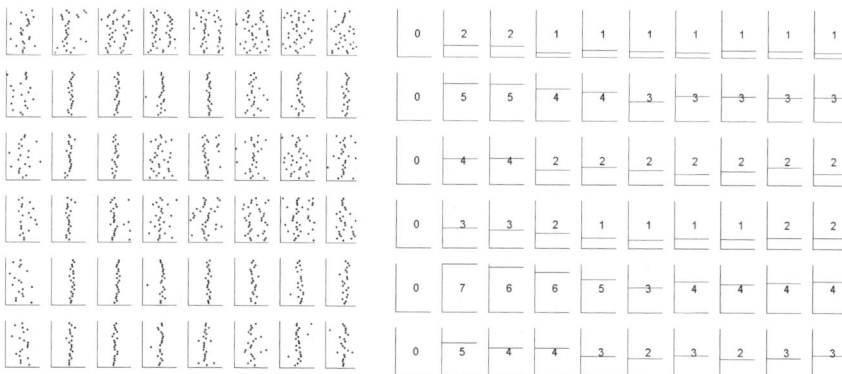

Figure 8. Phase plot (left) and Inhibitory drive (right) for stimulus 2. For details, see the legends of Figures 6 and 7.

6.   DYNAMICAL MECHANISMS RESPONSIBLE OF THE NETWORK BEHAVIOUR AND CODE

Previous simulation results have enlightened two major observations: first, the PNs and LNs which do not receive the stimulus remain silent while the others fire in a globally quasi-synchronized fashion, leading to LFP oscillations; second, on a finer temporal scale, some of the PNs are robustly and tightly phase-locked with the LFP at a particular oscillation while others are not and this subset of phase-locked PNs change in time in a stimulus-specific way.

In the last decade, there have been many studies about the emergence of synchronous activity in spiking neural networks [16]. In a recent paper, Börgers and Kopell [17] have addressed the case of a network with sparse and random connectivity between two populations of theta neurons, one excitatory and the other inhibitory. However, their network is limited to excitatory to inhibitory and inhibitory to excitatory connections and the dynamical behaviour is studied at the population level only. In contrast, we address here a more complex situation in which spike adaptation and inhibitory to inhibitory connectivity are likely to play a role in the dynamics and we address both population and individual neuron levels. In order to understand how the firings of PNs and LNs get synchronized, we will first study two simple cases: a PN which has received inhibition and a LN which has received excitation. While unraveling progressively the behaviour of individual neurons, we will be able to explain the dynamical construction of the neural code.

6.1   Influence of LN inhibition on the next PN firing time

Due to the sparseness of the connectivity, the number of afferent connections and by consequence the resulting inhibitory input strength varies from neuron to neuron. Furthermore, when they receive inhibition, PNs are in different states because of their different past temporal evolutions. Thus, we will consider the problem of a PN receiving, at time $t=0$, an exponential inhibitory current plus an external current i^{ext} due to the stimulus and will study the influence of the strength g of the inhibition and of the initial state of the neuron on its next firing time. The total current this theta neuron receives is given by Equation (7),

$$i(t) = i^{ext} - g\, e^{-t/\tau_I} \tag{7}$$

Examples of the time evolution of $\theta(t)$ are given in Figure 9 for different initial states $\theta(0)$ and for a given inhibition strength g. For all the negative $\theta(0)$, we can see that any two neighboring trajectories get closer in time leading to similar firing times T obtained when $\theta(T) = \pi$. We note that for the majority of positive $\theta(0)$, the neuron fires earlier, but as its state is just after reinitialized to $\theta = -\pi$, we are led to the previous situation where $\theta(0)$ is negative. The effect of the inhibition strength g can also be inferred from

Figure 9. On the left and the right, the value of g corresponds to the reception of one and three LN spikes, respectively. As expected, firing times T increase with g. However, the standard deviation of T decreases with g and thus PN synchronization is tighter when the inhibition is stronger (see also Figure 10). In Figure 9 are also indicated, as if the input current was constant over time, the 'instantaneous' resting and threshold states of the neuron when the total current is negative (or subthreshold): we can see that the neuron, when it is inhibited, tends to remain close and under the resting state.

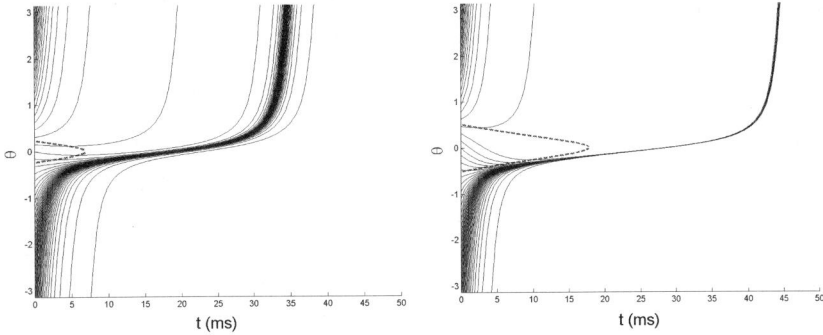

Figure 9. Trajectories of a PN receiving an inhibitory synaptic input at $t=0$ for different intial states $\theta(0)$. At the left, the PN receives a single LN spike and thus the inhibition strength is $g=-0.5$. At the right, the PN receives simultaneously three LN spikes and thus the inhibition strength is $g=-1.5$. The dashed curve in both figures is the locus of points where the time derivative of θ vanishes. (As θ is 2π-periodic, trajectories hitting the upper bound π are continued with the value $-\pi$).

To explain why PN synchronization gets tighter when the inhibitory drive gets stronger, let us consider the QIF version of the theta neuron (recall that $v = \tan \theta/2$) given by

$$\frac{dv}{dt} = F(v) + \alpha.i \tag{8}$$

where $F(v) = v^2$. It can easily be shown that, for two neighboring trajectories $v(t)$ and $v'(t) = v(t) + \delta v(t)$ where $\delta v(t)$ is small, we have the following linearized equation (9)

$$\frac{d\delta v}{dt} = \frac{dF(v)}{dv} \delta v = 2v\delta v \tag{9}$$

which is independent of the total current. Furthermore, if $dF/dv < 0$, that is when $v < 0$ (similarly $\theta < 0$), the two trajectories will get closer in time, and vice versa. This explains what was observed in Figure 9. A sufficiently inhibited PN will have a negative potential v (or θ) for some duration bringing together neighbouring trajectories as $d\delta v/dt < 0$. Note that this explanation does not depend on a particular choice of a model for the PN as soon as the function F is a decreasing function of v below some potential v_0 and then an increasing function above, which is generally the case when one wants to represent conductance based neurons with one variable models (Chapter 4 of [18]). Furthermore, a similar behaviour has been found for the original conductance based neuron models of the PNs, an observation which validates our choice of the theta neuron here [19].

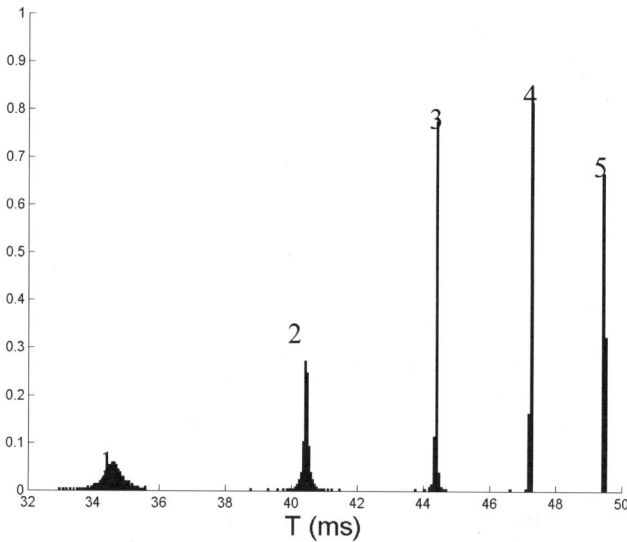

Figure 10. Superposition on the same plot of the theta neuron firing time T distributions relatively to its initial state and when it receives from 1 to 5 inhibitory synaptic inputs.

We now return to what happens in the network. Let us consider that a particular PN has received some inhibition coming from a given number n of LNs at the previous LFP oscillation, with $n > 0$. As these LNs are quasi-synchronized, this inhibition is approximately equivalent to a unique inhibitory drive of strength $g = n\, g^{syn}$ where g^{syn} is the strength of a single inhibitory LN-PN connection. Then, if this PN does not receive additional inhibition before its firing, it will fire at a time given by the distribution corresponding to the nth peak in Figure 10, the average time of these well separated peaks increasing with n. Due to the sparseness of the connectivity, this number n may not be the same for different PNs. As a consequence, the PNs that have received a smaller number n of LN spikes will be synchronized earlier than the PNs that have received a larger number n' of LN spikes. Using equation (4.11) in [17], it can be shown that the time difference Δt that separates the average firing time of these two sets of neurons is approximately given by

$$\Delta t = \tau_I \ln(n'/n) \tag{10}$$

where τ_I is the time constant of the inhibitory synapse. For example, in Figure 6 left at the 3rd peak of the LFP, the spikes emitted by PN2 and PN5 are clearly separated from those of PN1 and PN4. The phase difference between these two sets of neurons is 0.73 radians leading to a time difference of about 7 ms for 16 Hz LFP oscillations. Moreover, Figure 7 left indicates that the mean inhibitory drive received by the cluster (PN2,PN5) and (PN1,PN4) is $n=2$ and $n'=4$ LN spikes, respectively. This gives a theoretical time difference of $\Delta t = \tau_I \ln(2) = 6.93$ ms, where $\tau_I = 10$ ms here. This theoretical value is in good agreement with the experimental value derived above.

In conclusion, the firing of the PN population is divided into different quasi-synchronized clusters. Each cluster n is separated from the next one by a time difference of $\Delta t = \tau_I \ln(1 + 1/n)$ where n is the number of LN spikes the first set of neurons received at the previous LFP oscillation. The PNs which have not been inhibited at all, i.e. for which $n=0$, can fire freely at any time but their number is sufficiently low so that they do not to perturb the other neurons. Only the PNs that receive a sufficient but not a too large number of inhibitory synaptic inputs from the LNs will be phase-locked to the LFP.

To complete the above analysis, non-perfect synchronization of the LNs and noise in the input should be considered [19]. The former introduces some variability in the total inhibitory synaptic drive of the PNs. As a consequence, the firing time distributions observed in Figure 10 are more spread but the general tendency with increasing n is conserved. Noise in the input, although different in nature, plays a similar role. The above linearized equation for the evolution of two neighboring trajectories of a neuron contains in this case an additional term which is the difference of the noisy currents for the two trajectories: as this noise is relatively weak, the general stabilizing behaviour observed when $v < 0$ and vice versa remains dominant, and the influence of noise essentially accumulates for $v > 0$, which slightly modifies the firing times obtained without noise.

6.1.1 Influence of PN excitation on the LN next firing time

Previous simulations have shown that the silent period that separates consecutive bursts of network activity lasts for about 50 ms. Because this is much longer than the synaptic time constants (5 and 10 ms for excitation and inhibition, respectively), the total synaptic current previously received by the neurons has dramatically decreased at the end of the silent period. Thus, at that time, LNs are almost in the state they would have without considering synaptic entries, this state however depends strongly on their adaptation current. Then, right after the volley of PN spikes, each LN receives a given number of excitatory synaptic inputs which varies from neuron to neuron due to the sparseness of the connectivity. The first LNs to fire are those which receive the stimulus and which are the least adapted and the most excited by the PNs. This first set of LNs eventually inhibits the other LNs before they can fire because of the strong inhibitory connections that exist between them. This leads to a complex competition between LNs (see for example in Figure 4). As the PN spike volley is quasi-synchronized and as the mean excitatory synaptic input to a LN is high (it is approximately equal to the product between the probability of connection, the number of PNs that fire and the excitatory connection strength, *i.e.* about $0.5 \times 20 \times 0.05 = 0.5$), the LNs will fire just a few ms after the PNs in a quasi-synchronous fashion.

6.1.2 How is the code finally constructed?

In the above discussion, we have shown how neurons can fire at relatively precise times, but nothing has been said about the identity of the neurons that fire with respect to those receiving the stimulus. Nevertheless, section 5 has revealed that the identity of the neurons that fire precisely is given by the stimulus identity. From one firing period to the other, the neurons that fire are quite robustly determined by the connectivity. Actually, given an assembly $\mathcal{L}_{k-1}$ of emitting LNs at the $(k\text{-}1)^{\text{th}}$ oscillation of the LFP, the assembly $\mathcal{P}_k$ of emitting PNs at the next oscillation is robustly determined by $\mathcal{L}_{k-1}$ because of the mechanisms discussed above and because of the connectivity between these two assemblies. The link between $\mathcal{P}_k$ and $\mathcal{L}_k$ exists for the same reasons but depends also on previous firing assemblies $\mathcal{L}_{k-m}$, with $m > 0$ because of the LN adaptation. Of course, all these assemblies depend on the assemblies $\mathcal{P}_{stim}$ of PNs and $\mathcal{L}_{stim}$ of LNs that receive the stimulus because they determine the neurons that can fire. If we concentrate now on the PNs that fire precisely, as they are members of the $\mathcal{P}_k$ assemblies, they are also determined by the connectivity.

Therefore, the network exhibits a spatio-temporal code of the stimulus, in the sense that at any LFP oscillation, an assembly of precisely firing PNs is robustly determined by the stimulus, the connectivity and the past activity of the network.

7. TEMPORAL PROPERTIES OF THE CODE

In order to study the temporal properties of the neural code, we transform the output of each PN into a binary vector (0, 1, 1, 1, 0, …) where the kth bit 1 and 0 corresponds to a synchronization or a desynchronization at the kth peak of the LFP, respectively. This was done by computing the standard deviation of the phases of the PN spikes over the 20 runs for each peak of the LFP and assigning 0 or 1 when the standard deviation was respectively higher or lower to a given threshold equal here to 0.5 and which corresponds in time to about 5 ms. This value corresponds to experimental observations showing that PN spikes occur within a ± 5 ms window when they are phase-locked with the LFP [20]. Figure 11 shows the spatio-temporal code obtained this way for the stimulus 1 corresponding to Figure 6.

	LFP 1	2	3	4	5	6	7	8	9	10
PN 1	0	1	1	1	0	1	0	0	0	0
PN 2	0	1	0	0	0	0	0	0	0	0
PN 3	0	1	1	1	1	1	1	0	0	1
PN 4	0	1	1	1	0	0	0	0	0	0
PN 5	0	0	1	1	0	0	0	0	0	0
PN 6	0	1	1	1	1	0	1	0	0	0

Figure 11. Spatio-temporal binary code obtained for stimulus 1. Each row corresponds to a given PN (from 1 to 6) and each column corresponds to a given peak of the LFP (from 1 to 10). The bit 1 or 0 within a given box corresponds to a synchronization or a desynchronization of a given PN at a given peak of the LFP. For example, the bit 1 found in the box (LFP 4, PN 3) means that PN 3 is synchronized at the 4th peak of the LFP. This has to be compared to figure 6 left where the phases of the spikes fired by PN 3 at the 4th peak 4 of the LFP present a very small jitter over the 20 trials.

We now compare the binary spatio-temporal code obtained for stimulus 1 with the one obtained for a different stimulus for which we changed the identity of a single PN among all the neurons that receive the stimulus so that the two stimuli present a maximum overlap. Figure 12 represents the time evolution of the hamming distance between the two obtained spatio-temporal binary codes. When the network is intact (plain curve), this distance increases with time so that it would become easy to separate these two stimuli at the 10th peak of the LFP. Thus, time seems to play a role in decorrelating the AL representations of similar stimuli as it was experimentally shown in the olfactory bulb of the zebrafish [10]. However, this temporal decorrelation is lost when the same simulation was performed for a network without any frequency adaptation for the LNs (dotted curve in Figure 12). Indeed, the PNs were generally either synchronized or desynchronized at all peaks of the LFP (see Figure 13).

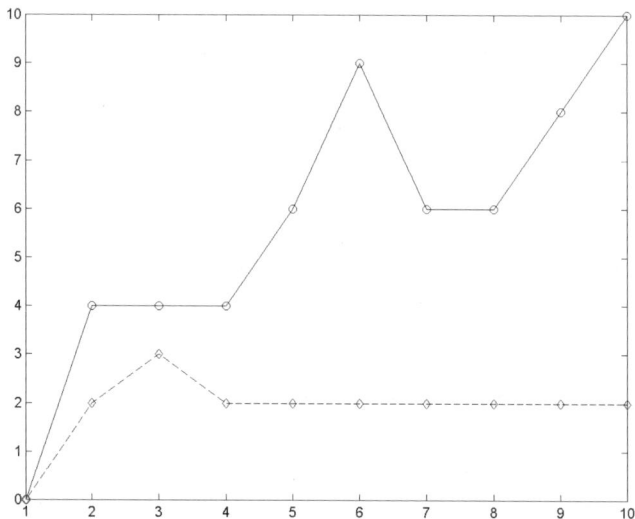

Figure 12. Time evolution of the hamming distance between the binary codes corresponding to stimulus 1 and 2 obtained at each peak of the LFP. The x and y axis is the peak of the LFP and the number of different bits between the 2 codewords. Plain curve is for an intact network and dashed curve is for a network without any frequency adaptation for the LNs ($g_j^{adapt} = 0$).

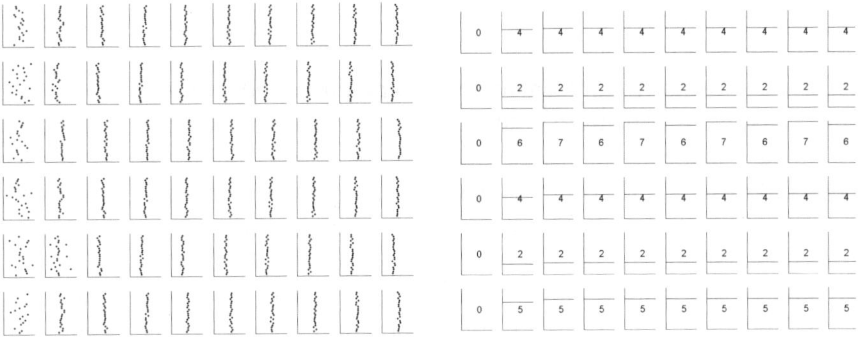

Figure 13. Phase plot (left) and Inhibitory drive (right) for stimulus 1 and for a network without any frequency adaptation for the LNs (g_j^{adapt} = 0). For more explanations, see the legends of figures 6 and 7.

8.   DISCUSSION

In this paper we have proposed a simplified model of the insect AL in order to explore the neural code in olfaction. A possible role of the AL is to transform a multidimensional input vector representing the odorant stimulus into a spatio-temporal code given by a sequence of quasi-synchronized assemblies of PNs, in which each PN is individually phase-locked to the LFP.

In contrast to the model of Bazhenov *et al.* 2001 [6] that consists of conductance based neurons and biologically detailed synapses, our model is a network of single variable neurons coupled via simple exponential synapses. This reduced complexity allows a deeper understanding of the mechanisms responsible of the network oscillatory behaviour and of the spatio-temporal coding of the stimulus. In particular, the network exhibits a repeated alternance of a quasi-synchronized PN spike volley followed right after by a similar LN spike volley in a way similar to the one described by Börgers and Kopell [17]. This phenomenon, referred as PING (pyramidal interneuronal network gamma) is responsible of the 16 Hz LFP oscillations obtained in the simulations of our model. Although these authors have studied a much simpler network without any input pattern and frequency adaptation, we found that the same reason is responsible of the same behaviour, that is strong connections induce the quasi-synchronization of the two populations of neurons. However, their analysis differs from ours in the sense that we have been able to explain the emergence of a spatio-temporal code by unraveling progressively the behaviour of individual neurons. In

230

particular, the LN frequency adaptation was shown to be responsible of the temporal evolution of the spatial code. Moreover, we have shown in simulations that this temporal aspect of the code is crucial in enhancing the distance between the representations of very similar odours.

Because all that matters in this coding is to know whether or not a given PN is phase-locked to the LFP, a given odour can be represented as a binary codeword of the same size that the number of PNs and where the kth bit 1 or 0 corresponds to a synchronization or a desynchronization of the kth projection neuron with respect to the current peak of the LFP. Recognizing a given odour then consists in discriminating the binary codeword at each peak of the LFP from any codeword representing a different odour. Because the dimension of the representation space is relatively small (~900 PNs for the locust), this discrimination problem is likely to be nonlinearly separable and therefore difficult. A possible role of the KCs, in the MB, the second stage of the olfactory system, is to transform this dense 900 dimensional binary code into a sparse code in the huge dimensional space defined by the 50 000 KCs such that the problem becomes linearly separable. Thus, the sparsening of the odour representation in the KC layer facilitates odour discrimination [7,21,22]. This shares striking similarities with kernel methods[3] like Support Vector Machines (SVM) in pattern recognition.

Work is ongoing for designing an electronic nose inspired from the biological principles detailed above. This will include the adaptation of our AL model so as to interface it with gas sensors and the development of an SVM type MB model for discriminating the binary codewords provided by the AL model.

---

[3] http://www.kernel-machines.org

APPENDIX

Reduction of the PN and LN conductance based neuron models

In order to use models of neurons that are biologically plausible, we have first simulated the conductance based models of the PNs and LNs used in Bazhenov *et al.* [6]. Both neurons were found to be type I neurons and we chose to model them with the theta neuron model [11]. We then fitted the parameters of the theta models so as to match the instantaneous firing frequency vs. applied current curves (see Figure A1). Note however that these instantaneous frequency curves do not take into account a possible frequency adaptation leading to a decrease of the frequency over time. Therefore, the two parameters involved in the adaptation current of the theta models have been fitted independently so that the time responses to applied constant current correspond to the ones obtained with the conductance based model. Figures A2 and A3 clearly indicate a close match between the time responses of the two models. In particular, the frequency adaptation seen in the conductance based model of the LN is similar to the one of the theta model (see Figure A3). The parameters for the fitted theta models are given below.

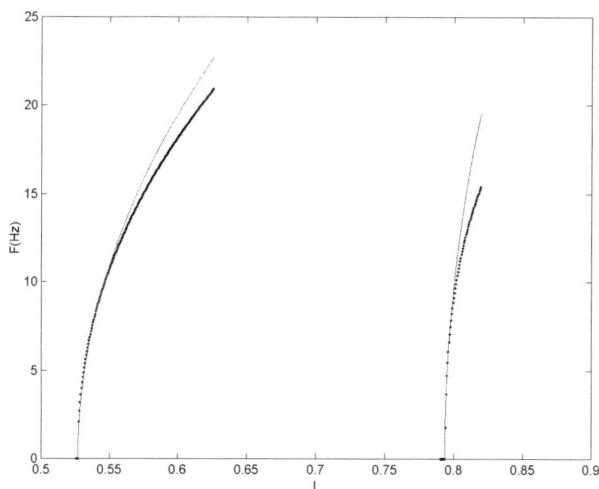

Figure A1. Instantaneous firing frequency vs. applied current for a PN (left) and a LN (right). Plain curves are for the simulations of the conductance based models from [6] and dotted curves are for the simulations of the corresponding fitted theta models.

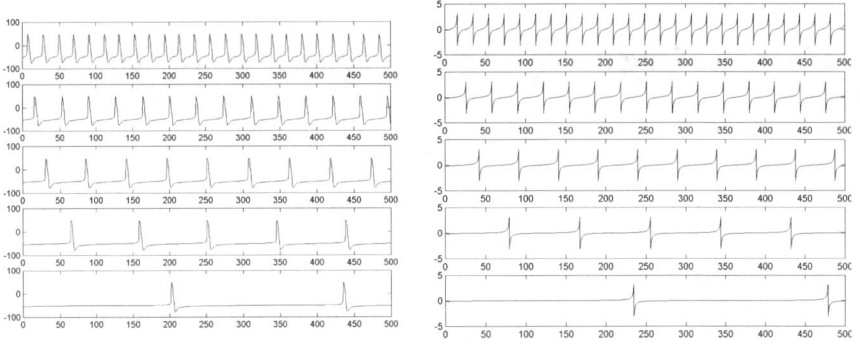

Figure A2. Temporal responses of the conductance based model of a PN (left) to a constant input current of different amplitudes (from top to bottom $I_j^{ext} = 1.0, 0.7, 0.6,$ 0.55, 0.53) compared with the fitted theta model (right). Time is in ms.

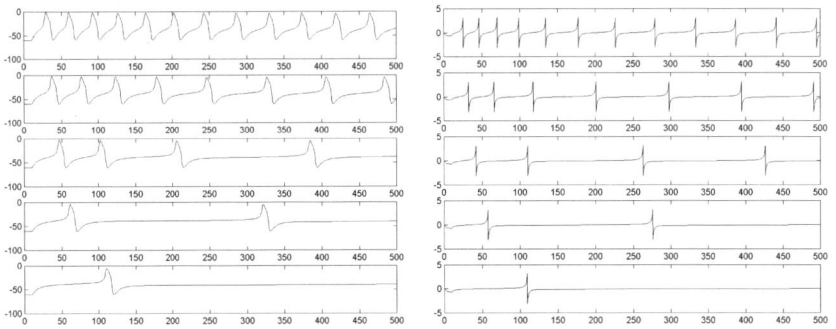

Figure A3. Temporal responses of the conductance based model of a PN (left) to a constant input current of different amplitudes (from top to bottom $I_j^{ext} = 1.0, 0.9,$ 0.85, 0.82 and 0.8) compared with the fitted theta model (right). Time is in ms.

Parameter values for the neurons:

$$I_j^{ext} = 0.75$$

$I_j^{th} = 0.527$ if neuron j is a PN and -0.7935 if neuron j is a LN.

$\alpha_j = 0.055255$ if neuron j is a PN and 0.14516 if neuron j is a LN.

$g_j^{adapt} = 0$ if neuron j is a PN and -0.05 if neuron j is a LN.

$\tau_j^{adapt} = 200$ ms.

Parameter values for the synapses:

Excitatory synapses: $g_{ij}^{syn} = 0$ if i and j are two PNs

$g_{ij}^{syn} = 0.05$ and $\tau_{ij}^{syn} = 5$ ms if i is a PN and j is a LN

Inhibitory synapses: $g_{ij}^{syn} = -0.1$ and $\tau_{ij}^{syn} = 10$ ms if i and j are two LNs

$g_{ij}^{syn} = -0.5$ and $\tau_{ij}^{syn} = 10$ ms if i is a LN and j is a PN

REFERENCES

1 R.A. Russel, (1999). Odor detection by mobile robots, World Scientific Series in Robotics and Intelligent Systems, World Scientific, 22 (1999).
2 H. Ishida, A. Kobayashi T. Nakamoto and T. Moriizumi, Three-dimensional odor compass, IEEE Trans. Robot. Autom., 15 (1999) 251-257.
3 T. Nakamoto H. Ishida and T. Moriizumi, An odor compass for localizing an odor source, Sens. Actuators B 35-36 (2001) 32-36.
4 E. Hugues, O. Rochel and D. Martinez, Navigation strategies for a robot in a turbulent odor plume using bilateral comparison, 11th International Conference on Advanced Robotics (ICAR), University of Coimbra, Coimbra, Portugal, 2003, 1, pp. 381-386.
5 P. Duchamp-Viret, M.A. Chaput and A. Duchamp, Odor response properties of rat olfactory receptor neurons, Science 284 (1999) 2171-2174.
6 M. Bazhenov, M. Stopfer, M. Rabinovich, R. Huerta, H.D. Abarbanel, T.J. Sejnowski and G. Laurent, Model of transient synchronization in the locust antennal lobe, Neuron 30 (2001) 553-567.
7 J. Perez-Orive, O. Mazor, G.C. Turner, S. Cassenaer, R.I. Wilson and G. Laurent, Oscillations and sparsening of odor representations in the mushroom body, Science 297 (2002) 359-365.
8 G. Laurent, Dynamical representation of odors by oscillating and evolving neural assemblies, Trends Neuroscience 19 (1996) 489-496.

234

9 M. Stopfer, S. Bhagavan, B.H. Smith and G. Laurent, Impaired odor discrimination on desynchronisation of odor-encoding neural assemblies, Nature 390 (1997) 70-74.

10 R. Friedrich and G. Laurent, Dynamic optimization of odor representation by slow temporal patterning of mitral cell activity, Science 291 (2001) 889-894

11 B. Ermentrout, Type I membranes, phase resetting curves, and synchrony, Neural Computation 8 (1996) 979-1001

12 F. Hoppenstead and E.M. Izhikevich, Canonical neural models, Brain Theory and Neural Networks, 2nd Ed., MIT Press, Cambridge, MA, (2002).

13 E.M. Izhikevich, Neural excitability, spiking and bursting, International Journal of Bifurcation and Chaos 10 (2000) 1171-1267.

14 E.M. Izhikevich, Which model to use for cortical spiking neurons?' IEEE Trans. Neural Networks, 2004 in press

15 G. Laurent, M. Wehr and D. Davidowitz, Temporal representations of odors in an olfactory network, The Journal of Neuroscience, 16 (1996) 3837-3847.

16 A.K. Sturm and P. König, Mechanisms to synchronize neuronal activity', Biol. Cybern. 84 (2001) 153-172

17 C. Börgers and N. Kopell, Synchronization in networks of excitatory and inhibitory neurons with sparse, random connectivity, Neural Computation 15 (2003) 509-538.

18 W. Gerstner and W.M. Kistler, Spiking neuron models – single neurons, populations, plasticity, Cambridge Univ. Press, (2002).

19 E. Hugues and D. Martinez D. (2004) in preparation

20 G.M. Laurent, R.W. Stopfer, M. Friedrich, A. Rabinovich, A. Volkovskii and H.D.I. Abarbanel, Odor coding as an active, dynamical process: experiments, computation and theory, Annu. Rev. Neuroscience 24 (2001) 263-297.

21 F.E. Theunissen, From synchrony to sparseness, Trends in Neurosciences 26 (2003) 61-64.

22 R. Huerta, T. Nowotny, M. García-Sanchez, H.D. Abarbanel and Rabinovich, Learning classification in the olfactory system of insects, (2004) submitted

Chapter 15

DETECTING CHEMICAL VAPOURS FROM EXPLOSIVES USING THE ZNOSE®, AN ULTRA-HIGH SPEED GAS CHROMATOGRAPH

Edward J. Staples
Electronic Sensor Technology, 1077 Business Center Circle, Newbury Park, California, USA

Abstract: Gas chromatography is the preferred analytical method for analysis of odours, fragrances, and other chemical vapours. A handheld electronic nose, called the zNose®, incorporating an ultra-high speed GC column, solid-state sensor, programmable gate array (PGA) processor, and an integrated preconcentrator can now provide near real time analysis of odours from explosives. Odours from Primacord, Detasheet, C4, and TATP were analyzed and their chemical composition quantified in less than 10 seconds with pictogram sensitivity. An expandable chemical/odour library based upon Kovats indices provides an effective method of recognizing odours from explosives and allows users to distribute and share odour signatures from chemical or biological threats as well. The zNose® is able to create an almost unlimited number of virtual chemical sensors for monitoring the concentration of target compounds within odours or fragrances. Virtual chemical sensors provide compound specific data for principal component analysis and neural network learning algorithms.

Key words: Gas chromatograph, explosives detector, zNose.

1.   ELECTRONIC NOSES AND BOMB DETECTORS

Conventional trace detectors are designed to respond to only energetic materials, e.g. nitrates, and not to detect other background chemicals in odours. Conversely electronic noses are designed to respond to all chemicals within an odour. Based upon this distinction an electronic nose might not be considered a good bomb detector where there are strong background odours.

J.W. Gardner and J.Yinon (eds.),
Electronic Noses & Sensors for the Detection of Explosives, 235-248.
© 2004 *Kluwer Academic Publishers. Printed in the Netherlands.*

236

On the other hand, trace detectors are not good electronic noses because they are blind to many important environmental and olfactory chemicals. However, the diversity of today's terrorist threats (e.g. explosive, chemical, and biological) makes it increasingly apparent that there is a complementary role for electronic noses with the ability to quickly learn and recognize threat vapours of all kinds.

Sensitivity is not the issue since electronic noses like the zNose® and trace detectors like ion-mobility spectrometers (IonTrack IMS) have essentially equal speed and sensitivity to explosive compounds. Specificity is the issue and users should understand the different role each type of instrument can undertake as part of an integrated security, force protection, or general law enforcement screening or investigative missions.

2.   NEW ANALYTICAL TOOL FOR ODOUR MEASUREMENTS, THE ZNOSE™

A new type of ultra-fast gas chromatograph, the zNose®, is able to perform analytical measurements of volatile organic vapours and odours in near real time with ppt sensitivity. Because of its picogram sensitivity it is a useful tool for detecting energetic materials (explosives) or volatile organics of any kind. The zNose® separates and quantifies the organic chemistry of odours through ultra-high speed chromatography in 10 seconds. Using a patented solid-state mass-sensitive detector, picogram sensitivity, universal non-polar selectivity, and electronically variable sensitivity have been achieved. An integrated vapour preconcentrator coupled with the electronically variable detector, allow the instrument to measure vapour concentrations spanning 6+ orders of magnitude. A portable zNose®, shown in Figure 1, is a useful tool for onsite odour and ambient air measurements.

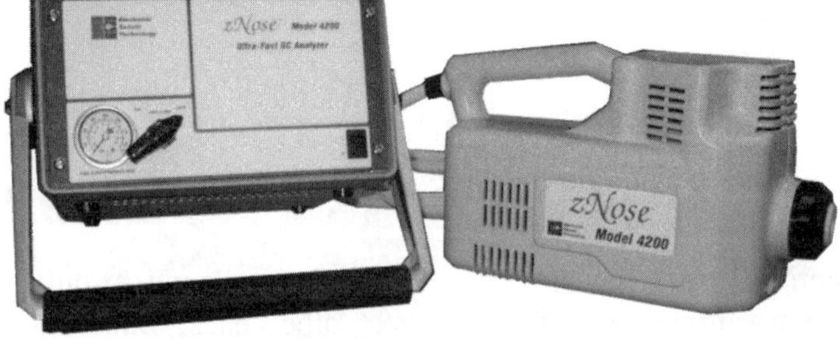

Figure 1. zNose™ technology has been incorporated into a portable handheld instrument with wireless connectivity and GPS interfacing.

3. HOW THE ZNOSE™ QUANTIFIES THE CHEMISTRY OF AN ODOUR

A simplified diagram of the zNose® system shown in Figure 2 consists of two sections. One section uses helium gas, a capillary tube (GC column) and a solid-state detector. The other section consists of a heated inlet and pump that draws ambient air into the instrument. Linking the two sections is a 'loop' trap, which acts as a preconcentrator when placed in the air section (sample position) and as an injector when placed in the helium section (inject position). Operation is a two-step process. Ambient air (odour) is first sampled and organic vapours collected (preconcentrated) on the trap. After sampling the trap is switched to the helium section where the collected organic compounds are injected into the helium flow. The organic compounds pass through a GC column with different velocities and exit the column at characteristic times. As they exit the column they are detected and quantified by a solid-state detector.

A high-speed programmable gate array (PGA) controls the processing of odour samples and includes electronic flow control, timing, electronic injection, and temperature control for the column, inlet, detector, and other parts of the instrument. The user interface can be a laptop or remote computer using a wireless modem (1.6 km range). A software program allows users to select appropriate measurement methods, interface with GPS receivers, and to identify specific energetic compounds found in explosives from a library of Kovats indices.

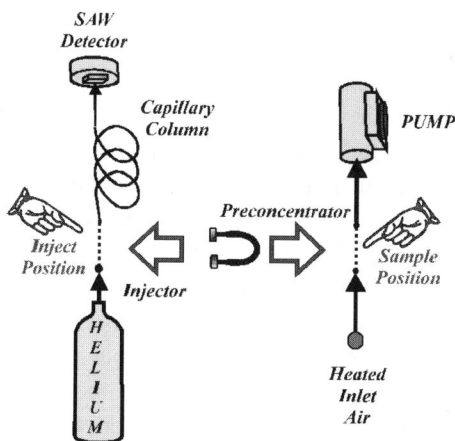

Figure 2. Simplified diagram of zNose® showing air section on the right and helium section on the left. A loop-trap preconcentrates organics from ambient air when in the sample position and injects collected organics into the helium section when in the inject position.

4. ENERGETIC COMPOUNDS

The chemistry of explosives involves what are called energetic compounds because they readily decompose with shock or high temperature. Some important characteristics of six common energetic compounds found in explosives are listed in Table 1.

Table 1. Energetic compounds commonly found in explosives.

Compound	CAS No.	Formula	Mol. Wt.	Density	Vapour pressure mmHg	Decomposition temp. (°C)
NG	55-63-0	$C_3H_5N_3O_9$	227.0872	1.6	4.0×10^{-3}	120
DNT	121-14-2	$C_7H_6N_2O_4$	182.1354	1.521	1.5×10^{-4}	300
TNT	118-96-7	$C_7H_5N_3O_6$	227.133	1.654	5.5×10^{-8}	240
PETN	78-11-5	$C_5H_8N_4O_{12}$	318.1378	1.77	1.2×10^{-8}	141
RDX	121-82-4	$C_3H_6N_6O_6$	222.117	1.82	4.1×10^{-9}	170
Tetryl	479-45-8	$C_7H_5N_5O_6$	287.1452	1.73	4.0×10^{-10}	220

The molecular structures of the six compounds are shown in Figure 3. Structures can be open or closed (aromatic) and molecular weights are typically above 200 Da. For detection purposes perhaps the most important characteristics are the vapour pressure and decomposition temperature. Low vapour pressure compounds tend to adhere to cool surfaces and require careful control of instrument temperatures. However, if temperatures are too high compounds like NG and PETN will decompose before they can be detected.

The lower vapour pressure of these compounds means their concentration in ambient air will also be low. If explosives are contained in an enclosure with cool surfaces the vapour concentration may be even lower than saturated values due to partitioning effects. The saturated equilibrium ambient air concentration of TNT, RDX, and PETN as a function of ambient temperature is shown in Figure 4. At room temperature there are approximately 100 picograms of TNT per mL available for detection. NG and DNT have even higher concentrations available for detection. However, PETN and RDX produce less than 1 picogram per mL and hence are much more difficult to detect as vapours. For these compounds it is easier to use a wipe to extract material from surfaces and then to desorb thermally the compounds as vapours into the detection system.

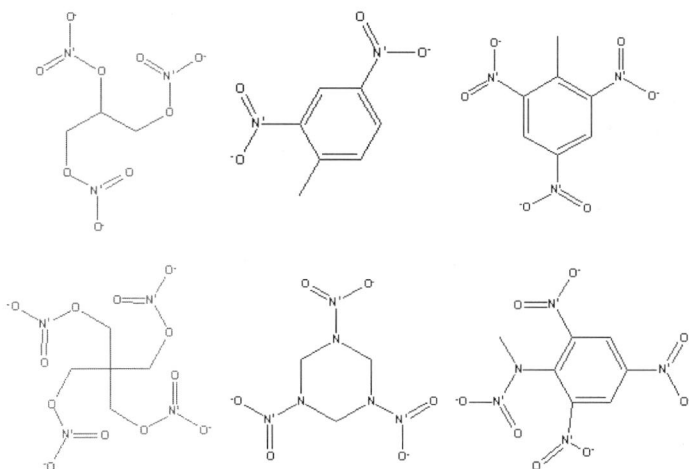

Figure 3. Chemical structure of nitroglycerine (NG), dinitrotoluene (DNT), trinitrotoluene (TNT), pentanerythritoltetranitrate (PETN), cyclotrimethylenetrinitramine (RDX), and trinitrophenyl-n-methylnitramine (Tetryl). Open structures (two on left) are more likely to decompose compared with closed aromatic ring structures.

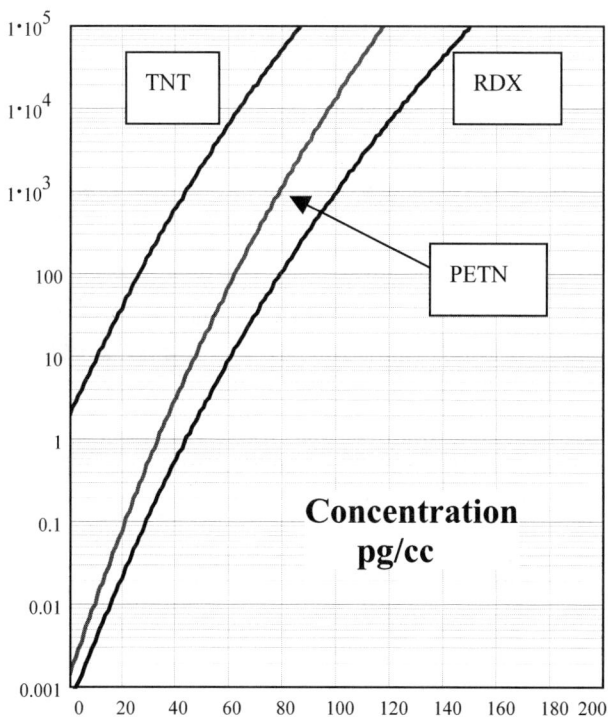

Figure 4. Saturated vapour concentration vs temperature (°C) of TNT, PETN, and RDX.

5.   CALIBRATION USING N-ALKANE ODOUR STANDARDS

An odour standard of n-alkane vapours is used to calibrate sensitivity and specificity of the zNose®. Specificity is what allows the instrument to recognize known chemicals and/or chemical groups (odour signatures) and to deliver the appropriate alarms. The zNose™ is an ultra-fast GC which separates and measures the concentration of the individual chemicals of an odour directly, typically in 10 seconds. Individual chemicals are recognized by their retention time relative to the retention times of the linear-chain alkanes. Tabulating the retention times and detector counts (cts) provides a complete and quantitative measure of any odour or fragrance.

To calibrate the instrument requires only that a known amount of each alkane be introduced into the instrument. For alkane numbers above 14 (and explosive compounds) this is best done by the three step process depicted in Figure 5. A glass tube is attached to the inlet and a known amount of alkanes dissolved in methanol are injected. A short drying (step 2) removes the volatile solvent vapours but leaves the semi-volatile alkanes (or explosives) attached to the walls of the tube. The final step is to heat the tube to vapourize the semi-volatiles and then collect them in the preconcentrator trap of the zNose®.

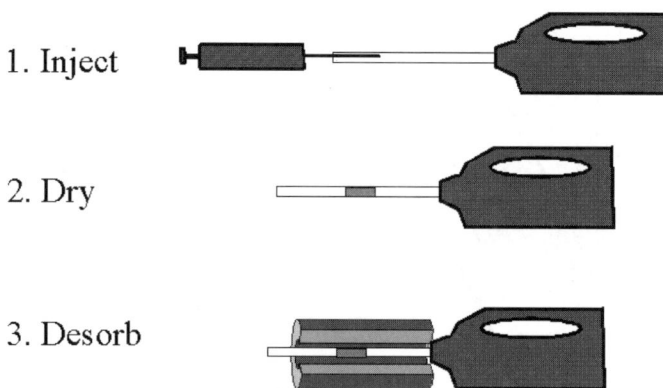

Figure 5. zNose® calibration process using n-alkanes.

Odours from explosives create chromatogram peaks whose retention times are indexed to the retention times of a standard vapour mixture of n-alkanes. As an example, shown in Figure 6 is the standard vapour response obtained using a calibration vapour containing C11 through C22 alkanes as well as DNT and TNT. Retention times are expressed as indices relative to the n-alkane peaks. In the chromatogram DNT has an index of 1537 (between C15 and C16) and TNT has an index of 1707 (slightly above C17).

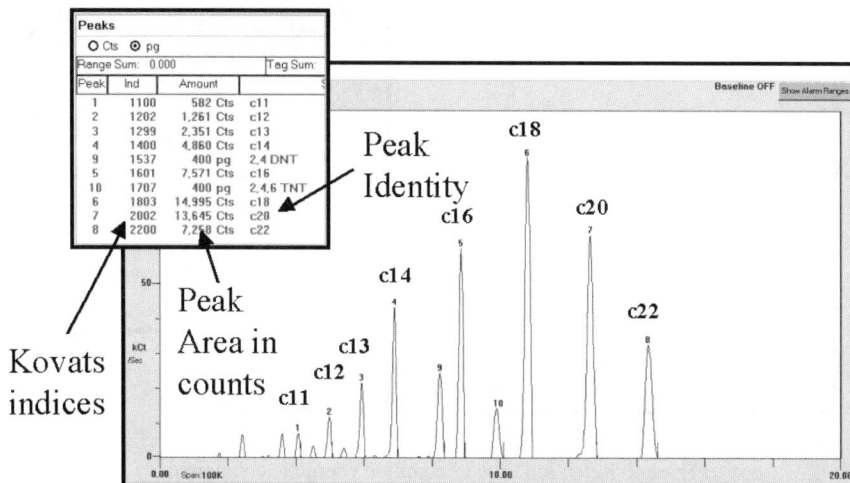

Figure 6. A 20 second chromatogram showing retention time indices of DNT, TNT, and n-alkanes.

6.   KOVATS INDICES OF COMMON EXPLOSIVES

The retention times and Kovats indices of energetic compounds were obtained by direct desorbtion with a methanol solution containing known concentrations and measuring the resulting odour chemistry with a zNose™ as shown in Figure 7. The Kovats indices for each of the 6 common explosives are given in Table 2.

242

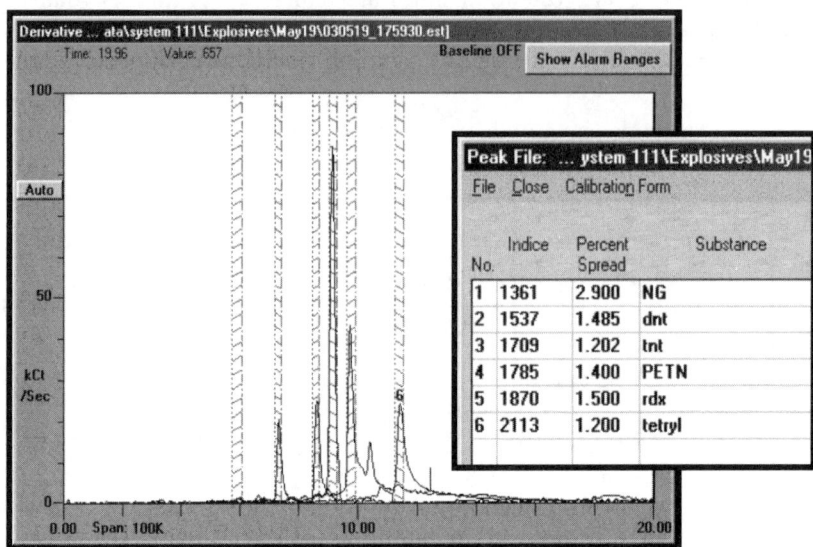

Figure 7. Retention time of common energetic compounds.

Table 2. Kovats indices of energetic materials.

Compound	CAS No.	Chemical Formula	Kovats Indices
NG	55-63-0	$C_3H_5N_3O_9$	1356
DNT	121-14-2	$C_7H_6N_2O_4$	1537
TNT	118-96-7	$C_7H_5N_3O_6$	1704
PETN	78-11-5	$C_5H_8N_4O_{12}$	1791
RDX	121-82-4	$C_3H_6N_6O_6$	1870
Tetryl	479-45-8	$C_7H_5N_5O_6$	2100

7. VIRTUAL CHEMICAL SENSORS WITH ALARMS

The indices, tabulated in Table 2, provide the basis for creating alarms or virtual sensors for each of the compounds. Because the retention times are relative to n-alkane vapours they are machine independent and only require knowledge of the n-alkane retention times to create a library of explosives, which applies to all zNose™ instruments. Tabulated chromatogram peak information, as shown in Figure 8, are used by the software to define retention times of peaks by index rather than time in seconds. Response factors of each sensor and alarm window width are also defined by the listing. Once defined, chromatographic measurements are reduced to a simple user display of six virtual sensors for the common energetic compounds together with their user selected alarm levels.

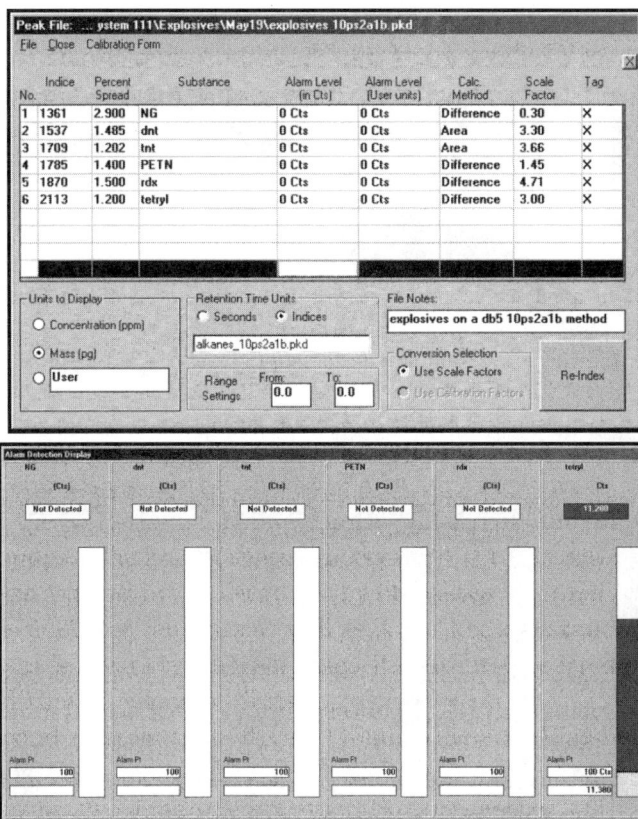

Figure 8. Once retention times and alarm levels are defined, chromatograms can be replaced by an array of virtual chemical sensors.

8. ODOUR CHEMICAL LIBRARIES

Computer processing of olfactory images quantifies the individual chemicals and allows the aggregate odour response to be recognized relative to a known odour standard, e.g. n-alkanes. The Kovats indices for known chemicals are stored in a library together with their odour description or perception. When an unknown odour is analyzed the retention time of peaks are converted to Kovats indices and clicking on individual peaks with a mouse pointer brings up the nearest library entry. The lookup process is illustrated in Figure 9 using the PETN 'peak' which has an index of 1776.

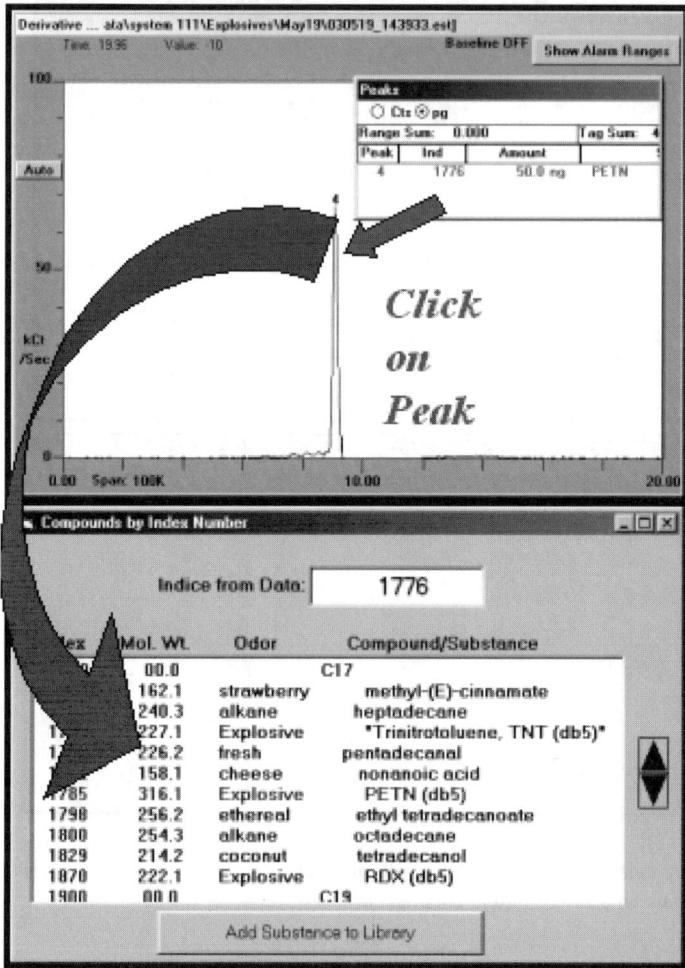

Figure 9. Machine independent library of Kovats indices assist in identification of peaks and odours.

9.   ODOURS FROM REAL WORLD EXPLOSIVES

Two common explosive materials are Primacord and Detasheet. PETN is the explosive core of Primacord, where it develops a velocity rate of 6,400 metres per second. Primacord is insensitive to friction and ordinary shock, but may be exploded by rifle fire. It also detonates sympathetically with the detonation of an adjacent high explosive. Odours from Primacord thermostated at 60° C were tested and the organic compounds detected confirmed the presence of only PETN as shown in Figure 10.

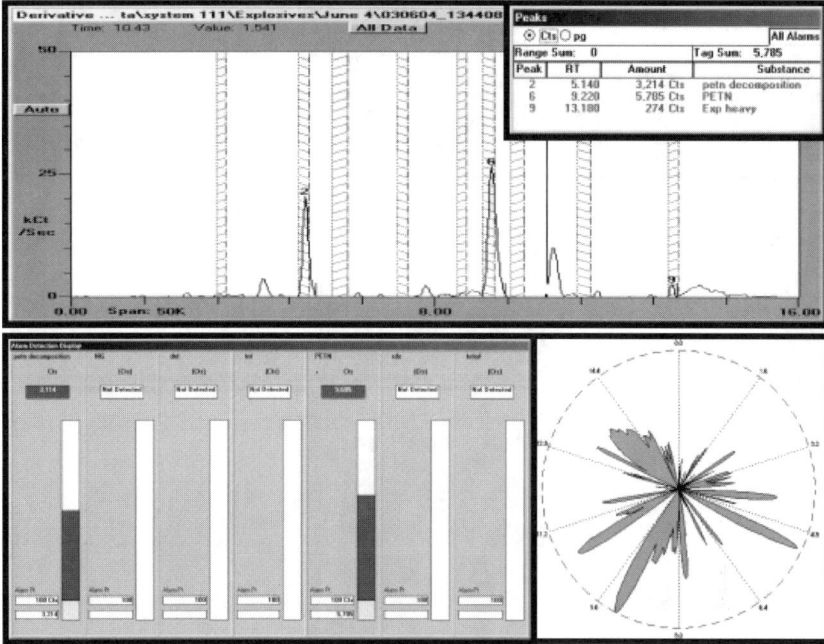

Figure 10. Odour chromatogram, Virtual sensor array, and olfactory image for Primacord.

Detasheet is a moulded and flexible explosive consisting of RDX, PETN and plasticizing wax. Volatile organics vapours from Detasheet confirmed the presence of both as shown in Figure 11. RDX is usually used in mixtures with other explosives, oils, or waxes; it is rarely used alone. It has a high degree of stability in storage and is considered the most powerful and brisant of the military high explosives. Incorporated with other explosives or inert material, RDX forms the basis for many common military explosives.

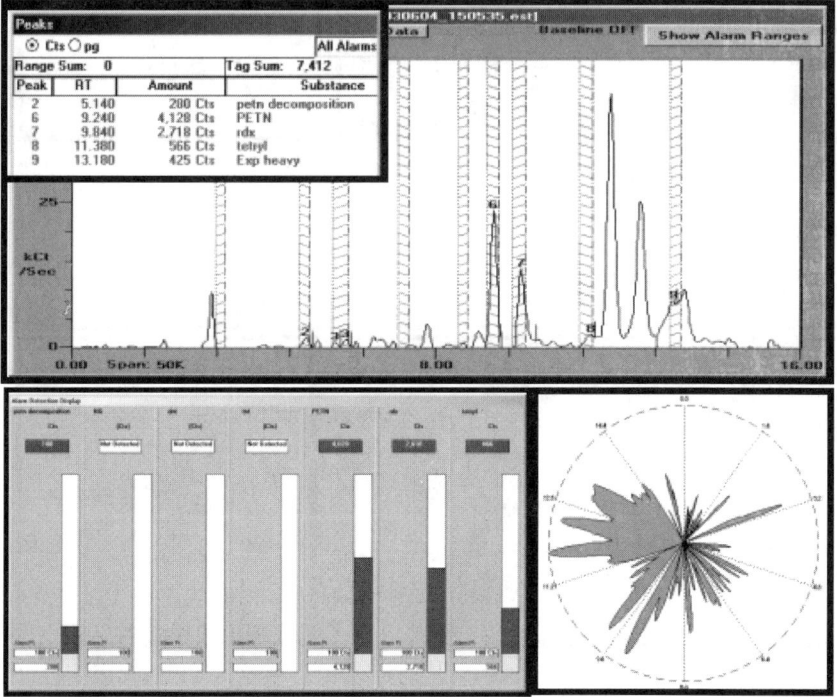

Figure 11. Odour chromatogram, Virtual sensor array, and olfactory image for Detasheet.

C4 is another well known explosive consisting of RDX, other explosives, and plasticizers. It can be moulded by hand for use in demolition work and packed into shaped charge devices. The odour of C4 explosive, shown in Figure 12, contains RDX, PETN, TNT, Tetryl and plasticizing wax. At room temperature RDX has a distinct odour consisting mainly of a very volatile taggant compound with an index of 1020.

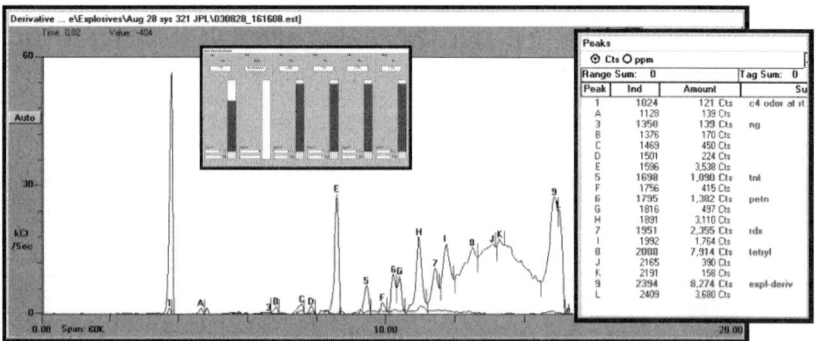

Figure 12. Odour chromatogram, quantification, and virtual sensor response for odours of C4.

In recent times because of restrictions placed upon conventional military explosives there has been an increase in the use of 'home made' bomb materials. Perhaps none is more notorious that triacetone triperoxide (TATP). This crystalline material can be produced from common acetone, peroxide, and citric acid, yet it has the explosive power of RDX. The most publicized account was that of Richard Reid, the shoe bomber, but TATP is often used by Hamas 'human' bombers in Israel.

TATP is a relatively volatile explosive and readily produces vapours that can be detected. Shown in Figure 13 is a typical chromatogram comparison which establishes the Kovats index for TATP of 1115 using a zNose® equipped with a db5 column.

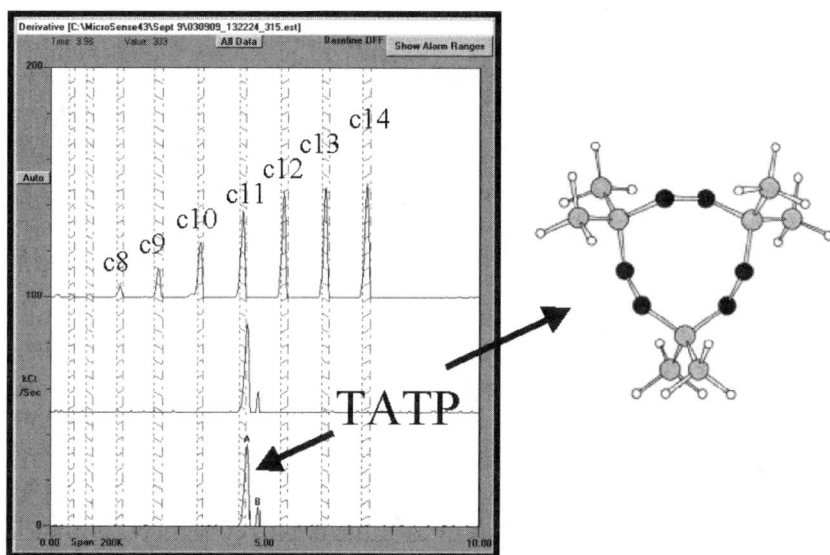

Figure 13. Replicate chromatograms from TATP odours compared with n-alkane response (top trace). This explosive contains no nitrates and often cannot be detected by trace detectors.

10. SUMMARY OF RESULTS

Gas chromatography is no longer considered 'slow'. A new type of electronic nose based upon ultra high-speed gas chromatography and a new solid state GC detector, now allows the chemistry of odours to be quantified in near real time with high precision, accuracy, and parts per trillion sensitivity. Odours from explosives were analyzed and compared using chromatograms and virtual chemical sensors for common energetic compounds. The sensitivity of the instrument allows compound concentrations at parts per trillion (pg/mL) levels to be made. Identification of explosive compounds is greatly simplified by indexing retention time to a single n-alkane odour standard. A library of indices allows unknown odours to be quickly analyzed and compared to known odour signatures.

Because the electronic nose is based upon the science of gas chromatography, odour measurements can be easily confirmed and validated by independent laboratory measurements taken on quality control samples. The ability to rapidly perform analytical measurements on odours of all kinds in real time will provide first responders with a cost effective new tool for identifying explosive, chemical, or biological threat odours.

Chapter 16

EXPLOSIVE VAPOUR DETECTION USING MICROMECHANICAL SENSORS

Thomas Thundat, Lal Pinnaduwage, and Richard Lareau[*]
Life Sciences Division, Oak Ridge National Laboratory, Oak Ridge, Tennessee, USA
*Transportation Security Administration, US Department of Homeland Security, Atlantic City, New Jersey, USA

Abstract: MEMS-based microcantilever platforms have been used to develop extremely sensitive explosive vapour sensors. Two unique approaches of detecting of explosive vapours are demonstrated. In the first approach a cantilever beam coated with a selective layer undergoes bending and resonance frequency variation due to explosive vapour adsorption. The resonance frequency variation is due to mass loading while adsorption-induced cantilever bending is due to a differential stress due molecular adsorption. In the second approach that does not utilize selective coatings for speciation, detection is achieved by deflagration of adsorbed explosive molecules. Deflagration of adsorbed explosive molecules causes the cantilever to bend due to released heat while its resonance frequency decreases due to mass unloading.

Keywords: Cantilever sensors, explosive vapour detection, resonance frequency, cantilever bending, adsorption-induced stress, nanodeflagration

1.   INTRODUCTION

Real-time detection of explosives is important for practical applications ranging from passenger baggage-screening to the disarming of landmines [1-7]. Explosive detection is one of the primary areas where the war on terrorism urgently requires new developments in sensor technology. Since the vapour pressures of many explosive substances are very low, a chemical detection technique with sub-parts per trillion is needed to detect explosives in real-time. Up to now, attaining this level of sensitivity required techniques

J.W. Gardner and J.Yinon (eds.),
Electronic Noses & Sensors for the Detection of Explosives, 249-266.
© 2004 *Kluwer Academic Publishers. Printed in the Netherlands.*

such as mass spectroscopy, nuclear magnetic or quadrupole resonance, that employed large and complex equipment. None of the technologies currently available offer a clear path to the development of an extremely sensitive, hand held, battery operated explosive detector that can be mass-produced.

Recently Micro Electro-Mechanical Systems (MEMS) have been emerging as sensor platform for the development of sensors with extreme high sensitivity [8-14]. Micromachined silicon cantilevers are the simplest MEMS sensors that can be micromachined and mass-produced. Microcantilever sensor technology is an upcoming sensing technique with broad applications in chemical, physical, and biological detection. With their compactness and potential low cost, detection techniques based on silicon-based cantilevers provide a path for the development of miniaturized sensors.

Microcantilever sensors offer many orders of magnitude better sensitivity compared to other sensors such as quartz crystal microbalances (QCM), flexural plate wave oscillators (FPW), and surface acoustic wave devices (SAW). There are several distinct advantages of the microcantilever sensors compared to the above mentioned and other MEMS sensors:

(i) Microcantilevers are quite small having surface area of the order of 10^{-5} cm^2 that is orders of magnitude smaller compared to the above devices,

(ii) Even though the other sensors have only one mode of detection (mass loading or gravimetric), the microcantilevers can be used in several modes, mass loading and bending being the most commonly used modes,

(iii) Silicon microcantilevers can be mass produced at relatively low cost using standard semiconductor manufacturing processes,

(iv) Studies conducted up to now have illustrated that microcantilever sensors have superior detection sensitivities for a number of chemical and biological species [15,16].

The typical dimensions of these microcantilevers are 50-200 μm long, 10-40 μm wide and 0.3-2 μm thick. The mass of the microcantilever is a few nanograms. The primary advantage of the microcantilever method originates from its sensitivity that is based on the ability to detect cantilever motion with sub-nanometer precision as well as the ease with which it may be fabricated into a multi-element sensor array.

Microcantilever sensors do not offer any intrinsic chemical selectivity. Chemical speciation is accomplished by coating the cantilevers with chemically selective layers. A number of chemically selective polymeric coatings have been developed for achieving chemical selectivity in SAW and QCM [17,18]. Chemical selectivity using self-assembled monolayers (SAM) has been reported for microcantilever sensors [19]. We will discuss only SAM-based chemical speciation in this chapter.

2. MODES OF OPERATION

Microcantilever sensors can be operated in many different modes. One of the operational modes of the microcantilever sensor is very similar to that of QCM and SAW devices where the adsorption produces changes in resonance frequency. However, unlike QCM and SAW devices, microcantilevers can also detect adsorption-induced forces by monitoring cantilever bending which is many orders of magnitude more sensitive than approaches based on resonance frequency variation for the cantilevers with resonance frequency in tens of kHz. Microcantilevers that are metal-coated on one side are very sensitive to temperature and undergo static bending as a result of slight variations in temperature due to the bimetallic effect. Gimzewski *et al.* have reported the use of microcantilever bending for calorimetric detection of chemical reactions at the pico- and femto-Joule level, corresponding to a 10^{-6} °C temperature variation of the cantilever [10].

2.1 Resonance frequency approach

Microcantilevers have a fundamental resonance frequency of vibration, v, that is dependent on its spring constant, k, and effective mass, m^*, as shown in equation (1).

$$v = \frac{1}{2\pi}\sqrt{\frac{k}{m^*}} \tag{1}$$

If the spring constant remains constant, shifts in resonance frequency can provide information on the mass adsorbed. The spring constant for a rectangular cantilever is given by equation (1).

$$k = \frac{Ewt}{4l^3} \tag{2}$$

where E is Young's modulus, w the width, t the thickness and l the length of the cantilever. Shorter cantilevers can give better sensitivity for frequency shifts.

2.2   Cantilever bending approach

In addition to resonance frequency variation, cantilevers undergo bending due to molecular adsorption when adsorption is confined to a single side of the cantilever. Microcantilever deflection varies sensitively as a function of adsorbate coverage. The exact mechanism of adsorbate-induced bending on 'real surfaces' still remains to be solved. Microcantilevers, such as those used in our experiments, have an intrinsic deflection due to unbalanced stresses on the opposing surfaces. For example, cantilevers with thin films (many atomic layers) of another material show bending. Although bending can be expected for films of many atomic layers due to differences in physical parameters such as elastic and lattice constants, bending due to submonolayer coverage as small as 10^{-3} monolayers (ML) is not intuitive. One monolayer of gold surface has 1.5×10^{15} atoms/cm^2 while on Si(111) surface the density is 7.4×10^{14} atoms/cm^2. Therefore, 10^{-3} ML corresponds to approximately 10^{12} atoms on the surface of the cantilever.

One way to explain the cantilever bending is by using free energy changes involved in adsorption. Change in free energy is equivalent to change in the surface stress. Surface stress is one of the most significant properties of a solid surface. However, measuring surface stress is extremely difficult. In addition, measuring variation in surface stress is even more challenging. Due to these technical challenges, accurately measuring the variation in surface stress still remains unexplored.

Microcantilever deflection changes as a function of adsorbate coverage when adsorption is confined to a single side of a cantilever (or when there is differential adsorption on opposite sides of the cantilever). Using Stoney's formula and equations of bending of a cantilever, a relation can be derived between the cantilever bending and changes in surface stress. Since we do not know the absolute value of the initial surface stress, we can only measure the variation in surface stress. The surface stress variation between top and bottom surface of a cantilever can be written as:

$$\Delta\sigma_1 - \Delta\sigma_2 = \frac{zEt^2}{4L^2(1-v)} \tag{3}$$

where, z is the cantilever deflection, E is the Young's modulus, L is the cantilever length, t is the thickness and v is the Poisson's ratio. Since all the quantities on the right hand side can be measured (or known *a priori*), the changes in surface stress due to adsorption can be calculated.

Surface stress, σ, and surface free energy, γ, can be related using the Shuttleworth equation [20],

$$\sigma = \gamma + \frac{\partial \gamma}{\partial \varepsilon} \qquad (4)$$

where σ is the surface stress. The surface strain $\partial \varepsilon$ is defined as ratio of change in surface area, $\partial \varepsilon = dA / A$. Since the bending of the cantilever is very small compared to the length of the cantilever, the strain contribution is only in the ppm (10^{-6}) range while the surface free energy changes are in the 10^{-3} range. Therefore, one can easily neglect the contribution from surface strain effects and equate the free energy change to surface stress variation [21].

2.3   Calorimetric approach (deflagration)

If adsorption or desorption of molecules involves variation in temperature, it can be detected by the bending of a bi-material cantilever. Since the thermal mass of a cantilever is extremely small they can be heated to hundreds of degrees centigrade in a millisecond. Rapid heating of a cantilever with adsorbed explosive molecules results in deflagration of adsorbed molecules [22]. It should be stated that scanning force microscopy and optical microscopy have shown that adsorbed explosive molecules form islands. The sizes of these circular islands (drops) increase with increased adsorption. Deflagration of these explosive droplets results in cantilever bending that can be monitored using an optical beam deflection method.

3.   APPARATUS

Detection of explosive vapours using resonance frequency and bending approaches requires coating the cantilevers with chemically selective coatings. The deflagration approach, however, does not require any chemically selective coatings. All the experiments were conducted using a modified atomic force microscope head where an optical beam deflection was used to monitor cantilever motion. The experimental setup used for the measurements is shown in Figure 1. A modified atomic force microscope (AFM) head measured the bending response of the microcantilever. The light from a laser diode was focused at the apex of the cantilever. The reflected laser beam was allowed to fall on a position-sensitive detector (PSD).

The key to achieving chemical selectivity using microcantilevers is the ability to functionalize one surface of the silicon microcantilever so that explosive molecules will be preferentially bound to that surface upon exposure to vapour. Choosing coatings that can provide highest affinity, therefore, can control the selectivity of detection. Another important requirement for a sensor system is fast recovery (sensor reversibility), so that the sensor can be used repetitively.

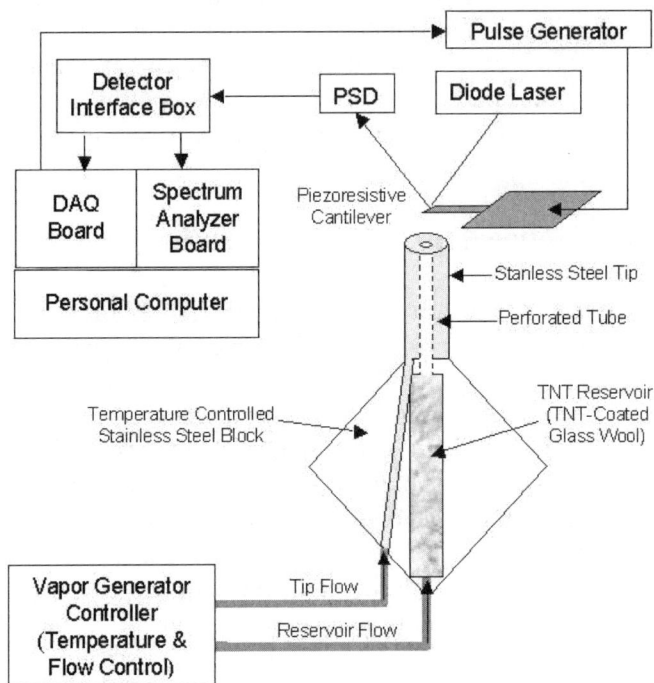

Figure 1. Schematic diagram of the experimental apparatus used in the present experiments.

One way to achieve chemical selectivity is modifying the cantilever surface with self-assembled monolayers (SAM) with functional head groups that will bind to explosive molecules. Significant acidity (pK_a of 5–7) has been measured for carboxyl-terminated SAMs on gold substrates [23]. Such acidic surfaces were expected to strongly bind basic nitro-substituted molecules of explosive vapours via hydrogen bonding [24]. SAMs of 4-mercaptobenzoic acid (4-MBA; also known as thiosalicylic acid) on gold are stable and efficiently provide surface carboxyl (-COOH) groups for conjugation to enzymes, antibodies, or antigens [25-26]. The conformational rigidity and specific surface orientation of 4-MBA aromatic SAMs prevent intermolecular hydrogen bonding and dimerization. In addition, earlier scanning tunneling microscopic studies of 4-MBA SAMs have found domains with periodic row structure and no hole defects [27]. All these features of 4-MBA SAMs make them suitable for highly sensitive detection of explosives [28].

To detect explosive molecules, we functionalized silicon cantilevers with 4-MBA SAMs. The cantilevers had typical dimensions of 180 μm length, 25 μm width, and 1 μm thickness (Park Scientific Instruments, Inc.) and a spring constant of 0.26 N/m. One side of the cantilever had 30 nm thick gold film with a 3 nm titanium adhesion layer. The formation of a 4-MBA SAM on the gold surface of the cantilever was achieved by immersing the cantilever into a 6×10^{-3} M solution of 4-MBA (97%, from Aldrich Chemical Company) in absolute ethanol for two days. Upon removal from the solution, the cantilever was rinsed with ethanol and then dried before use in the experiments. The monolayer coating was shown to be quite stable for several months under normal operating conditions.

The experimental arrangement used in the present experiments is shown in Figure 1. The microcantilever was held in place in a vacuum-tight glass flow cell by a spring-loaded wire. A modified atomic force microscope head measured the bending response of the microcantilever. The light from a laser diode was focused at the apex of the cantilever (the gold-coated side on which the monolayer was deposited). The reflected laser beam was allowed to fall on a position-sensitive detector (PSD). The output from the PSD was amplified and normalized through a homemade electronics box and fed to a Stanford Research System model SR 760 FFT spectrum analyzer (for resonance frequency measurements) and a Hewlett Packard model 34970A data logger (for bending measurements). This allowed simultaneous measurement of bending and resonance frequency. The flow rate of gas around the cantilever was kept constant to minimize bending due to gas flow.

A vapour generator developed at Idaho National Engineering and Environmental Laboratory (INEEL) was used to generate the PETN, RDX, and TNT vapour streams. Flowing ambient air through a reservoir containing PETN TNT or RDX generated the vapour stream. The reservoir consisted of 0.1 g of PETN, TNT or RDX dissolved in acetone and deposited on glass wool contained in a stainless steel block. The temperature of the reservoir was controlled using two thermoelectric elements that cooled or heated the reservoir, generating a saturation vapour pressure within the reservoir. When the explosive vapour stream was turned off, the same carrier stream was redirected to bypass the reservoir and was sent through the cantilever flow cell. This allowed maintaining a constant flow rate through the cantilever chamber and eliminated any cantilever bending due to changes in flow rate. Thus the cantilever was always subjected to a stream of ambient air, in this case at a flow rate of 100 standard cubic centimetres per minute (sccm).

Even though the data presented here were taken with the vapour generator tip sealed at the cantilever flow cell, we observed similar results even when the vapour generator tip was several millimeters below the flow cell. For the experiments involving RDX and PETN, both vapour generators were operated at 50°C. At this temperature, the vapour concentrations for PETN and RDX were 1400 ppt and 290 ppt, respectively.

Deflagration experiments were conducted only for TNT vapours. For deflagration experiments we have used a piezoresistive microcantilver mounted in the AFM head using a home made receptacle that provided electrical connections to the conducting channel in the cantilever. The piezoresistive track on the cantilever was used for passing the current to heat the cantilever. No attempt was made to measure the cantilever deflection using piezoresistive effect. Instead, optical beam deflection was used to determine the bending response of the cantilever. The output from the PSD was amplified and normalized through a homemade electronics box and fed to a PCI-MIO-16-E1 data acquisition board and a MI 4551 dynamic signal acquisition card (both installed in the personal computer) from National Instruments (NI). This allowed simultaneous measurement of bending and resonance frequency. The explosive vapour generator described before was used for producing controlled amounts of TNT vapour. The tip of the vapour generator was placed about 5 mm below the cantilever for optimum TNT loading. Commercially available piezoresistive microcantilevers from Thermomicroscopes, Inc. (model PLCT-NTSA; resistance $\approx$ 2.2 kΩ), were used for these experiments. Stanford Research model DS 345 function generator provided the 10V heating pulse with $\approx$ 50ns rise time; however, quantitatively similar data were later acquired by using a 10V pulse (with 100 µs rise time) from the PCI-MIO-16-E1 data acquisition board.

Exposing cantilevers to TNT vapours at ambient conditions led to adsorption of TNT onto the untreated, uncoated cantilever surfaces. Of particular advantage to the proposed sensor is the unique, intrinsic 'stickiness' of the TNT molecule in the presence of humidity. From optical and AFM microscopy it was observed that TNT forms circular, 'pancake like' islands on the cantilever surface. The thickness of the islands increased with increasing time of exposure to TNT vapour. The mass of adsorbed TNT was calculated by the corresponding decrease in the cantilever's resonance frequency. After the desired amount of TNT is adsorbed on the cantilever, a voltage pulse from the NI card (rise and fall times of the order of 100 µs) of 10 Volts and 10 ms duration was applied across the piezoresistive cantilever, which led to rapid heating of the cantilever and subsequent deflagration of TNT deposited on the cantilever. The deflagration event was captured on video with a Hitachi model WKC 150 MOS colour video camera.

4. RESULTS AND DISCUSSIONS

4.1 Cantilever bending response

Figure 2 shows the cantilever bending and resonance frequency variation when a SAM coated cantilever is exposed to PETN. As seen from Figure 2, the bending response of the cantilever to the PETN exposure is extremely sensitive and fast. Since the noise level of the bending response in these experiments is ≈ 2 nm (3× standard deviation of the noise level), the detection sensitivity corresponding to Figure 2 is ≈ 14 ppt. Maximum bending of the cantilever is achieved within 20 s. The amount of PETN delivered by the generator in 20 s is ≈ 660 pg. However, the mass of PETN exposed to the cantilever is much smaller, since the cross-sectional area of the hole in the delivery tube of the vapour generator is ≈ 0.07 cm^2 and the cantilever surface area is ≈ 8×10^{-5} cm^2. Allowing for several wall bounces, it can be estimated that a few picograms of PETN impinging on the cantilever in 20 s is sufficient to yield a 200 nm deflection of the cantilever. Since the minimum detection level above the noise level is a few nanometers, a low-femtogram (10^{-15} g) level of LOD is implied. It should be noted that with a vapour preconcentrator, the detection capabilities of a given sensor are better described by LOD (based on the amount of explosive material) than by vapour concentration.

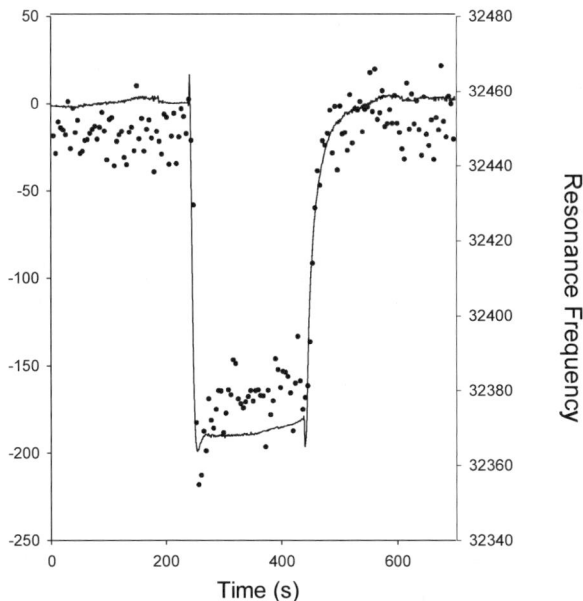

Figure 2. The response of a 4-mercaptobenzoic acid (4-MBA)-coated silicon cantilever to a PETN vapours of 1.4 ppb concentration in ambient air.

The solid curve depicts the bending response, and the dots depict the resonance frequency of the cantilever. The frequency shift due to the adsorption of PETN vapour corresponds to a mass loading of 15 pg on the cantilever.

The bending and frequency responses of the SAM-coated cantilever to an RDX vapour stream with a concentration of 290 ppt are shown in Figure 3. In this case the frequency response, and thus the mass loading, is quite small. Yet the bending response is quite clear, with a 'direct' detection sensitivity of $\approx$ 30 ppt. The maximum cantilever bending is achieved within $\approx$ 25 s, and the mass of RDX delivered by the generator during this time is $\approx$ 96 pg. Again, if we compare the cantilever cross-sectional area with that of the vapour generator, this corresponds to $\approx$ 0.1 pg delivered to the cantilever, for a total cantilever deflection of 20 nm, and thus an implied LOD of a few femtograms.

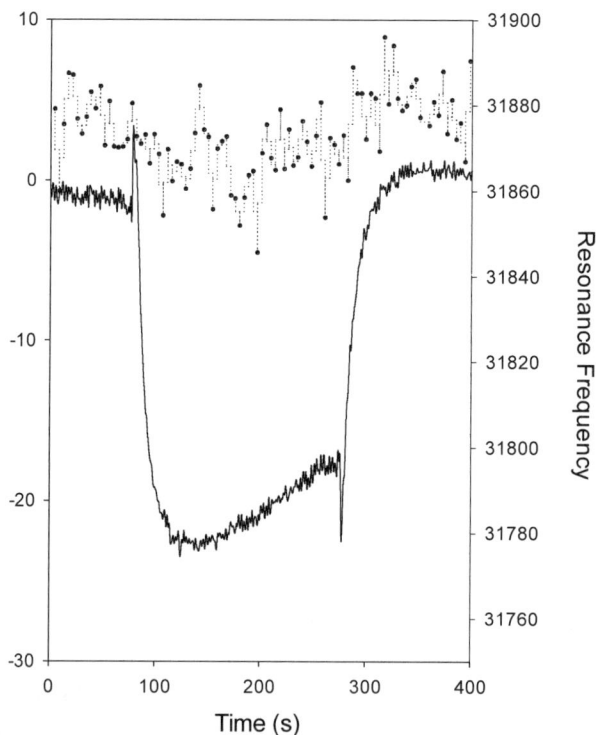

Figure 3. The response of a 4-mercaptobenzoic acid (4-MBA)-coated silicon cantilever to a RDX vapours of 290 ppt concentration in ambient air. The solid curve depicts the bending response, and the dots depict the resonance frequency of the cantilever. The frequency shift due to the adsorption of RDX vapour is barely discernible.

The rapidity with which explosive vapours can be detected and the relatively fast relaxation of the cantilever when the vapour stream is turned off is shown in Figure 4. When the PETN stream is turned on for 10 s, a 40 nm deflection signal is observed. When the vapour stream is turned off, the cantilever is relaxed back almost to the original position within 60 s. Another important observation from the data shown in Figure 4 is that the resonance frequency of the cantilever does not change significantly as a result of the small amount of PETN deposited in 10 s. The bending of the cantilever is still quite easily detected.

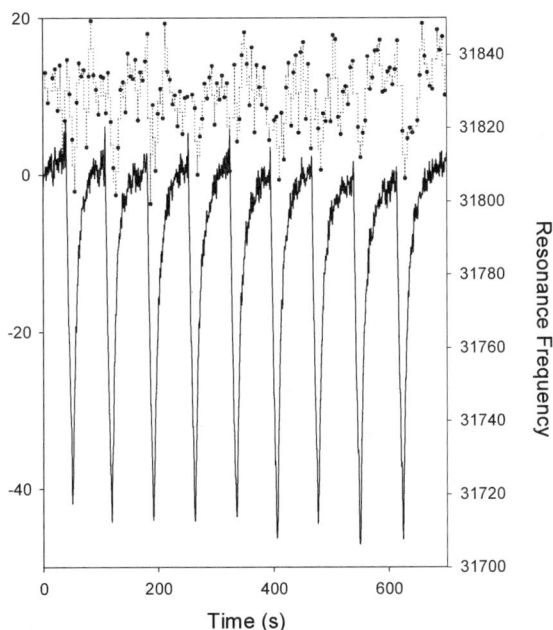

Figure 4. The response of a 4-mercaptobenzoic acid (4-MBA)-coated silicon cantilever to the periodic turning on (10 s) and off (60 s) of a PETN vapours of 1.4 ppb concentration in ambient air. The solid curve depicts the bending response, and the dots connected by dashed lines depict the resonance frequency of the cantilever.

The mechanism for cantilever bending is assumed to be adsorption-induced stress. The adsorption decreases the surface free energy and surface free energy is density is surface stress. It is speculated that the hydrogen bonding between the nitro groups of the explosives molecules and the hydroxyl group of 4-MBA is responsible for the easily reversible adsorption of explosive vapours on the SAM-coated top surface of the cantilever. This hydrogen bonding creates a differential surface stress since hydrogen bonding is confined to only one of the surfaces.

4.2 Deflagration

For deflagration experiments the cantilevers were first exposed to controlled amounts of TNT vapour from the vapour generator. The amount of TNT adsorbed on the cantilever was calculated from the resonance frequency shift using Equation (1). The variation in the resonance frequency of the cantilever due to TNT adsorption is shown in Figure 5. The initial mass of the cantilever was calculated to be approximately 130 ng.

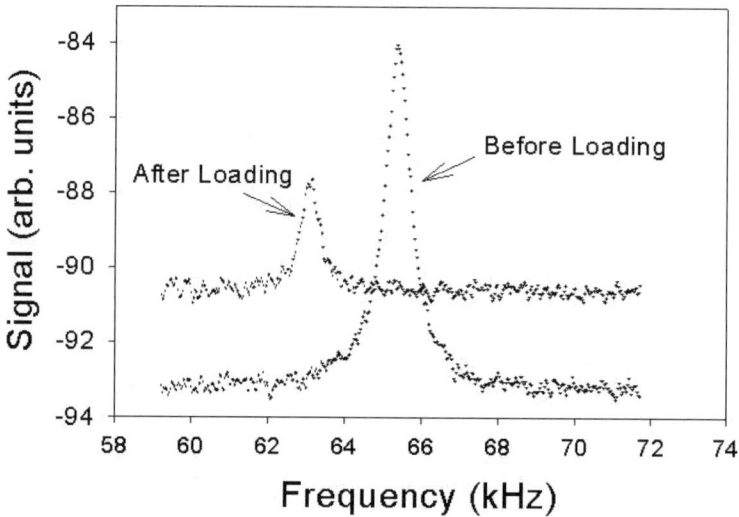

Figure 5. The resonance curves of the cantilever before and after loading with TNT. With TNT loading, the resonance frequency shifted to a lower value due to the increased mass. After deflagration, the resonance frequency returned to the original value.

Applying a voltage pulse of 10 V across the piezoresistive cantilever (corresponding to a current of $\approx$ 5 mA) heated the cantilever to $\approx$ 500^{0}C. This corresponds to temporal-time gradient of 5×10^6 $^{\circ}$C/s. This rapid heating resulted in cantilever bending, presumably due to the difference in expansion coefficients of the doped and un-doped regions. Typical deflection signals from the cantilever, with and without TNT deposited on it, are shown in Figure 6. The solid curve of Figure 6 represents the bending of the cantilever in the absence of TNT on it. From Figure 6 it is clear that the cantilever comes to thermal equilibrium during heating within about a milliseconds, and the bending signal levels off; when the voltage pulse is ended, the cantilever comes back to room temperature within about a millisecond and the bending relaxes back to its original position. (The rise time of the voltage pulse is ≈ 100 µs).

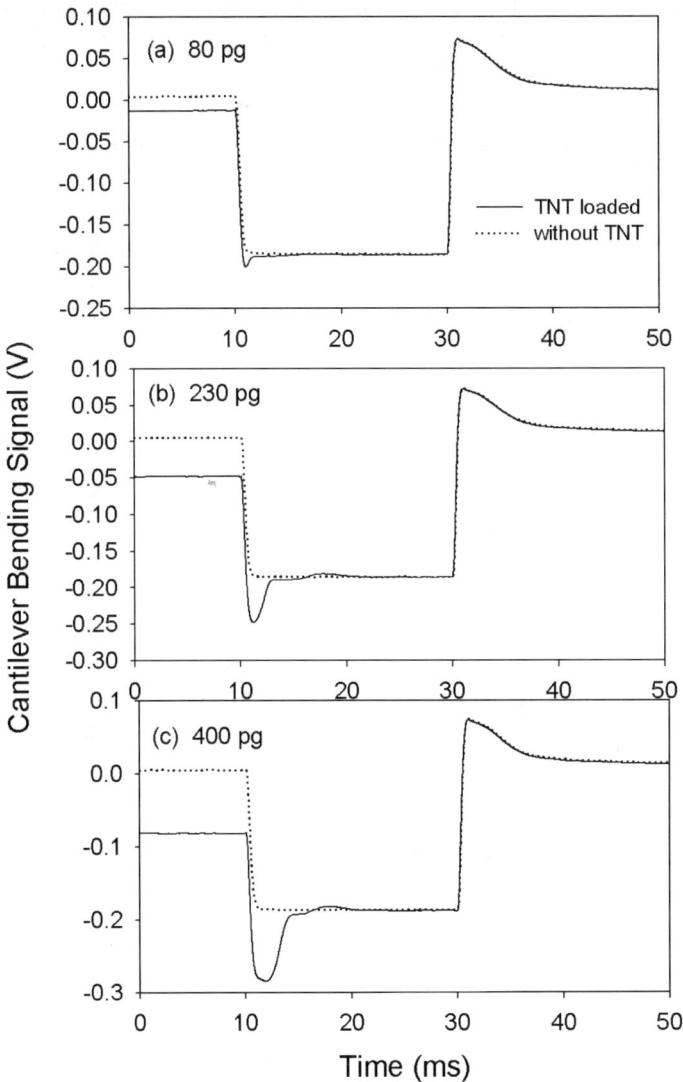

Figure 6. Bending response of a cantilever (as measured by the voltage output from a position sensitive detector) to applied voltage pulse with and without TNT adsorbed on the surfaces. The bending of the uncoated cantilever follows the time profile of the applied voltage pulse (except the lengthening of the rise and fall times) and is presumably due to the difference between the thermal expansion coefficients of silicon and the doping material. The exothermic nature of the TNT deflagration event is clear due to the enhancement in bending of the cantilever.

When a voltage pulse is applied to the cantilever loaded with TNT, the rapid rise in temperature of the cantilever leads to the deflagration of the TNT. This deflagration results in a rapid, exothermic reaction. The heat released during deflagration induces an *additional bending* of the cantilever, which appears as a 'bump' during the fast heating regime. This additional bending is shown as a dotted curve in Figure 6 [29]. Since the amount of heat released can be expected to increase with the mass of TNT deposited on the cantilever, we can expect the area of the 'bump' to increase linearly with the mass of TNT deposited on the cantilever. As described earlier, the mass of deposited TNT on the cantilever can be deduced from the shift in resonance frequency. A plot of the area of the 'bump' versus TNT mass measurement is shown in Figure 7 [29]. The linear behaviour shown in Figure 7 is consistent with the notion that adsorbed TNT is responsible for the observed bump.

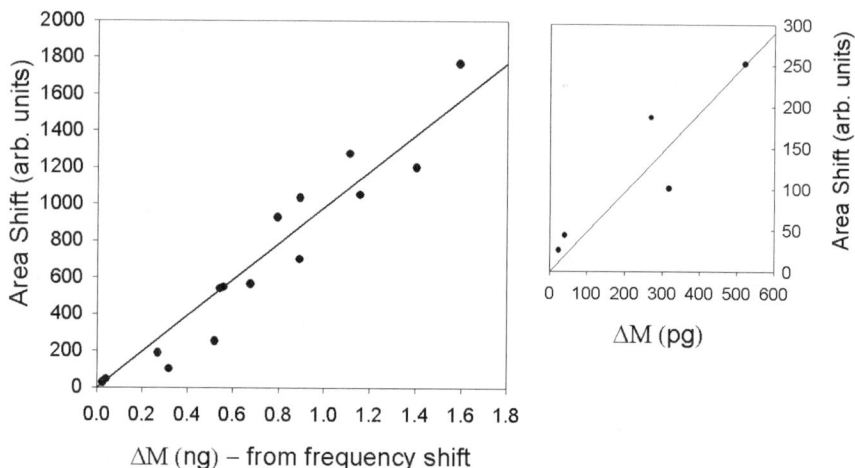

Figure 7. Relationship between the area under the curve obtained from the peaks shown in figure 2 and the mass of adsorbed TNT calculated from the observed changes in resonance frequency of the piezolever.

Various reactions are possible on the cantilever surface due to the presence of TNT. These include melting, vapourization and decomposition of TNT. All these reactions are endothermic except deflagration. Since the resonance frequency of the cantilever returns to its original value after deflagration it can be assumed that all the adsorbed TNT vanished from the surface. This rules out melting which occurs at 83°C for TNT as the sole cause of the observed effect. This conclusion is supported by images captured with a high-speed camera that correspond to the point in time that increased bending is observed and the change in resonant frequency takes place.

We can compare the heat expected to be released by the deflagration of 6.9 ng of TNT in Figure 5 to the heat generated by the voltage pulse during the 'heating phase'. At the rising edge of the voltage pulse, it takes about 1 ms for the cantilever to reach the steady state. Since the amplitude of the voltage pulse is 10V and its resistance is ≈ 2.2 kΩ, the heat generated during 1 ms is ≈ 45 μJ. Since the heat of explosion for TNT is ≈ 4560 J/g, the heat released by the deflagration of 6.9 ng of TNT is ≈ 31 μJ. These numbers are in reasonable agreement with the bending signal generated by the deflagration event and that due the voltage pulse in Figure 3. Therefore, the 'additional bending' of the cantilever with TNT deposited on it can be most certainly attributed to the deflagration event.

There are two important factors that determine the usefulness of a detection technique: sensitivity and selectivity. The sensitivity of the present technique can be estimated as follows. In the experiments that have been conducted so far, the minimum amount of TNT that we were able detect on the cantilever is ≈ 50 pg; see Figure 4. Since the sticking coefficient for TNT on the cantilever surface at low TNT concentrations is ≈ 0.1, the cantilever needs to be exposed to 500 pg or so of TNT to be detected. Therefore, the current Limit of Detection (LOD) for this technique is better than a nanogram. In the actual experiment, we probably had 700 times more TNT that came out of the generator but the cantilever did not 'see' most of it (the delivery tube had a cross section of ≈ 0.07 cm^2, compared to a cross sectional area of $\approx 10^{-4}$ cm^2 for the cantilever).

In the present vapour detection scheme, some possible interferences that pertains – for example to conditions at an airport terminal – are water vapour, gasoline vapour, and common solvents such as acetone and alcohols. It was found that none of these interfere with our detection method. We have conducted a comprehensive study of the desorption characteristics of TNT and several other explosive vapours and the above-mentioned common interferences from cantilever surfaces. This study showed that nanogram quantities of TNT takes tens of minutes to desorb from cantilever surfaces, while much larger amounts of the above interferences are desorbed within seconds. Therefore, such interferences will not lead to a 'signal' in the present technique. In addition, during rapid heating these interferents showed an 'endothermic bump' corresponding to removal of heat from the cantilever. This is consistent with the notion that, since there is no exothermic reaction involved in this case, and heat is lost from the cantilever in the vapourization of water, thus producing a signal that is opposite in direction with that of TNT.

Important factors to be considered in using this technique for detecting TNT are sensitivity and selectivity. Ability to detect about 50 picograms of TNT present on the piezolever has already been demonstrated while improvement in this value can be achieved by designing a small spring constant cantilever that can be heated to a temperature high enough for initiating deflagration.

5.   CONCLUSIONS

We have demonstrated two novel techniques of detecting explosives using microcantilevers. The first method utilizes cantilevers with selective coatings. Experiments conducted using a selective coating of a self-assembled monolayer of mercaptobenzoic acid show sensitivity in the parts per trillion ranges. The second method is based on deflagration of adsorbed explosive molecules by rapid heating of the cantilever. The amplitude of deflagration signal depends on the amount of explosive molecules adsorbed on the cantilever. For TNT adsorption on a cantilever with a spring constant of 30 N/m, a deflagration signal can be discerned when a 50 pg of TNT is adsorbed on the cantilever. Higher sensitivity could be achieved using cantilevers with smaller spring constants.

ACKNOWLEDGMENTS

The experiments reported here were conducted with the help of Drs. V. Boiadjiev, F. Tian, G. Muralidharan, and D. Hedden, J. E.Hawk, T. Gehl, and D. Yi. This work was supported by the National Safe Skies Alliance, the Department of Homeland Security, and the Bureau of Alcohol Tobacco, and Firearms (ATF), and ORNL. Oak Ridge National Laboratory is managed by UT-Battelle, LLC, for the U. S. Dept. of Energy under contract DE-AC05-00OR22725.

REFERENCES

1 R.J. Colton and J.N. Russell, Making the world a safer place, Science 299 (2003) 1324-1325.
2 R.G. Ewing and C.J. Miller, Detection of volatile vapours emitted from explosives with a handheld ion mobility spectrometer, Field Analytical Chemistry and Technology, 5 (2001) 215-221.
3 A. Fainberg, Explosive detection for aviation security, Science 255 (1992) 1531-1537.

4   K.G. Furton and L.J. Myers, The scientific foundation and efficacy of the use of canines as chemical detectors for explosives, Talanta 54 (2001) 487-500.

5   A.N. Garroway, M.L. Buess, J.B. Miller, B.H. Suits, A.D. Hibbs, G.A. Barrall, R. Matthews and L.J. Burnett, Remote sensing by nuclear quadrupole resonance, IEEE Trans. on Geoscience and Remote Sensing 39 (2001) 1108-1118.

6   C.M. Harris, The science of detecting terror, Analytical Chemistry, 74 (2002) 127A.

7   J.I. Steinfeld and J. Wormhoudt, Explosives detection: a challenge for physical chemistry, Annual Review of Physical Chemistry, 49 (1998) 203-232.

8   R. Berger, F. Delamarche, H.P. Lang, C. Gerber, J.K. Gimzewski, E. Meyer and H.J. Guntherodt, Surface stress in the self-assembly of alkanethiols on gold, Science 276 (1997) 2021-2024.

9   G.Y. Chen, T. Thundat, E.A. Wachter and R.J. Warmack, Adsorption-induced surface stress and its effects on resonance frequency of microcantilevers, Journal of Applied Physics 77 (1995) 3618-3622.

10 J.K. Gimzewski, C. Gerber, E. Meyer and R.R. Schlittler, Observation of a chemical reaction using a micromechanical sensor, Chemical Physics Letters 217 (1994) 589-594.

11 T. Thundat, R.J. Warmack, G.Y. Chen and D.P. Allison, Thermal and ambient-induced deflections of scanning force microcope cantilevers, Applied Physics Letters 64 (1994) 2894-2896.

12 T. Thundat, G.Y. Chen, R.J. Warmack, D.P. Allison and E.A. Wachter, Vapour detection using resonating microcantilevers, Analytical Chemistry, 67 (1995) 519-521.

13 T. Thundat, P.I. Oden and R.J. Warmack, Microcantilever sensors, Microscale Thermophysical Engineering 1 (1997) 185-199.

14 E.A. Wachter and T. Thundat, Micromechanical sensors for chemical and physical measurements, Review of Scientific Instruments, 66 (1995) 3662-3667.

15 J. Fritz, M.K. Baller, H.P. Lang, T. Strunz, E. Meyer, H.J. Guntherodt, E. Delamarche, C. Gerber and J.K. Gimzewski, Translating bimolecular recognition into nanomechanics, Science 288 (2004) 316-318.

16 G.H. Wu, R.H. Datar, K.M. Hansen, T. Thundat, R.J. Cote and A. Majumdar, Bioassay of prostate-specific antigen (PSA) using microcantilevers, Nature Biotechnology 19 (2001) 856-860.

17 J.W. Grate and M.H. Abraham, Solubility interactions and the design of chemically selective sorbent coatings for chemical sensors and arrays, Sens. Actuators B 3 (1991) 85-111.

18 R.A. McGill, M.H. Abraham and J.W. Grate, Choosing polymer coatings for chemical sensors, Chemtech 24 (1994) 27.

19 Y.M. Yang, H.F. Ji and T. Thundat, Nerve agents detection using a Cu^{2+}/L-cysteine bilayer-coated microcantilever, Journal of the American Chemical Society, 125 (2003) 1124-1125.

20 R. Shuttleworth, The surface tension of solids, Proc. Phys. Soc.(London) 63A (1950) 444-457.

21 H.J. Butt, A sensitive method to measure changes in the surface stress of solids, Journal of Colloid and Interface Science 180 (1996) 251-260.

22 L.A. Pinnaduwage, A. Gehl, D.L. Hedden, G. Muralidharan, T. Thundat, R.T. Lareau, T. Sulchek, L. Manning, B. Rogers, M. Jones and J.D. Adams, A microsensor for trinitrotoluene vapour, Nature 425 (2003) 474.

23 Z. Dai and H.X. Ju, Effect of chain length on the surface properties of omega-carboxy alkanethiol self-assembled monolayers, Physical Chemistry Chemical Physics, 3 (2001) 3769-3773.

24 E.J. Houser, T.E. Mlsna, V.K. Nguyen, R. Chung, R.L. Mowery and R.A. McGill, Rational materials design of sorbent coatings for explosives: applications with chemical sensors, Talanta 54 (2002) 469-485.

25 S.E. Creager and C.M. Steiger, Conformational rigidity in a self-assembled monolayer of 4-mercaptobenzoic acid on gold, Langmuir 11 (1995) 1852-1854.

26 S. Susmel, C.K. O'Sullivan and G.G. Guilbault, Human cytomegalovirus detection by a quartz crystal microbalance immunosensor, Enzyme and Microbial Technology 27 (2000) 639-645.

27 A. Ulman, An Introduction to Ultra-thin Organic Films, Academic Press: New York, 1991.

28 L.A. Pinnaduwage, V. Boiadjiev, J.E. Hawk and T. Thundat, Sensitive detection of plastic explosives with self-assembled monolayer-coated microcantilevers, Applied Physics Letters 83 (2003) 1471-1473.

29 L.A. Pinnaduwage, A. Wig, D.L. Hedden, A. Gehl, D. Yi and T. Thundat, Detection of trinitrotoluene via deflagration on a microcantilever, Journal of Applied Physics, (2004) submitted.

Chapter 17

SOLID PHASE MICROEXTRACTION FOR FIELD SAMPLING

Krishna C. Persaud, Peter Wareham and Anna Maria Pisanelli
Department of Instrumentation and Analytical Science, UMIST, Manchester, UK

Abstract: Solid phase micro extraction sampling methods have been evaluated and adopted to develop a robust method of sampling trace volatiles that can be applied to an array of gas sensors, for use in the field to recognize a condition.

Key words: Solid phase micro extraction, field sampling, volatiles, gas sensing, electronic nose

1. INTRODUCTION

Solid phase micro extraction (SPME) fibres have improved in quality over the last few years and are now commonly used for introducing analytes into a gas chromatograph [1,2]. The technique utilises a one cm length of fused silica coated with an adsorbent or absorbent phase. This allows preconcentration of organic compounds from a volatile or liquid sample that are adsorbed or absorbed in the fibre. The fibre is inserted into a heated injector port of a gas chromatograph where the analytes are thermally desorbed directly onto the column. The fibres are made up of a variety of coatings, of which the most common ones are shown in Table 1.

SPME is a multiphase equilibration process. Analytes, from the matrix in which the fibre is exposed to, will be transported into the coating of the SPME fibre as soon as the fibre is exposed. The extraction time can be considered complete when the analyte concentration has reached equilibrium between the sample matrix and the fibre coating. When using this in experimental conditions, it means that when the equilibrium is reached the

267

J.W. Gardner and J.Yinon (eds.),
Electronic Noses & Sensors for the Detection of Explosives, 267-277.

amount of the extracted analytes is constant and independent of an increase in extraction time.

We are interested in trace volatile sampling and measurement with emphasis on odour measurement. Commercially about 11 SPME fibre types are available such as those from Supelco™, but in the case of odour (trace volatile) detection in air, we have found that three fibre types are most applicable. These comprise 100 μm polydimethylsiloxane (PDMS), 65 μm polydimethylsiloxane/divinylbenzene (PDMS/DVB) and 75 μm carboxen/polydimethylsiloxane (CAR/PDMS) types. Our previous work has investigated the use of both 100 μm PDMS and 65 μm PDMS/DVB fibre types, and we have adopted the latter after initial screening studies.

Table 1. Common SPME fibre coatings.

Coating Material	Acronym
Polydimethylsiloxane	PDMS
Polydimethylsiloxane/divinylbenzene	PDMS/DVB
Polyacrylate	PA
Carbowax/divinylbenzene	CW/DVB
Polydimethylsiloxane/carboxen	PDMS/CAR
Polydimethylsiloxane/divinylbenzene/carboxen1006	PDMS/DVB/CAR
Carbowax/Templated resin	CW/TPR

As part of a project aimed at producing an 'electronic nose' instrument that can be used in the field for odour detection and measurement, we have developed a fully portable prototype system that has been tested in the field.

Electronic nose instruments are well documented [3,4]. The core technology is based on an array of gas sensors that are broadly selective (sensitive) to a range of odour volatiles. Different odours have differing volatile constituents, to which the sensors react producing differing response patterns. By comparing unknown response patterns to those obtained from known odours, an 'electronic nose' instrument may be trained to recognise particular odours. The instrument we have devised is based on an array of metal oxide semiconductor gas sensors coupled to electronics for data acquisition, using an external portable computer for data analysis.

In an environmental application the consistent collection and delivery of odour volatiles into an electronic nose becomes problematic. A number of factors contribute to this namely, inherent environmental variables such as temperature and humidity and physical parameters such as site accessibility to the target sampling area and airborne particulates. In addition, site-specific background interference odours and low concentration target volatiles also add to the complexity of the problem.

As a method to circumvent many of these sampling problems we have exploited solid phase micro extraction (SPME) techniques to produce a novel approach to instrument sample delivery. Use of a hand-held sampling device and commercial SPME fibres allows a user to track volatiles from relatively inaccessible site locations.

Our aims are to be able to carry out field based measurements, identify and discriminate low concentrations of volatiles in varying background environments, using an array of gas sensors used as detectors.

2. METHOD DEVELOPMENT

2.1 Investigation of SPME fibre characteristics

For initial work we utilised a conventional SPME fibre holder manufactured by Supelco[TM] that is conventionally used for headspace sampling for gas chromatography (Figure 1) that consists of two parts. At the heart of the device is a fibre, housed within a hollow needle sheath. At the tip of the fibre is a length of bonded coating material that absorbs/adsorbs volatile species, enabling a sample of a headspace (or aqueous phase) to be retained. The fibre retracts within the needle for mechanical protection. The fibre is exposed and retracted for sampling and sample delivery by means of a simple plunger arrangement, into which the fibre and needle are mounted.

In particular we were interested in parameters that affect sampling methodology such as temperature variation, humidity variation, sampling time, interference volatiles, as well as SPME types.

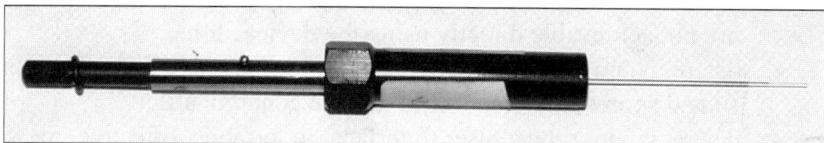

Figure 1. Commercial SPME fibre holder (Supelco[TM]).

Inter and intra-sample temperature variation is potentially the biggest interference to achieving desired reproducibility whilst using the SPME sampling protocol. Inter-sample temperature variation has a marked affect on the sampling sensitivity of SPME, due to the temperature-dependant nature of distribution coefficients (ratio of concentrations across the fibre/headspace boundary) for target analytes. Thus, sensitivity (extraction efficiency) towards a particular target analyte is likely to vary and thus cause differences in sample presentation to the sensor array. This can be compensated for using a set of predictive equations, but given the inherent multi-component nature of a field-based sample matrix, such compensation is likely to be of limited use. We found that the sampling sensitivity of SPME devices is generally independent of humidity variation. In the case of PDMS, sufficient water absorption to change polymer polarity producing a slight change in an analyte distribution coefficient is only apparent at humidity values approaching 100%.

While SPME sampling of volatiles in a controlled environment is quite reproducible, and numerous papers confirm this [3], we were more interested in the variability that may occur in a field sampling regime. We investigated the repeatability of measurements in a situation that is not well controlled – the combustion products of a fire. Figure 2 shows replicate gas chromatograms of combustion volatiles from three paper fires, sampled with a polydimethylsiloxane/divinylbenzene (PDMS/DVD) SPME fibre. For clarity the chromatograms are offset by 10 minutes. It can be seen that the peaks are quite repeatable, though the quantitative adsorption of the volatiles may vary somewhat from measurement to measurement. This gives some idea of the repeatable performance of such fibres in uncontrolled environments.

2.2 Adapting SPME sampling to field use

Commercial SPME sampling devices have a number of inherent problems when deployed in a hostile sampling environment:

- SPME devices are inherently fragile and susceptible to mechanical damage during deployment.
- There is a need to ensure repeatable sampling conditions, which are not achievable directly using the device alone.
- There is no facility for filtering airborne contaminants (dust, fungal spores) that could lead to fibre contamination
- There is no robust user interface to provide effective SPME deployment by semi-skilled operatives inherent with the device.
- There is no easy traceability for fibre deployment, fibre identification and tracking.

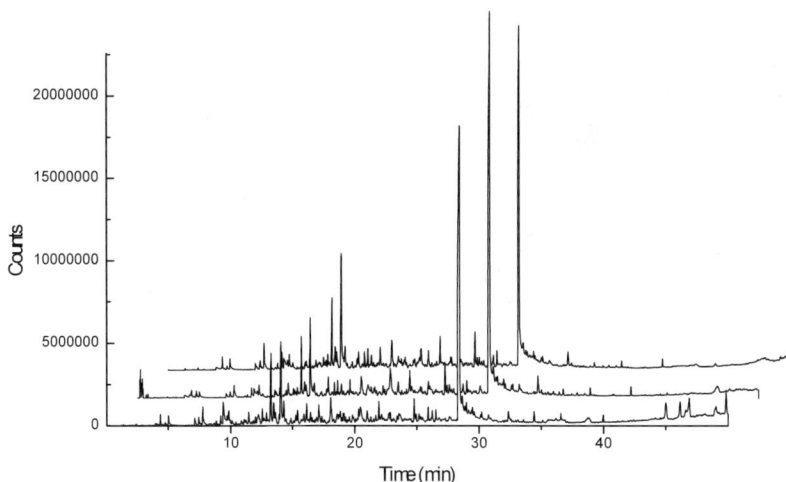

Figure 2. Replicate gas chromatograms of volatiles from combustion of paper sampled with a polydimethylsiloxane/divinylbenzene (PDMS/DVD) SPME fibre.

We adapted the sampling system of Figure 1 so that it could be used for controlled sampling in a field environment. Figure 3 shows the addition of a particulate filter together with a small sampling pump that could be used to control the time to which the fibre was exposed to the environment. The operational mode of the sampler is as follows: The probe end of the assembled sampler is placed in the immediate location of interest. The SPME device is loaded into the sampler and the pump (powered by a rechargeable battery pack) started. At the beginning of the required sampling time, the SPME fibre is exposed within the sampler to trap odour components from the sample air stream. At the end of the required sampling time, the SPME fibre is retracted and the SPME device removed from the sampler. The sampler can remain *in situ* for repeat samples or be transferred to any other location desired.

For a nominally air/trace volatile headspace, extraction equilibration times are generally short, of the order 20 to 100 seconds, the latter time corresponding to compounds with relatively low volatility. These times apply if the sampling system is dynamic (agitated system), which is the case with the modified field sampling device. Increasing air flow rates over the fibre has the effect of decreasing the equilibration time of low volatility compounds only, for a given sampling time of 300 seconds; it is likely that all target volatiles will be at equilibrium.

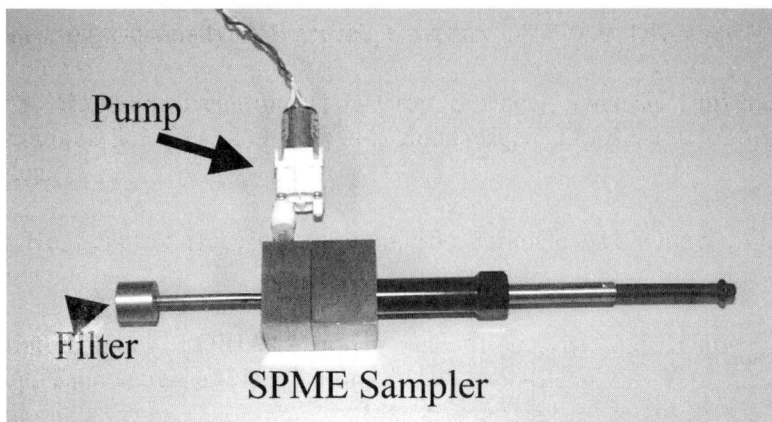

Figure 3. Modified field sampler.

2.3   Interface to an electronic nose

The volatile loaded SPME fibre requires heating so that desorption can occur for subsequent analysis. SPME devices are normally used in combination with bench-top analytical instruments with integrated desorption interfaces such as the injector port of a gas chromatograph. For sample delivery into an 'electronic nose', we have taken a different approach. The metal oxide semiconductor sensors used within the instrument operate at elevated temperatures, powered by individual heater elements on each sensor. We have mounted the sensors in a symmetrical 3-dimensional array within a stainless-steel header block. The array encloses a small headspace volume that is rapidly heated to elevated temperatures by the sensors themselves. The header block allows the introduction of the SPME fibre directly into this volume. The sample volatiles are thus desorbed directly into the enclosed region between the sensors as shown in Figure 4. The electronics used within the instrument for interrogation of the sensors are custom designed, based on a resistance measurement circuit. We have integrated both sensor measurement and sensor heater control circuits on a single printed circuit board (PCB). Additionally a micro-controller is used to provide the interface to an external computer system. Data is acquired at a rate of 1 measurement (of all 8 sensors) per second. This gives sufficient resolution to enable detailed analysis of sensor response profiles.

Parameters that are important to achieve reproducible results include the condition of fibre coating (damage, contamination by high molecular weight semi-volatiles), sampling temperature, sampling time (if non-equilibrium conditions apply), time between extraction and analysis (volatile bleed from fibre), volatile adsorption on intermediate surfaces, fibre positioning within header, condition within header (physical contaminants, debris etc).

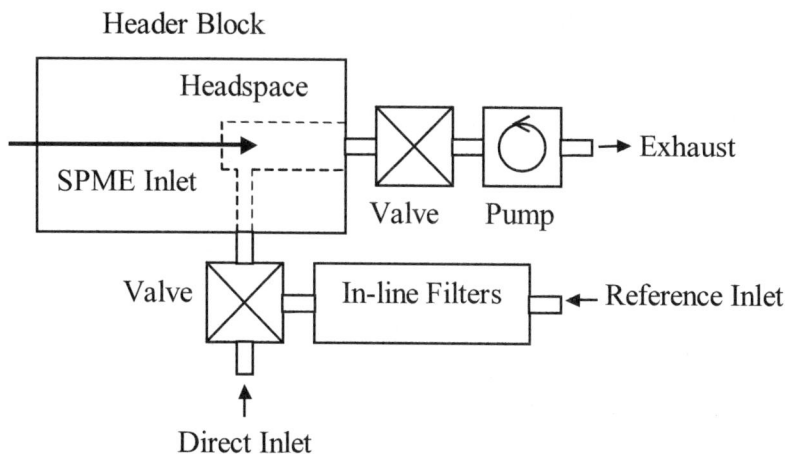

Header Block

Headspace

SPME Inlet

Valve   Pump   → Exhaust

Valve   In-line Filters   ← Reference Inlet

↑
Direct Inlet

Figure 4. Interface to an electronic nose.

3. RESULTS

3.1 Sensor responses

Figure 5 shows the dynamic response of each sensor in the array as it senses the volatiles desorbed from the fibre. The general profile reflects a combination of factors inherent within the measurement system namely; desorption characteristics of the volatile species from the SPME fibre, diffusion of the volatiles through the static headspace between the fibre and the sensors and the response characteristics of the sensors to the volatiles. Different portions of the general response profiles are potentially of use in terms of resolving differences between samples and thus classifying sample types.

274

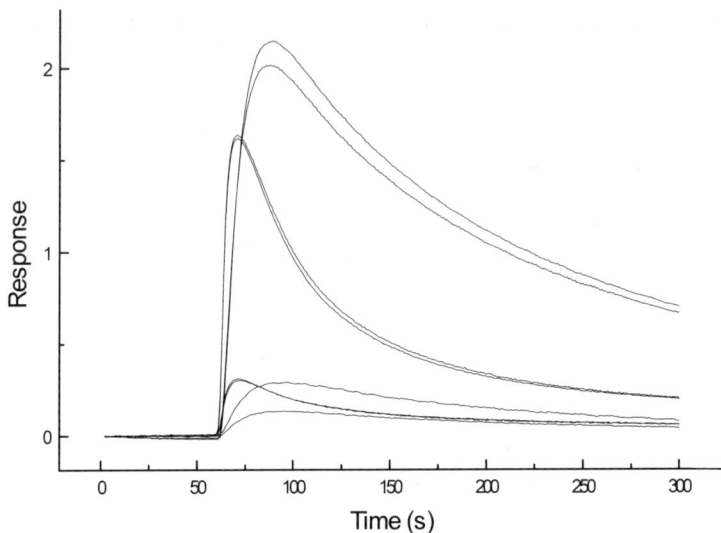

Figure 5. Sensor response profiles.

3.2 Detection of mouldy odour (dry rot)

For field use we were interested in detection of mouldy odour in buildings where the presence of dry rot is suspected caused by the fungus *Serpula lacrymans.* In this case the chemical concentration levels in the vapour phase may be in ppb range, and it is essential to be able to preconcentrate samples before analysis.

3.2.1 Identification of the chemical components

Both samples of cultured fungus, as well as samples of taken from a building were analysed using gas chromatography/mass spectrometry using different SPME fibre types. Samples were placed in a 20 ml headspace vial. In the case of cultured samples, samples were placed in a larger wide-mouthed container, with the headspace extracted via a pre-drilled septum. The Teflon–lined septum covering the vial was pierced and the fibre was exposed to the sample headspace for 30 min. The fibre was then retracted into the needle and immediately transferred to the gas chromatograph and desorbed for 5 min in the GC injection port. Compounds were resolved on a ZB-1701 column (30 m × 0.25 mm × 0.25 µm) under the following conditions: injection port temperature 250°C; helium flow 2.5 ml/min; oven temperature program 40°C (4 min hold) then 4 °C/min to 220°C.

The results are summarised in Table 2. The presence of 1-octen-3-ol (mushroom odour) and 3-octanone (resinous odour) appear to be common to all samples of dry rot, and to a slightly lesser extent, 3-octanol is also identified. Where dry rot fungus is present on pine wood substrate, additional compounds also appear in the headspace including alpha-pinene (pine odour), hexanal (cut grass odour) and carene (lemon odour), likely to be originating from the wood itself. Different fibres types had different characteristics and were able to adsorb some compounds more selectively. CAR/PDMS yielded noticeably larger peaks for most components.

Table 2. Serpula lacrymans analysis

SPME Fibre type	CAR/PDMS	PDMS/DVB	PDMS
Cultured *Serpula lacrymans*	3-Octanone 1-Octen-3-ol	3-Heptanone 3-Octanone 1-Octen-3-ol 3-Octanol	3-Octanone 1-Octen-3-ol Octenal Octanol 2-Octen-1-ol
Cultured *Serpula lacrymans* on pine wood	3-Cycloheptanone Hexanal alpha-Pinene 2-Heptanone Carene 3-Octanone 1-Octen-3-ol 3-Octanol	alpha-Pinene 3-Octanone 1-Octen-3-ol 3-Octanol	
Dry rot sample recovered from an infected building	alpha-Pinene Carene 3-Octanone 1-Octen-3-ol 3-Octanol	3-Octanone 1-Octen-3-ol	
Dry rot sample recovered from a second infected building	3-Octanone 1-Octen-3-ol 3-Octanol	3-Octanone 1-Octen-3-ol 3-Octanol	

3.2.2 Differentiation of field collected samples using an electronic nose

Volatile samples were collected and analysed in the field using the sampling system in conjunction with an array of metal oxide gas sensors. The sensors were calibrated using vapour from an ethanol/water solution and this was used as a reference. A sample of room air was used as a control – background odour. The SPME sampler was introduced to the suspect area, and after sampling the fibre was desorbed into the heated sensor block. When many response patterns are compared it is difficult to visualise differences between sample types. Principal components analysis (PCA) is a useful way of visualizing multidimensional data [5]. It is a method of reducing multidimensional data to lower dimensions based on the variance between individual patterns. What is observed is a general clustering or grouping of data points in areas of the graph according to the type of sample. This indicates that defined differences in sensor response patterns to each sample type are present, which may be used as a basis for recognition and classification of sample type. Thus, if a sensor response pattern for an unknown sample were to lie in an area of the graph bounded by points of a known sample, we can predict with a certain probability what that unknown sample is likely to be, based on previous knowledge.

Figure 6 shows a PCA analysis representing the results of one such field visit. The plot shows three separate clusters representing the reference, background and the volatile sample with a mouldy odour identified as dry rot. These data have been replicated in many sites, and show promise that a robust instrument for field use can be achieved.

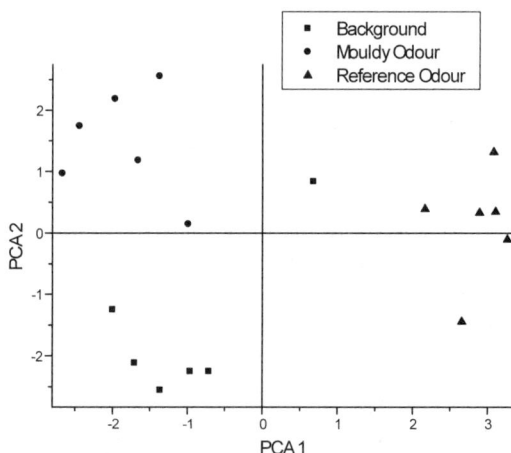

Figure 6. Principal components analysis of results from field sampling of volatiles.

4. CONCLUSION

An investigation of several SPME fibres using gas chromatography/mass spectrometry have shown them to be repeatable when used appropriately for sampling ill-controlled volatiles in the environment. To remediate the mechanical fragility of the fibres and to achieve reliable sampling in the field, an effective and portable sampling and sensing system for volatile chemicals in trace quantities has been developed and tested. SPME sampling of volatiles from otherwise inaccessible locations in the field proved to be a viable and successful method. The techniques developed can be used to improve sampling and sensing for a variety of environmental monitoring applications apart from that described; these may also include narcotics, explosives or contraband where trace quantities of volatile substances need to be detected and analysed, but where traditional analytical systems lack the necessary sensitivity.

ACKNOWLEDGEMENTS

Peter Wareham was funded through a DTI Link project and EPSRC UK, in collaboration with: ECOLAB (Terminix), Stockport, UK. Anna Maria Pisanelli was funded through IST-2001-38404 project IMOS, through the European Commission, Brussels. We are extremely grateful to Kevin Masterson of ECOLAB for his support throughout the project.

REFERENCES

1 T. Gorecki, A. BoydBoland, Z.Y. Zhang and J. Pawliszyn, 1995 McBryde Medal Award Lecture: Solid phase microextraction - A unique tool for chemical measurements, Canadian J. Chem. Revue Canadienne de Chimie 74 (1996) 1297-1308.
2 K.J. James, The determination of volatile organic compounds in soils using solid phase microextraction with gas chromatography mass spectrometry, HRC – J. High Resolution Chromatography 19 (1996)515-519.
3 T.C. Pearce, S.S. Schiffman, H.T. Nagle, and J.W. Gardner (eds.), Handbook of Machine Olfaction, Wiley-VCH: Weinheim, (2003), 592pp.
4 K.C. Persaud, Olfactory System Cybernetics: Artificial Noses, in Handbook of Olfaction and Gustation, 2nd edition, Richard L. Doty (ed.), Marcel Dekker (2001) ISBN 0-8247-0719-2.
5 I. Borg and P. Groenen, Modern Multidimensional Scaling: Theory and Applications, Springer-Verlag: New York, (1997).

Chapter 18

A DIGITAL MICROFLUIDICS PLATFORM FOR MULTIPLEXED EXPLOSIVE DETECTION
Colorimetric Approach

Vamsee K. Pamula
Department of Electrical and Computer Engineering, Duke University, Durham, North Carolina, USA

Abstract: Portable and automated field screening equipment would be very effective in assessing contamination due to explosives at many defense sites. A droplet based microfluidic lab-on-a-chip utilizing electrowetting has been presented for fully automated detection of TNT. Microliter droplets of TNT in DMSO and KOH in water are reacted on a chip in a programmed way. The same platform has integrated colorimetric detection. The detection of TNT is linear in the range of 12.5–50 µg/mL.

Key words: TNT, electrowetting, microfluidics, colorimetric

1.   INTRODUCTION

Contamination of soil, groundwater, and seawater with explosives at formerly used defense sites (FUDS) poses a serious environmental threat. Many of these sites were involved in munitions manufacturing and storage. Such contamination is a result of a variety of disposal practices due to manufacturing spills, unexploded ordnance (UXO), lagoon disposal of explosive-contaminated wastewater, and excess propellants among other causes. Although a number of munitions facilities have discontinued on-site waste disposal, they still possess high levels of soil and groundwater contamination. Even at very low parts per billion concentrations, explosives pose a health risk to humans and the environment [1]. The most common explosive, 2,4,6-trinitrotoluene (TNT), is suspected to be a carcinogen besides being highly toxic for humans, plants, and animals.

J.W. Gardner and J.Yinon (eds.),
Electronic Noses & Sensors for the Detection of Explosives, 279-288.

These contaminated environments need to be remediated due to their toxicity to both humans and other species. The cleanup is a very expensive affair involving costly and time-consuming laboratory analysis. There is a need for on-site testing mainly to define the extent of contamination and to define, in real-time, the sampling of the areas of contamination to ensure that the remediation efforts meet US Environmental Protection Agency (EPA) requirements. In addition, on-site field-screening methods can yield large savings in the remediation costs and reduce the time for analysis as the samples need not be transported to an off-site laboratory.

The most commonly used explosives can be classified into nitroaromatics (TNT, DNT, TNB), nitramines (RDX, HMX), and nitrate esters (PETN). Of these, TNT is the most widely used explosive. Currently, the laboratory method of TNT analysis is SW-846 Method 8330, which is performed in a high performance liquid chromatography system (HPLC) in an off-site laboratory [2]. This method is a highly selective method and sensitive to parts per billion concentrations (ppb) however the average turnaround times for the results of a single test are 3 days and 1 month with costs of $1,000 and $250, respectively [3].

Several field screening test kits are available for the detection of TNT in soil and water. The most common methods can be categorized into immunoassay and colorimetric based methods. Strategic Diagnostics [4] has two immunoassay based methods commercially available in the market:

1. DTECH is semiquantitative.
2. RaPID assays are quantitative and use a slightly laborious magnetic-bead based approach. The reagents used in immunoassay methods may be sensitive to temperature and have shorter shelf lives.

On the other hand, colorimetric methods are quantitative, utilize stable and inexpensive reagents. Field screening colorimetric methods for quantitative detection of TNT in soil has been reported by Jenkins [5] and Medary [6]. In Jenkins' method, TNT was extracted from undried soils with acetone and reacted with potassium hydroxide and sodium sulfite to form the highly coloured Janowsky anion with absorbance at 540 nm. Medary's method is similar but utilized methanol for extraction and 10% aqueous solution of sodium hydroxide to form a coloured anion with absorbance at 516 nm. Strategic Diagnostics also has a commercial colorimetric kit field EnSys which is based on Jenkins' method. In the EnSys method, the only difference is that the soil is dried before extraction. In a comparison of various commercially available methods for the detection of TNT and RDX, CRREL-EnSys colorimetric method was found to be the most suitable for on-site detection based on the accuracy, detection limits, precision, ease of use, cost per sample, and the ability to detect classes of explosives [7].

Acetone or acetonitrile are commonly used as solvents to extract TNT from soils, although a mixture of DMSO/Ethanol was used as a solvent to extract TNT in a report on toxicity studies [8]. These colorimetric methods can be used to detect a class of explosives, such as nitroaromatics, nitramines, and nitrate esters. However, explosives within a class cannot be distinguished very well. The most preferred on-site method relies on colorimetric detection as explained in the EPA Method 8515 for TNT. These methods require manual sample extraction for soil samples and preconcentration for water samples. Further, calibration with a control solution needs to be performed *manually*, the sample and reagents have to be mixed *manually*, and the absorbance from a spectrophotometer is noted *manually*.

An ideal on-site detection system would be inexpensive, sensitive, fully *automated*, reliable, multiplex sample handling, and detect a broad range of explosives. The advent of microfluidic lab-on-a-chip technology might offer such a detection system. Microfluidic capillary electrophoresis chips have been utilized for the detection of nitroaromatics such as TNT, DNT, NT, and DNB [9-12]. Due to the good redox properties of nitroaromatics and the inherent suitability for miniaturization, most of the microfluidic methods so far used electrochemical methods for detection. The individual components of nitroaromatics can be detected in the capillary electrophoresis chips (analyte-specific) unlike the colorimetric methods (class-specific) where nitroaromatics are detected broadly.

2.   LAB-ON-A-CHIP ARCHITECTURE

In this section, the concept of 'digital microfluidics' and the technique of electrowetting actuation, as it applies to the automation of on-site explosive detection, are explained. The fabrication details of the electrowetting microfluidic chips are also presented. Finally, the colorimetric reaction for the detection of TNT is presented.

2.1   Digital microfluidics

Currently, most microfluidic devices are based on continuous fluid flow, primarily using electrokinetic phenomena for actuation. An alternative approach towards microfluidics is to manipulate the liquid as unit-sized discrete microdroplets. Due to the architectural similarities with digital microelectronic systems, we refer to this approach as 'digital' microfluidics.

Digital microfluidic systems have several advantages over continuous-flow based devices, the most important being reconfigurability and scalability of architecture which makes it a very good choice for implementing a lab-on-a-chip. Electrowetting is one of the few microfluidic techniques that has been used successfully to implement a digital microfluidic lab-on-a-chip integrating all the operations, such as transport, mixing, splitting, dispensing, and disposal of liquids onto a single platform. Electrowetting refers to the modulation of interfacial tension between a conducting liquid and a solid electrode, by the application of an electric field between the two. We have utilized electrowetting to demonstrate an automated and chip-based colorimetric enzyme-kinetic glucose assay using standard glucose solutions [13]. In this chapter, we extend the use of our electrowetting platform to analyze TNT-laded droplets particularly automating the on-chip storage of reagents, dispensing of reagents, mixing of the sample and the reagents, and colorimetric detection. We have not addressed the sample extraction as all the experiments were performed on TNT samples constituted in DMSO (dimethlysulfoxide) in our lab.

2.2   Chip fabrication

The electrowetting actuation system consists of two parallel electrode plates, a continuous ground plate on top and an addressable electrode array as the bottom plate as shown in Figure 1. Both the top and bottom plates were fabricated on glass substrates with all the electrodes patterned in indium-tin-oxide (ITO) due to its transparent nature enabling easy integration of optical measurement techniques with the electrowetting system. Using standard microfabrication techniques, an array of independently addressable control electrodes were patterned in ITO on the bottom plates. It is further coated with Parylene C (800 nm) for insulation. The top glass plate is coated with a layer of ITO to form a continuous ground electrode. Both top and bottom plates are coated with a thin hydrophobic layer of Teflon AF 1600 (50 nm). A spacer is used to separate the top and bottom plates, yielding a fixed gap. The droplet is sandwiched between the two plates, and is surrounded by immiscible silicone oil. Silicone oil prevents evapouration of the droplets and reduces the voltages required for driving the droplets. In the experiments reported in this paper, we have used electrowetting chips with an electrode pitch of 1.5 mm and a gap spacing of 500 μm. A custom electronic controller was built to address and switch each electrode independently.

These electrowetting chips are used to dispense the sample and the reagent droplets, transport and mixed them for chemical reactions to occur.

An optical absorbance measurement system is also integrated to form a lab-on-a-chip.

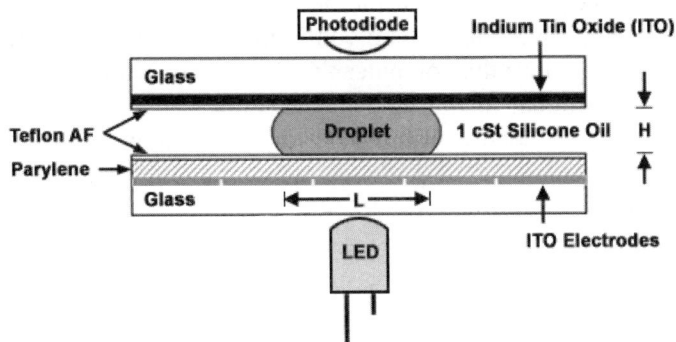

Figure 1. Schematic of electrowetting actuation device with optical detection.

2.3 TNT assay

A polynitroaromatic compound such as TNT reacts with an alkali such as potassium hydroxide to give an intensely coloured Jackson-Meisenheimer reaction product. When this reaction takes place in a ketone solution such as acetone, it is known as Janowsky reaction. In this paper, we slightly modify the approach by replacing acetone with DMSO. We have chosen DMSO as the solvent in our system because the DMSO droplets would be immiscible with the surrounding silicone oil while acetone would mix with the silicone oil. The resulting Jackson-Meisenheimer coloured product from the reaction of TNT in DMSO with KOH in water has an absorbance peak around 540 nm. The absorbance is related to the concentration of TNT by Beer's law,

$$A = \varepsilon \times C \times l \tag{1}$$

where A is the absorbance, ε is the extinction coefficient of the reaction product under the assay conditions, C is the concentration of TNT, and l is the optical path length. The concentration of TNT can be calculated directly from the absorbance measured at the end of the reaction.

2.4 Optical detection

As shown in Figure 1, optical detection is performed in a plane perpendicular to that of the electrowetting chip. The setup consists of a green

LED and a photodiode at 180° to each other. The photodiode (TSL257, obtained from Texas Advanced Optoelectronic Solutions) is a light to voltage converter that combines a photodiode and an amplifier on the same monolithic device. The output of the photodiode is externally amplified 10 times using a non-inverting op-amp configuration and the amplified voltage signal $V(t)$ is logged by the computer through a data acquisition board. $V(t)$ is directly proportional to the light intensity measured by the photodiode. The absorbance is calculated from $V(t)$ from the equation:

$$A(t) = \ln\left[\frac{V_0 - V_{dark}}{V(t) - V_{dark}}\right] \qquad (2)$$

where V_0 corresponds to zero absorbance (or 100% transmittance), and V_{dark} corresponds to the dark voltage of the photodiode.

2.5   Chemicals

Commercial grade 2,4,6-trinitrotoluene (TNT) was provided by Sandia National Labs. KOH and DMSO were reagent grade. A stock solution of TNT was prepared in DMSO at a concentration of 1 mg/mL and diluted to obtain other concentrations. 7 mM KOH solution was prepared in DI water.

3.   RESULTS AND DISCUSSION

In the results presented below, the TNT assay was performed on the chip in three steps: manual dispensing, electrowetting-enabled mixing, and colorimetric detection. First, a droplet of TNT in DMSO and a droplet of KOH in water are dispensed manually by a pipette on the electrowetting chip. By applying appropriate voltages to the electrodes, these two droplets are merged. The merged droplet is further mixed by shuttling it across three electrodes for 10 seconds at an actuation voltage of 50 V. Our earlier results indicate that we should achieve complete mixing in less than 5 seconds for this pattern of mixing [14]. At the completion of mixing, the absorbance is logged by the LED/Photodiode setup described in the previous section. 1 cSt silicone oil is used a filler medium in all the experiments reported in this paper.

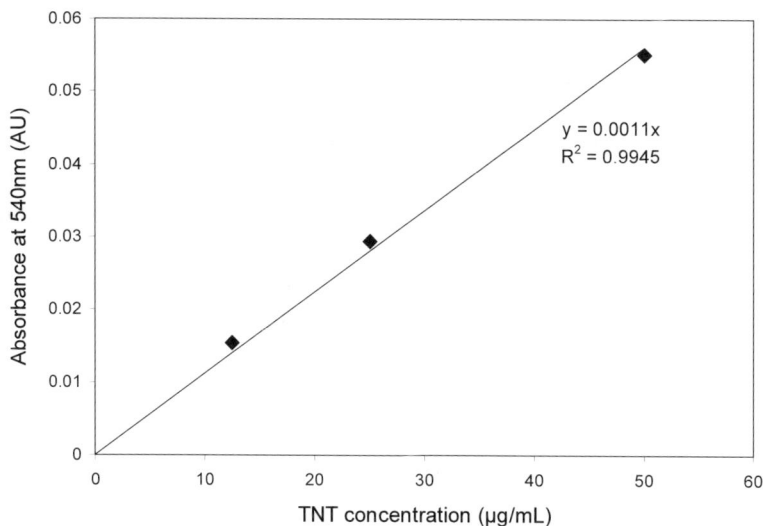

Figure 2. Absorbance of the reaction product with respect to the initial concentration of TNT on the electrowetting chip.

A stock solution of 1mg/mL TNT in DMSO was diluted 20×, 40×, and 80× to yield concentrations of 50 µg/mL, 25 µg/mL, and 12.5 µg/mL. A 1.3 µL droplet of each of these solutions was dispensed on the chip. A 1.3 µL droplet of 7 mM KOH in DI water was also dispensed on the chip. Photodiode readings were taken after 3 minutes of initial merging and mixing of the droplet on the chip. The absorbance results are shown in Figure 2. By mixing a droplet of TNT and a droplet of KOH with equal volume, the concentration of TNT will reduce by a factor of 2 due to dilution. In Figure 2 only the original concentrations of the TNT droplets are shown. The absorbance is linear among the three concentrations presented. We have also demonstrated the multiplexing capability to perform 3 simultaneous TNT droplet reactions as shown in Figure 3. These results are only preliminary and no optimization of the reactions or the optics was performed to improve the detection limits. Below a concentration of 10 µg/mL (which amounts to 5 µg/mL after dilution on the chip), the change in absorbance is not discernible with the current detection setup. The path length of ∼ 500 µm in our system is very short compared to the 1 cm used in the standard spectrophotometers. In order to address the problem of short path lengths: the gap height of the droplet can be increased; detection can be performed on a sideways elongated droplet with the LED and the photodiode setup also sideways; the reaction can be modified to yield an intense colour for the reaction products.

Stability of the reaction: We observed that over various concentrations of TNT in DMSO the absorbance increases initially for 3 minutes after merging with the KOH droplet on the chip. However, after 3 minutes it stabilizes and remains stable for 5 more minutes. It was reported elsewhere that the same reaction between TNT and KOH in acetone was found to be stable over a range (10% to 75%) of moisture content of the soil from which TNT was extracted (Ref.5). In our experiments the water content is 50% and the reaction seems to be stable even though in our case the solvent was DMSO. We did not perform experiments to evaluate the long term stability of the reaction since the measurements on our platform would be finished well within the 5 minutes of stability.

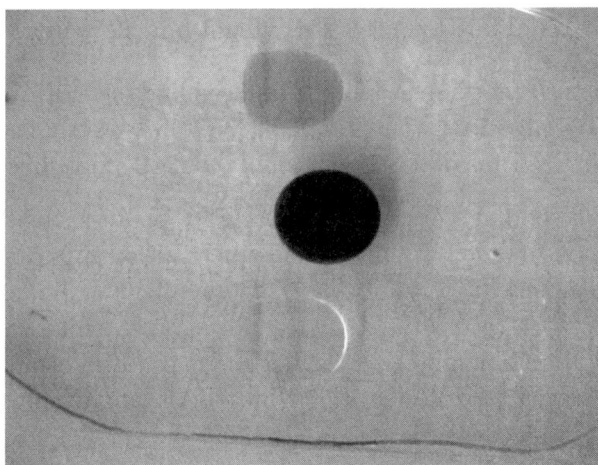

Figure 3. Top view of three droplets of TNT in DMSO + KOH in water on-chip with different concentrations of TNT yielding different intensities in colour.

We have also performed a colorimetric experiment with DNT in DMSO and KOH in DI water on the same electrowetting chip. However, we have not performed quantitative measurements nor determined the stability of the reaction. Further experiments are needed to assess the stability of the reaction and establish the linear range of detection for DNT. Also, experiments need to be performed on nitroaromatic samples extracted from soil and for samples in water to establish the suitability of a digital microfluidic system for field screening of nitroaromatics.

4. CONCLUSIONS

A digital microfluidic lab-on-a-chip based on electrowetting actuation of droplets has been demonstrated for the detection of 2,4,6-trinitrotoluene (TNT). We have used DMSO as a solvent instead of acetone or acetonitrile which is commonly used. We have demonstrated a linear range of detection for TNT between 12.5 µg/mL – 50 µg/mL in less than 5 minutes. We have also demonstrated simultaneous detection of various concentrations of TNT and feasibility of detection of 2,4-dinitrotoluene (DNT). Future work would involve increasing the detection limits of the system from µg/mL to ng/mL and simultaneous detection of other nitroaromatics such as DNT and TNB.

ACKNOWLEDGMENTS

I would like to acknowledge Vijay Srinivasan, Department of Electrical and Computer Engineering, Duke University, for helping with the experiments and Harinath Chakrapani, Department of Chemistry, Duke University for technical discussions. Also, I would like to thank Dr. Phil Rodacy of Sandia National Labs for providing the TNT samples.

REFERENCES

1 J.C. Bart, L.L. Judd, K.E. Hoffman, A.M. Wilkins and A.W. Kusterbeck, Application of a portable immunosensor to detect the explosives TNT and RDX in groundwater samples, Environmental Science and Technology 31 (1997) 1505-1511.
2 United States Environmental Protection Agency, SW-846 Method 8330.
3 A.B. Crockett, H.D. Craig, T.F. Jenkins and W.E. Sisk, Field sampling and selecting on-site analytical methods for explosives in soil, EPA Federal Facilities Forum Issue, EPA/540/R-97/501, November, (1996).
4 http://www.sdix.com/ProductSpecs.asp?nProductID=21
5 T.F. Jenkins and M.E. Walsh, Development of field screening methods for TNT, 2,4-DNT and RDX in soil, Talanta 39 (1992) 419-428.
6 R.T. Medary, Inexpensive rapid field screening test for 2,4,6-trinitrotoluene in soil, Analytica Chimica Acta 258 (1992) 341-346.
7 H. Craig, G. Ferguson, A. Markos, A. Kesterbeck, L. Shriver-Lake, T. Jenkins and P. Thorne, Field demonstration of on-site analytical methods for TNT and RDX in ground water, Proceedings of the HSRC/WERC Joint Conference on the Environment, http://www.engg.ksu.edu/HSRC/Proceedings.html, May, (1996).
8 L. Berthe-Corti, H. Jacobi, S. Kleihauer and I. Witte, Cytotoxicity and mutagenicity of a 2,4,6-trinitrotoluene (TNT) and hexogen contaminated soils in S. Typhimurium and mammalian cells, Chemosphere 37 (1998) 209-218.
9 J. Wang, B. Tian and E. Sahlin, Micromachined electrophoresis chips with thick-film electrochemical detectors, Analytical Chemistry 71 (1999) 5436-5440.

10 J. Wang, R. Polsky, B. Tian and M.P. Chatrathi, Voltammetry on microfluidic chip platforms, Analytical Chemistry 72 (2000) 5285-5289.

11 A. Hilmi and J.H.T. Luong, Electrochemical detectors prepared by electroless deposition for microfabricated electrophoresis chips, Analytical Chemistry 72 (2000) 677-682.

12 S.R. Wallenborg and C.G. Bailey, Separation and detection of explosives on a microchip using micellar electrokinetic chromatography and indirect laser-induced fluorescence, Analytical Chemistry 72 (2000) 1872-1878.

13 V. Srinivasan, V.K. Pamula, M.G. Pollack and R.B. Fair, A digital microfluidic biosensor for multianalyte detection, Proc. IEEE MEMS (2003) 327-330.

14 P. Paik, V.K. Pamula and R.B. Fair, Electrowetting-based droplet mixers for microfluidic systems, Lab on a chip 3 (2003) 28-33.

Chapter 19

NEXT GENERATION TRACE EXPLOSIVES DETECTION SYSTEMS
Sample Collection and Novel Electronic Sensors

Richard Lareau

Transportation Security Research and Development Laboratory, US Department of Homeland Security, William J.Hughes Technical Center, Atlantic CityInternational Airport, New Jersey, USA

Abstract: Trace-based explosives detection (ETD) systems currently in use have limitations in selectivity, sensitivity, size, and certainly cost (approximately $25,000 – $100,000$^+$). Miniaturization of systems to bench top or even handheld levels has great potential, but these systems will still be limited in operation and perhaps only a quarter of the price of present day systems. The use of electronic-based sensors, based on microelectromechanical systems, MEMS, and even further in the future, nanotechnology based systems (or nano-electromechanical systems, NEMS), shows promise for large improvements in sensitivity, selectivity, improved alarm rates, size and cost (from approximately $1,000 to $25 each unit). With current events and emphasis in antiterrorism, there is a need to quickly develop and deploy systems to all airports, along with numerous other transportation modes, protection of government facilities, public facilities, schools, etc. This could be realized with low cost miniaturized electronic chemical detection systems, with the capability of detecting trace explosives, but also could include chemical and biological agent detection systems, as well as, hazardous chemical detection systems. This paper will provide an emphasis on the analytical challenges of trace explosives detection, including contamination studies, sample collection, preconcentration, and sensors for low level detection. ETD systems operate via the collection of small residues of particles and/or vapours that indicate larger quantities of explosives present in the object or environment. To appropriately analyze for traces of explosives, improvised explosive devices (IED) are fabricated in various manners and the contamination spread from these IEDs are measured by trace analytical instruments. This assists in establishing our specification or threshold levels, while at the same time permits the development of trace standards which simulate real IEDs in particle composition and calibrated levels. The current types of trace detection techniques presently deployed include ion mobility

J.W. Gardner and J.Yinon (eds.),
Electronic Noses & Sensors for the Detection of Explosives, 289-299.

spectrometry, chemiluminescence and canines. Strengths and weaknesses of these 'systems' provide directional guides for future development of electronic sensors.

Keywords: Trace detection, explosive detection, MEMS, nanotechnology, trace chemical sampling, explosive vapour or particle collection

1.   INTRODUCTION

New technologies in trace analysis of explosives include trace walk through portals, miniaturized systems (e.g., hand-held systems), and next-generation sensor arrays (e.g., MicroElectroMechanical Systems (MEMS) and Nanotechnology based sensor platforms). The later next-generation sensor arrays/systems, currently in the R&D or prototype stages, are based on microcantilever, micro thermal analysis, lab-on-a-chip, optical detection, and other miniaturized platforms. Each shows promise for enhanced sensitivity, selectivity, and improved alarm rates, and also have the potential for lower cost, weight, and size. However, like much of the past CW/BW detector development work, many of the smaller size electronic explosives sensors/detectors lack a well developed front end sampling system for collection and transport of the trace analyte (vapour and/or particulate) to the advanced sensor system.

The ideal electronic sensor system will encompass a non-contact surface or air-volume sampler, in which both particle and vapour are collected, the analyte then thermally desorbed or thermally transported, perhaps pre-concentrated, and efficiently delivered to the sensor array. Careful attention to high volume/surface area collection and thermal flow conductance must be included in the engineered designs. This front-end collection and delivery system is as critical to the overall electronic sensor system as is the sensor detection specifications itself.

Advanced ETD systems could find use in numerous applications, including, transportation (e.g., airport checkpoint, checked bag, cargo, aircraft, train stations, bus terminals, etc.), protection of government facilities (e.g., embassy, government offices, court houses, national museum, etc.), customs (e.g., port-of-entry/boarder crossing, etc.), prisons, schools, and so forth. The systems can be designed to 'sense' many types of chemical species, including CW/BW agents, illegal drugs, and explosives.

2. NEXT-GENERATION TRACE EXPLOSIVES DETECTION PLATFORMS

MEMS systems are physically small, exhibit electrical and mechanical properties, and are normally low power devices. MEMS devices typically employ existing silicon-based semiconductor fabrication technology, although more recent fabrication based on alternate materials and fabrication schemes have been successfully explored. The device size features are on the micron to millimeter sizes; large for current semiconductor computer based components. The following classes can describe current MEMS applications: Sens. Actuators (including liquid and gas sensors, radiation detectors, *chemical* detectors, etc.), resonators (to sense changes in motion, environmental properties, etc.), and motors/gears, etc. [1]. One additional attribute of the MEMS devices is the ease of integration with on-chip complementary metal-oxide-silicon (CMOS) technology for computer integration – providing 'smart based', real time detection and ease for array detection, data fusion, and system integration.

Several MEMS devices are mass-produced and utilized in commercial products today. These include a MEMS-based accelerometer that is the key sensing electronic component in today's automotive air bag deployment system, inkjet-printer cartridges, systems to monitor air pressure and temperature in automotive tires, and chemical and environmental monitoring of processes and/or conditions in laboratories, etc. The US Department of Defense (DOD)/Defense Advanced Research Projects Agency (DARPA), as well as many other international government agencies have high dollar investments in R&D programs for the development of MEMS-based inertia, guidance, RF communication based systems, and Chemical/Biological detectors, however, currently, little investment in MEMS-based explosives detection systems.

Nanotechnology based systems are at the next scale down from MEMS devices, however, MEMS devices can be platforms for a nano-event to be studied or measured. The US National Nanotechnology Initiative, Interagency Working Group on Nanoscience [2], defines Nanotechnology as:

'Research and technology development at the atomic, molecular or macromolecular levels, in the length scale of approximately 1-100 nanometer range, to create structures, devices and systems that have novel properties and functions because of their small and/or intermediate size......Nanotechnology R&D includes integration of nanoscale structures into larger material components, systems and architectures....'

Systems that utilize sensors at this scale can be true Nanoscale systems, like carbon nano-tube (CNT) devices, or MEMS based platforms that measure Nano-events, like measuring clusters of molecules utilizing a microcantilever sensor system.

Future MEMS/Nano based trace explosives detection systems have several key attributes, including:

1. Small size – with integration of the MEMS chip, computer system, display, etc., portable, hand-held, or miniaturized non-operator based systems can be mass produced.
2. These detectors can be passive (like household smoke detectors) or active detectors (constant read-out to an integrated central security system).
3. Improved sensitivity (with enhanced collection/preconcentration) and selectivity (with novel polymeric film coatings/collection).
4. Dual or multiple sensor technology to improve alarm rates (multiple detector arrays with data fusion for detection confirmation).
5. Improved reliability realized from proven semi-conductor manufacturing (standard silicon based chips).
6. Lower cost (due to inexpensive mass produced manufacturing processes; system cost of $1,000 - $25 ea., with availability to all anti-terrorism applications).

The success of these systems will be realized with the full integration into the airport security sectors, including, sensors embedded in luggage tags (for checked and carry-on luggage), at the checkpoint carry-on and personnel screening site, on cargo loading containers, as well as, within critical locations in the airport and airplane itself. Use of these systems in other transportation security areas will be just as critical, and may add the additional requirement of portability (i.e., battery operated).

R&D efforts at the US Department of Homeland Security, Transportation Security Administration, R & D Division include near- and long-range programs for the development of novel MEMS/NEMS based trace explosives detection systems. These efforts include the development of novel miniaturized front-end collection systems, for the collection of explosive particles and/or vapour, followed by preconcentration and delivery to several MEMS/NEMS based sensors. A few examples of sensor projects currently under development include:

1. R&D and testing of a hand-held miniaturized detection system based on multiple MEMS-based sensors for improved alarm verification (see discussions below on the Sandia National Laboratory (SNL) μChemLab on a chip).

2. R&D and testing of microcantilever arrays for nanodetection of clusters of explosive molecules (see discussions below on ORNL/NSSA nanoexplosives detector).
3. R&D and testing of a polymer coated microcantilever trace explosives detection system (see discussions below on the Bureau of Alcohol Tabacoo and Firearms (ATF)/Naval Research Laboratory (NRL)/Oak Ridge National Laboratory (ORNL) system).
4. R&D and testing of a mass spectrograph (MS) on a chip (see discussions below on the DARPA-TSA funded, Army Research Laboratory/Northrup Grumman (ARL/NG) MS project).
5. R&D and testing of a carbon nano-tube (CNT) trace based explosives sensor (see discussion below on the NASA CNT project).

Additionally, three front-end collection systems are under development:

1. A miniaturized, polymer coated pre-concentrator system (in conjunction with the SNL Micro Chem Lab on a chip project).
2. A MEMS based thermal collection/separator system (in conjunction with the ORNL micro cantilever systems).
3. A MEMS based plate array/front-end collection system (being developed independently by NRL as a universal small sensor front-end collection systems; referred to as CASPAR).

3. EXAMPLES OF TRACE EXPLOSIVES DETECTION R&D PROGRAMS

An example of one of TSA/TSL's R&D funded MEMS based project is the Sandia National Laboratories (SNL) MicroHound project. This is based on the SNL 'Micro Chem Lab on a Chip', illustrated in Figure 1. The original prototype system from SNL was developed for high vapour pressure, chemical weapons (CW) detection, which utilized a MEMS GC separator, with miniature surface acoustic wave (SAW's) based sensors. The system included an inlet, coated pre-concentrators, detectors, and pumps. To make this useful for trace explosives detection, the addition of an alternate front-end sample collection/macro-preconcentrator and MEMS based coated-preconcentrator is necessary, along with the option to utilize or exclude the MEMS GC separator followed by detection by either, or both, SAW's and miniaturized IMS detectors.

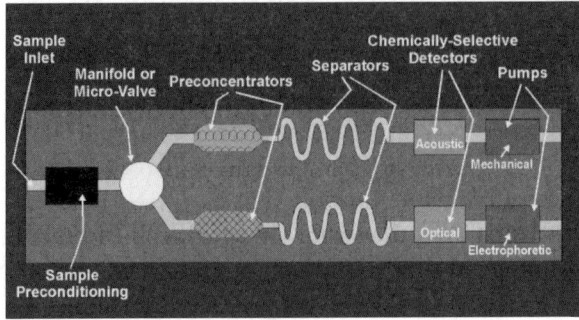

Figure 1. SNL's MicroChemLab on a Chip.

Figure 2 illustrates the components of the MicroHound system for trace explosives detection. The goal for this system is to provide low parts-per-trillion detection sensitivities, with collection times of less than 10 seconds, while at the same time having sufficient sensitivity and reliable confirmation to differentiate between different types of explosive signatures. The components will be integrated into a hand-held system, which will be operated by battery pack or direct AC power (see Figure 3 for an illustration of the proposed SNL MicroHound™ System).

Explosives Vapours/Particles

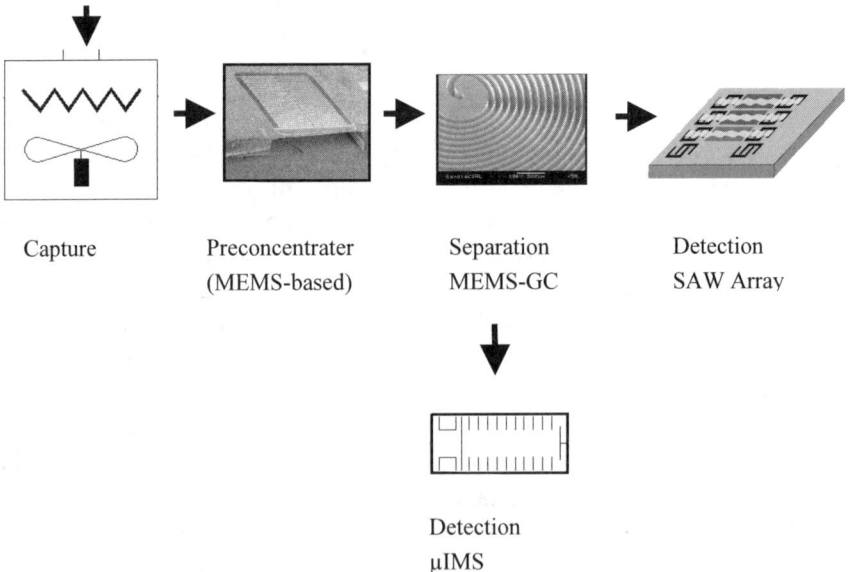

| Capture | Preconcentrater (MEMS-based) | Separation MEMS-GC | Detection SAW Array |

Detection
μIMS

Figure 2. Components of SNL MicroHound trace explosives detection system.

MicroHound^{TM} Concept

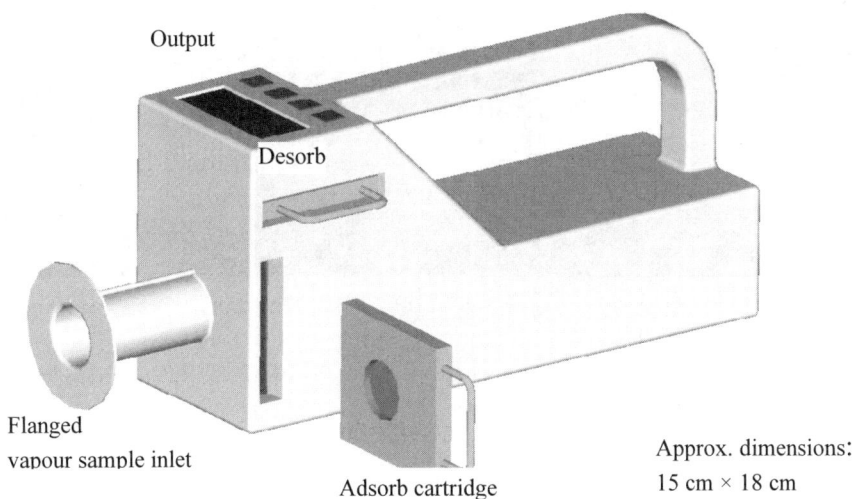

Figure 3. SNL MicroHound TM concept.

Another funded MEMS effort involves the use of microcantilevers for the detection of explosive vapours. Two R&D programs are currently funded at Oak Ridge National Laboratory (ORNL), US Department of Energy, in microcantilever technology for trace explosives detection. The first microcantilever project involves changes in resistivity, or frequency shifts, of a bimetallic silicon based microcantilever, whereby the temperature of the cantilever tip is rapidly heated till a deflection of the cantilever is observed, thus signifying the presence of a cluster of explosive molecules. Figure 4a presents a scanning electron micrograph (SEM) of one type of bimetallic silicon cantilever and Figure 4b is a similar SEM of alternate cantilever designs, compared to a strand of human hair, for size comparison. In this case, each explosive molecule gives a specific temperature for deflagration, and hence provides the selectivity necessary. Low ppt sensitivities in a few second analysis times are expected with this vapour detection system. An efficient front-end collection system is necessary in this case, in order to collect vapour and particle evidence; the particles would be thermally desorbed on a heated collection or 'trapping' surface to form the vapour for the cantilever collection surface.

296

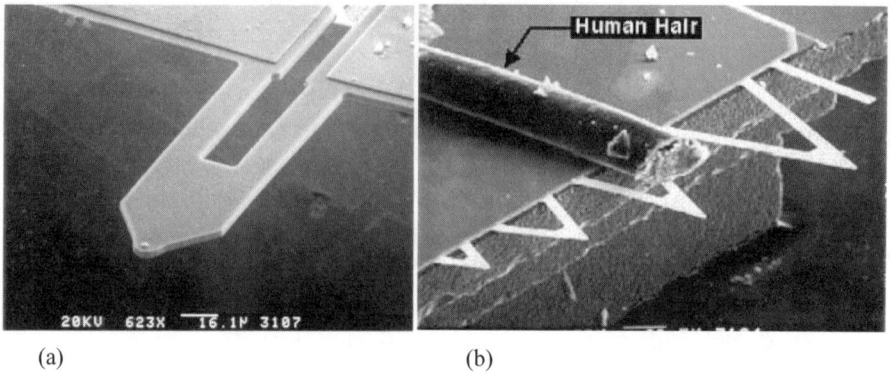

(a)   (b)

Figure 4. (a) SEM of bimetallic Si cantilever, and (b) SEM of alternate designed cantilevers, with human hair as size reference.

Figure 5 is an illustration of the cantilever deflection detector, which utilizes a laser diode light system for precise measurement of the cantilever 'bending' that occurs when the cluster of explosive molecules undergo deflagration from the surface.

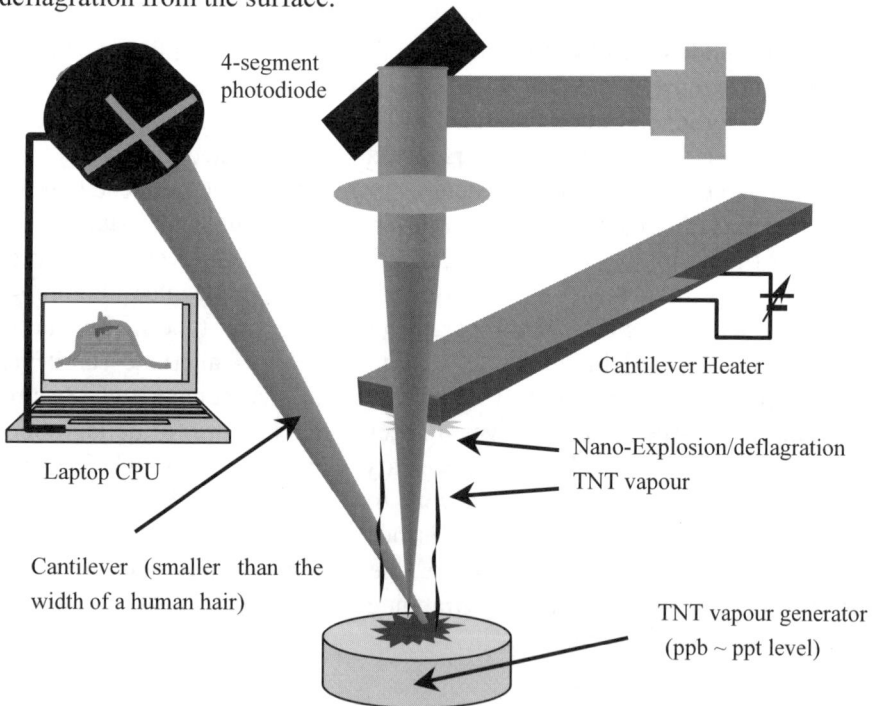

Figure 5. Illustration of the cantilever deflection detector – test assembly.

The second ORNL microcantilever project, funded by the ATF and technically co-directed by TSA, involves polymer-coated micro cantilevers. In this case, the explosive molecule would be adsorbed on the surface of the polymer, and the 'swelling' or change in physical properties would be measured indicating the presence of an explosive. Several polymers are currently being developed to provide the necessary selectivity for the different types of explosives signatures. The sensitivity is similar to the non-coated cantilever; estimated limit of detection of low parts-per-billion (with proper collection and preconcentration system, LOD should approach low parts-per-trillion). Figure 6 illustrates two modes of detection for the coated cantilevers: beam stress response and beam resonant frequency response.

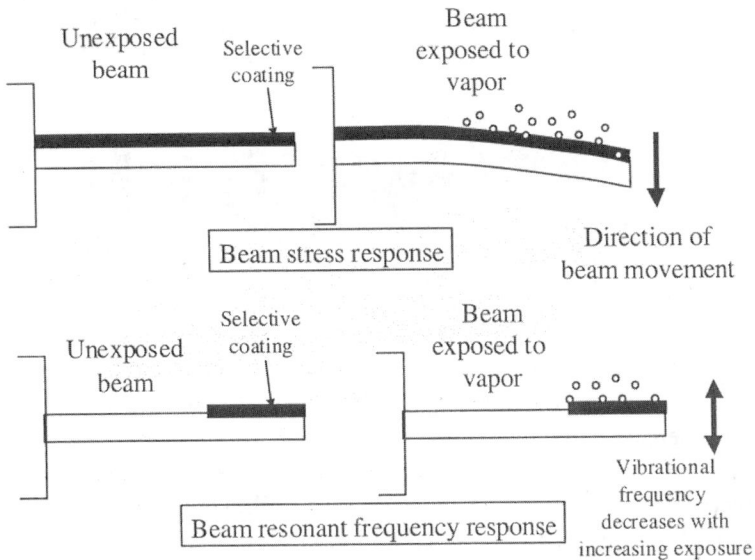

Figure 6. Illustration of the two modes of detection for the coated cantilevers: beam stress response and beam resonant frequency response

In the best-case situation, both types of microcantilever sensors would be grouped in an array to provide a cross platform of sensitive and selective explosive detection system. Additional co-funded (TSA/ATF) work is currently on going in materials development for novel microcantilevers. This involves R&D of silicon carbide (SiC) based cantilevers, for improvements in material properties (e.g., less fragile compared to silicon) and to provide a platform for wide band gap type materials, like aluminum nitride (AlN). With an AlN/SiC based cantilever, the sensor can now work in the piezoelectric resonator mode, providing enhance response and hence sensitivity to the analyte, along with a direct measurement by frequency/resistance response, versus the more complex optical detection measurement mode. Figure 7 is a SEM of a SiC test cantilever developed by Boston Micro Systems, Inc. and tested with and without coatings by the US Naval Research Laboratory.

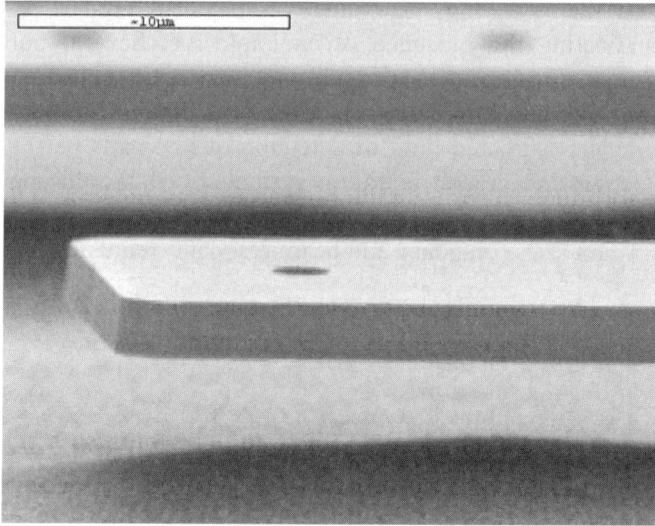

Figure 7. *SEM* of a SiC test cantilever developed by Boston Micro Systems, Inc.

The last example of these small-scale trace detectors is one based on nanotechnology platform and principles. This TSA funded project with NASA Ames Research Center involves the use of carbon nanotubes (CNT) as surface sensitive trace explosives sensors. In this case, the individual molecules of an explosive vapour interact on the atomic scale with a molecule(s) from the CNT surface. Figure 8 illustrates the NASA Ames 'conductive CNT explosives detector'; the CNT bridges between the source and drain of the transistor, and the change in conductance of the CNT bridge is measured as the molecular interaction with its' surface (an array of CNT wires).

Figure 8. Illustration of NASA Ames 'conductive CNT explosives detector.' CNT not to scale (actual CNT is a surface array of single wire CNT strands).

The key to the above mentioned microelectronic detectors will be the success of both the micro sensor platform itself and an efficient front-end vapour and particle collection system. Currently, several initiatives are moving forward for front-end collection systems. This is key for collecting both the vapour present in the environment, but also to collect the particles present in the air or on surrounding surfaces.

4. CONCLUSIONS

It is clear that both miniaturized and MEMS/Nano based systems are real time goals. If these small, microelectronic based, trace explosives sensor are developed successfully, and mass production adds superior reliability and low cost to the end product, then these 'smaller' MEMS/Nano sensors due indeed 'promise' to be better overall. The development, test and deployment of these micro systems are not expected to be complete until approximately 5 years from now.

ACKNOWLEDGEMENTS

The authors would like to acknowledge the scientific collaboration of – Kevin Linker and colleagues at SNL/DOE, Drs. Andy McGill and Eric Houser at NRL/DOD, Dr. Rick Mylck at BMS, Inc., Dr. Thomas Thundat at ORNL/DOE, and Linda Deel, ATF.

REFERENCES

1. http://mems.isi.edu (see The MEMS Clearinghouse – and related links therein).
2. http://www.nano.gov (see vision statements and related links).

Chapter 20

SUMMARY OF THE WORKSHOP
Will Electronic Noses complement or replace existing technologies for vapour and trace detection of explosives?

Jehuda Yinon
National Center for Forensic Science, University of Central Florida, Orlando, Florida, USA

1.   OVERVIEW

Eighteen papers and two posters were presented by scientists from eight different countries during the two days of the NATO ARW.

The topics can be summarized as follows:

o   Basics of electronic nose technology
o   Polymers for electronic noses
o   Biological noses
o   Single-compound sensors
o   Sensor arrays and pattern recognition
o   Metal oxide sensors
o   Electrochemical sensors
o   Micromechanical sensors
o   Chemiluminescence sensor
o   GC-SAW sensors
o   Preconcentration
o   Application of electronic noses/sensors for detection of explosives

J.W. Gardner and J.Yinon (eds.),
Electronic Noses & Sensors for the Detection of Explosives, 301-303.
© 2004 *Kluwer Academic Publishers. Printed in the Netherlands.*

The applications of the various electronic noses and sensors for the detection of explosives can be summarized in Table 1.

Table 1. Summary of e-noses/sensors and their application.

Type of Sensor	Principle	Detected Explosive	Sensitivity	Application	Time
Amplifying fluorescent polymer arrays	Fluorescence quenching when target is adsorbed on polymer	Nitroaro-matic compounds	fg	Landmines	
Optical sensor microarrays	Multi-analyte microspheres with fibre-optic sensors				
Tin oxide and indium oxide thin film sensors	Chemical interaction between analyte and thin film	TNT			
Electroche-mical	Thermo-Redox	TNT, NG, EGDN, DMNB	ng	Commercial	15 sec
Electroche-mical	Square-wave voltammetry	TNT	ppb	In water	2 sec
Electroche-mical	Cyclic voltammetry	TNT	30 ppt	Vapour and liquid phase	
Micro-machined cantilever	Deflection of cantilever as function of molecular adsorption		10-30 ppt (fg range) PETN RDX		
Micro-machined cantilever	Heated canti-lever causes nanoexplosions detected by an optical system		70 pg		
SAW sensor	Acoustic wave confined to the surface of a piezoelectric substrate	NG, TNT, PETN, RDX, C4		Needs GC and precon-centration	10 sec
Digital microfluidics	Colorimetric detection	TNT	ppm		
Chemilumi-nescence	NO_2 + Luminol $\rightarrow$ light	TNT, RDX, EGDN, PETN, C4, Semtex, TATP, DMNB	ng	Vapours and particles	20 sec

A discussion followed regarding the question whether electronic noses will be able to complement or replace existing technologies for trace and vapour detection of explosives

2. CONCLUDING REMARKS

A number of conclusions may be drawn or points made from the NATO workshop on electronic noses and sensors for the detection of explosive materials:

- Electronic noses are more appropriate for sensing a range of different chemical compounds rather than detecting a single compound.
- Electronic noses/sensors for detection of explosives will be ready for use in the field in about 4-5 years, and will be able to complement/replace existing technologies.
- Problems to be solved by then:
  - Transport of the analyte to the detector.
  - Ruggedness of detector.
  - Capability of detecting the complete range of explosives. including improvised explosives.
  - Increasing the sensitivity.
- The requirements of electronic noses/sensors are:
  - Small and portable.
  - Sensitive.
  - Low cost.

Index